Peter W. Downs Kenneth J. Gregory

RIVER CHANNEL Management

Towards Sustainable Catchment Hydrosystems

Routledge
Taylor & Francis Group
LONDON AND NEW YORK

First published 2004 by Hodder Arnold, a member of the Hodder Headline Group

Published 2014 by Routledge
2 Park Square, Milton Park, Abingdon, Oxon OX14 4RN
711 Third Avenue, New York, NY, 10017, USA

Routledge is an imprint of the Taylor & Francis Group, an informa business

British Library Cataloguing in Publication Data
A catalogue record for this book is available from the British Library

Library of Congress Cataloging-in-Publication Data
A catalog record for this book is available from the Library of Congress

ISBN 13: 978-0-340-75969-1 (pbk)

Typeset in 10/13pt Gill Sans by Phoenix Photosetting, Chatham, Kent

Dedication

By Peter Downs:
To Renata Marson Teixeira de Andrade for love and inspiration

By Ken Gregory:
To the next generation: Victoria Woolf, Michael Woolf, Jaivin Raj, Kaisun Raj, George Gregory, James Robertson, Christy Gregory and Thomas Roberston; that they enjoy environment and river channels

Contents

Preface

The idea for this book arose when Kenneth Gregory received a Leverhulme Emeritus Fellowship after retiring from the post of Warden of Goldsmiths College, University of London. Peter Downs taught and researched in river channel management at the University of Nottingham at the inception of the book, but was a consultant scientist in California by its completion. Progress has inevitably been slowed by the distraction of other tasks in retirement for one author, and the demands of moving and changing career for the other. While relying heavily on the wonder of e-mail and time zones to maximize our productivity, we have striven to meet in person at every available opportunity in Liphook, Beeston and then Albany, and are grateful to various hotels and hostelries in Yosemite, Luton, Phoenix, Southsea and Reno that have unwittingly played their part. Those meetings involved intense discussion about the book's structure and content and we hope that the reader will think the result worthwhile. Intriguingly, the ethos in both higher education management and in river channel management has developed along similar paths in recent years – both becoming more holistic, user-friendly and catholic in approach, but we hope that the most recent disparate and disjointed trends in one will not be emulated by managers of river channels.

No one book has hitherto focused exclusively on river channel management or dealt comprehensively with the rapid evolution in the objectives, approaches and methods to the topic over the last four decades. It is worth noting that the quantitative study of river channel processes was only emphasized in the mid-1960s (Leopold *et al.*, 1964), that the magnitude of human impact on river channels began to be articulated only in the late 1960s (Wolman, 1967), and it was not until the early 1970s that the full potential of river channel change was appreciated (Schumm, 1971, noted that '… it appears likely that with time a channel may undergo a complete change of morphology (river metamorphosis) if changes in discharge and sediment load are of sufficient magnitude'). From these beginnings, geomorphological, ecological and engineering research into natural river channel processes and on the impacts of human activity progressed to such an extent that radically new approaches to management could be conceived so that a book such as this could be contemplated.

We have been increasingly aware during preparation of these chapters how impossible it is for any book to be comprehensive. This is not a manual for 'doing' river channel management but instead a vehicle designed for those who wish to *think* about how river channel management reached its contemporary state, the essential elements of that state and how it might change in the future. Starting from the viewpoint of geomorphology, we have tried to reflect a background to river channel management that is increasingly multidisciplinary in scope and subject to international influences and trends. We take an intentionally physical approach to the topic, based on the catchment hydrosystem, and acknowledge freely that a different story could be told starting from the viewpoint of management institutions, or from engineering or economics, etc.; but all would eventually return to the river channel itself as arbiter of whether they had been successful. If we can present to a broad audience a perspective on the way in which river channel management has changed and evolved and so engender further thinking about its future we will be satisfied. This book is completed during the UN Year of Freshwater, which reminds us of the responsibility of citizens to use water wisely and sustainably – and to safeguard the river channel. Confucius astutely said 'Study the past, if you would divine the future' – his statement still seems apposite for river channel management.

Peter Downs, Ken Gregory
Oakland, California, and Liphook, Hampshire
August 2003

Acknowledgements

We are grateful to many individuals who have given advice and encouragement and especially to friends who kindly reviewed chapters and provided comments including Andrew Brookes, Will Graf, Janet Hooke, Matt Kondolf, Malcolm Newson, Philip Soar, Kevin Skinner, Des Walling and Geraldene Wharton. Thanks also to colleagues who have provided photographs to illustrate the text including Tim Abbe, Andrew Brookes, Giles Brown, Martin Doyle, Matt Kondolf, Scott McBain and Kevin Skinner.

KJG acknowledges the provision of an Emeritus Fellowship by the Leverhulme Trust (1998–2001); PWD appreciates, first, the continued friendship and patience of KJG, second, the provision of sabbatical leave from the University of Nottingham (1999) spent as a Visiting Scholar at the University of California, Berkeley and, more recently, the forbearance of colleagues at Stillwater Sciences. We could not have completed this book without the great understanding shown by, and enormous support provided from, Renata Marson Teixeira de Andrade and Chris Gregory. They will be as relieved to see the book finished as we are!

We gratefully acknowledge the permissions given by publishers to reproduce figures which are acknowledged in the underlines and include:

John Wiley & Sons Limited (for Figure 1.3 B from Lawler, 1993; Figure 3.3 from Gregory and Maizels, 1991; Figure 3.4 from Gregory 1992b; Figure 3.6 after Brown, 1996; Table 3.4 from Bathurst, 1993; Table 3.7 modified from Brookes and Shields, 1996; Figure 4.1 from Braga and Gervasconi, 1989; Figure 4.3 from Petts 1984; Figure 4.5A from Whitlow and Gregory, 1989; Figure 4.5B from Roberts, 1989; Figure 4.7 based on Rutherfurd, 2000; Figure 5.1 from Panizza, 1987; Figure 5.4 from Simon, 1989; Figure 5.6 from Brookes, 1987a, 1987c; Figure 6.1 from Ning Chien, 1985; Figure 6.3A from Gardiner 1991; Figure 6.3B from Brookes and Shields 1996; Table 6.3 from Hooper and Margerum 2000; Table 6.6 from Newson, 1994; Table 7.3 from Brookes and Shields 1996a; Figure 8.1 from Downs and Brookes, 1994; Figure 8.2 from Thomson, Taylor, Fryirs, and Brierley, 2001; Figure 8.4 from Downs, 1995b; Figure 8.5 from Brown, 1996; Figure 8.6 from Sear, Newson and Brookes, 1995; Figure 8.8 from Richards and Lane, 1997; Table 8.2 based on Kondolf and Larsen, 1995; Figure 9.1 from Boon, 1992; Figure 9.2 from Sear, 1994; Figure 10.7 from Downs and Thorne 1998a; Figure 10.10 from Watson and Biedenharn, 1999; and Table 10.7 from Shields, 1996); Professor S. A. Schumm (for Figure 3.1A from Schumm, 1977); The Institute of British Geographers (for Figure 1.4 from Beaumont, 1978; Figure 5.2 from Brunsden and Thornes, 1979; Table 8.5 from Thorne, Allen and Simon, 1996); Dr Wayne

D.Erskine (for Table 4.8 from Erskine, 1998); Cambridge University Press (for Table 4.12 from Schumm, 1991; Figure 7.4 from Bradshaw, 2002; and Table 9.1 from Downs, Skinner and Kondolf, 2002a); Blackwell Publishing (for Figure 1.4 from Graf, 2001; Figure 4.4 from Wolman, 1967; Table 10.2 from Leuven and Poudevigne, 2002; and Table 10.5 from Buijse, Coops, Staras, Jans, Van Geest, Grift, Ibelings, Oosterberg and Roozen, 2002); Blackwell Publishers (for Figure 3.5 from Thomas and Goudie, 2000; Table 3.7 from Brookes, 1994; Figure 6.4 from Petts, 1994; Figure 7.2 from Petts and Maddock, 1996; Figure 10.4 from Large and Petts, 1994; Table 10.10 from Hey, 1994a; Figure 10.5 from Hey, 1994, 1996); Gebru Borntraeger http://www.schweizerbart.de (for Table 5.1 from Schumm, 1988); American Journal of Science (for Table 4.2 from Schumm and Lichty, 1965; Figure 4.8 from Graf, 1977); Professor W.L.Graf (Figure 4.8 from Graf, 1977); reprinted with permission from Elsevier (Figure 1.2 from Kondolf, 1994b; Figure 1.5 from Brookes, Gregory and Dawson, 1983; Figure 3.1C from Rosgen, 1994; Figure 3.6B after Nanson and Croke, 1992; Figure 4.6 from Knox, 2000; Figure 5.4 from Field, 2001; Figure 5.9 from Hupp and Simon, 1991); Doug Wilcox Editor of Wetlands (for Figure 7.1 from Mahoney and Rood, 1998); Professor Antony R. Orme Editor in Chief, Physical Geography (for Figure 8.7 based on Fryirs and Brierley, 2000 in Physical Geography, Volume 21, by courtesy of Bellwether Publishing); Hodder Arnold (for Figure 1.3 above and Figure 3.5 below from Knighton, 1998, reproduced by permission of Hodder Arnold; and Table 3.2 from Newson and Newson, 2000); The National Academies Press (Figure 4.2 reprinted with permission from New Strategies for America's Watersheds [1999] by the National Academy of Sciences, courtesy of the National Academies Press, Washington, D.C.); River Restoration Centre (Figures 10.7 and 10.10 from River Restoration: Manual of Techniques); Springer Verlag GmbH & Co.(Figure 7.3 from Huppert and Kantor, 1998; Figure 8.9 from Larson and Greco, 2002; Table 8.4 from Reid, 1998; Table 8.6 adapted from Downs and Kondolf, 2002; Figure 10.12 after Kondolf, Smeltzer and Railsback, 2001; Figure 10.14 from Downs and Kondolf, 2002); Dr Daniel D. Huppert (for Figure 7.3 from Huppert and Kantor, 1998); Dr Leslie Read (Table 8.4 from Reid, 1998); Professor G.E.Petts (for Figure 5.6 from Petts, 1982); Federal Highway Administration (for Figure 5.2 from Shen, Schumm, Nelson, Doehring, Skinner and Smith, 1981); to Center for Watershed Protection (for Figure 5.10 from Cararo, 2000); to Scott McBain (for Figure 10.4A); to Routledge Publishers (for Figure 1.1. from Newson, 1992; and Figure 6.2 from Newson, 1997); and to the University of Texas Press (for Figure 6.1A after Greer, 1979).

In addition to citing the source of all figures and tables every effort has been made to trace all copyright holders to secure their permission. Any rights not acknowledged here will be noted in subsequent printings if notice is given to the publisher.

1

Introduction

1
The Need for River Channel Management

Synopsis

This text considers river channel management in the context of the temporal adjustments and evolution of river channels and the desire for a sustainable approach. The context is provided by explaining the relation between the river, the river channel and the drainage basin (1.1), reviewing past and present uses of the river channel (1.2) as a basis for outlining approaches to river channel management (1.3) and showing how different disciplines have been involved in developing and applying management approaches (1.4). This background is used to justify the structure of the volume (1.5) including its key assumptions (1.6).

River channels are thought of in different ways by individuals according to a combination of personal perceptions, scientific training and professional experience. Furthermore, they are perceived in different ways in specific countries and were perceived differently in the past. The management of river channels is, therefore, intimately linked with individual and collective understanding. As such, one recurring theme throughout this book is that accomplishing the sustainable management of river channels requires the individuals involved to have a thorough understanding of the dynamic behaviour of river channel systems: this is the foundation for progress 'towards sustainable catchment hydrosystems'.

1.1 River, river channel and drainage basin

Definitions of river and river channel can be sought from legal practice. In the UK *Wisdom's Law of Watercourses* (5th edition, Howarth, 1992) concludes, following Woolrych's (1851) *A Treatise on the Law of Waters* (2nd Edition, 1851), that the meaning of 'river' includes all natural streams, however small, which have a definite and permanent course. Howarth also defines 'watercourse' to refer to a range of different kinds of moving waters, encompassing estuaries, rivers, streams and their tributaries above and below ground, which are commonly but loosely distinguished by characteristics of length, breadth and depth. Wisdom asserts that '... the principal characteristic of a watercourse at common law is that it consists of water flowing in a channel with reasonably well-defined banks, it is not essential that the flow of water should be maintained continuously in a particular channel' (Howarth, 1992: 7). In Hobday (1952) 'river' is also defined to include all natural streams that have a definite and permanent course, but excludes all bodies of water, however large, which are of a temporary character. From a common law legal perspective therefore, 'river' is defined very broadly to include all natural flowing water streams and courses, and in this book the term **river** applies to all defined flows of freshwater from source to tidal limit. It is also stated that 'For legal purposes the constituent parts of a watercourse have traditionally been the subject of a three fold distinction: (1) the bed or alveus; (2) the bank or shore; and (3) the water flowing therein' (Howarth, 1992: 11), a distinction that provides the basis for determining the *river channel* as explicitly concerned with the banks and bed of the river. As such, *river channel management* is concerned with the treatment of the river bed and banks and is a subset of *river management*, which should involve a concern for both the bed and banks *and* flows in the river. The collection, distribution and allocation of river flows, along with subsurface waters, commonly termed *water resources management*, is not considered explicitly in this book, although *accommodating* the implications of water resources management decisions is one of the primary challenges for river channel managers.

Although until recently there were differences in the reported dimensions of world rivers and their drainage basin areas these can now be resolved more exactly by remote sensing (Renssen and Knoop, 2000). The relation between river, river channel and drainage basin is fundamental in understanding how the river functions. The network of river channels provides the arteries along which water, sediment and solutes are transmitted through the drainage basin or catchment by the river flow. A number of components can be recognized between the level of the drainage basin and the river channel, including the channel reach, river corridor, floodplain and drainage network (Figure 1.1; Table 1.1), and these components provide the starting point for understanding and managing the river channel. Such components, and the relationships between them, can be understood according to their categorization and classification, their process dynamics, and the ways in

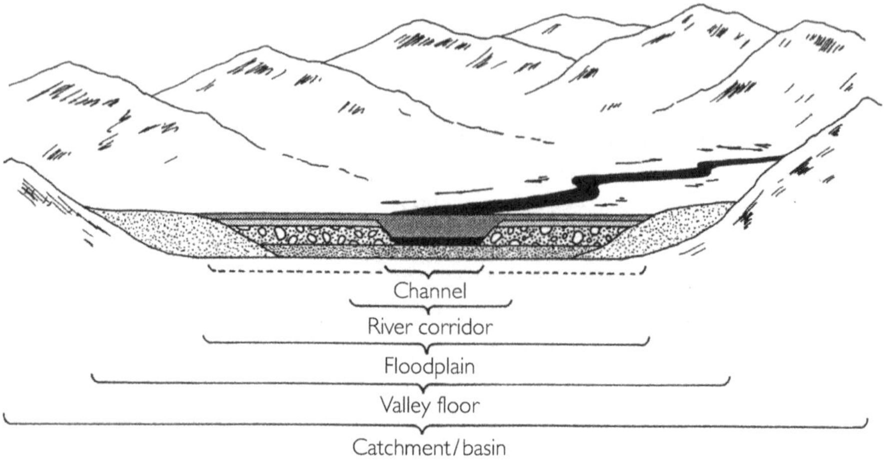

Figure 1.1 *Spatial scales in the drainage basin*
Source: from Newson (1992: 43). See also Table 1.1.

Table 1.1 Drainage basin components

Drainage basin component	Provisional definition
River channel	The linear feature along which surface water may flow, usually clearly differentiated from the adjacent floodplain or valley floor. There is no universal difference between stream and river channel although many languages have words for specific types of river or stream channel.
River reach	A homogeneous section of a river channel along which the controlling factors do not change significantly. Each reach may be viewed in terms of inputs, internal structural controls and process dynamics, and outputs.
Channel pattern or planform	The plan of the river channel from the air. It is considered over the length of a river channel reach, may be either single thread or multithread, and will vary according to the level of discharge.
Floodplain	The valley floor area adjacent to the river channel. A distinction may be made between the hydraulic floodplain, inundated at least once during a given return period, and the genetic floodplain which is the largely horizontally bedded landform composed of alluvial deposits adjacent to a channel.
River corridor	Linear features of the landscape bordering the river channel. The defining width of the corridor depends upon the topic under consideration and may encompass the full extent of the genetic floodplain, but is more likely to be constrained by human activities.
Drainage network	Composed of the network of stream and river channels that exist within a specific basin. The channels may be perennial, intermittent or ephemeral, the network may not be continuous and connected and the headwards extent of the network will change in response to storm events and human actions.
Drainage basin or catchment	Delimited by a topographic divide or watershed as the land area which collects all the surface runoff flowing in a network of channels to exit at a particular point on a river. The subsurface phreatic divide may not correspond exactly to the topographic divide. In the USA, the term 'watershed' is often applied to small and medium-sized drainage basins, and 'river basin' to large areas.
Catchment hydrosystem	A term used to encompass the unit area over which water and sediment drive the dynamics of the network of river channels. The term acknowledges that river channels are as dependent on water and sediment interactions on (and under) the land surface as in the clearly fluvial parts of the system. 'Hydrosystem' is used rather than 'ecosystem' to reflect our primary focus in this text on the physical rather than biological aspects of this system. In many aspects, the term requires a similar approach to study as do 'fluvial hydrosystems' (Petts and Amoros, 1996b), focusing on the four-dimensional connectivity of energy fluxes, and including the prospect of legacy effects over long timescales and the potentially significant impact of human activity.

which they change over time as the basis for placing the river channel in its evolutionary framework. It is also important to understand which components can be modified by the effects of human activity; indeed there has been considerable recent interest in attempting to decide what is 'natural' in landscapes because of the potential long history and longevity of human impacts. The process dynamics of the river channel should be viewed in relation to the water balance of the hydrological cycle and the hydrological response of the drainage basin (Gregory and Walling, 1973), because the river channel is part of a conveyor belt along which water, sediment, solutes and pollutants transfer through the hydrological system (Figure 1.2). River discharge, sediment and solute hydrographs, as recorded along the river at gauging stations, have a correlative expression with the other components listed in Table 1.1 and vary in time so that change is an integral feature of river channel systems. As such, even river channels 'in balance' with the prevailing forces upon them are in a condition of *dynamic* stability but this stability can be upset as a result of human activity or aspects of natural environmental change such as climatic forcing or tectonic processes. Whereas rivers and river channels were previously studied either in terms of contemporary processes or in terms of their evolution as an integral part of landscape history, there is now less distinction between the two strands (Gregory, 2000), facilitating impressive advances in the holistic understanding of the river channels, and requiring an appreciation that components of the drainage basin alter over different timescales (Figure 1.3). Therefore, river channel management must be implemented against the background of an understanding of temporal change in river channels. In contrast, until very recently, the majority of river channels were managed as definitive and static entities, rather than as dynamic components of a continuously evolving landscape.

1.2 A brief history of river use

River channel management is a constituent part of environmental management, that is, management that '... provides resources from the bioenvironmental systems of the planet but simultaneously tries to retain sanative, life-supporting ecosystems. It is therefore an attempt to harmonize and balance the various enterprises which humans have imposed on natural environments for their own benefit' (Thomas and Goudie, 2000: 176). Without human activity in close proximity to river channels, often requiring direct modification of the channel, there would be no need for river channel management. However, human interaction with river channel systems is highly varied in character and very long-established. It has been described as taking place in three major stages prior to the present (Cosgrove, 1990) which can be elaborated (Gregory, 2003b) to six broadly chronological but overlapping phases (Table 1.2), although not all phases may apply in any one geographical region.

Rivers have been important since the time of the first agricultural communities which naturally gravitated to flat fertile lands adjacent to rivers in regions with warm climates (Vallentine, 1967). However, with expanding demands in more arid areas, there was a need, *firstly*, to control and divert river flows. It is generally believed that King Menes, founder of Dynasty I in 3100 BC, dammed the Nile somewhere near Memphis to prevent the river overflowing and to protect the city (Said, 1983). So-called hydraulic civilizations developed in several areas (Wittfogel, 1956) involving the development of irrigation

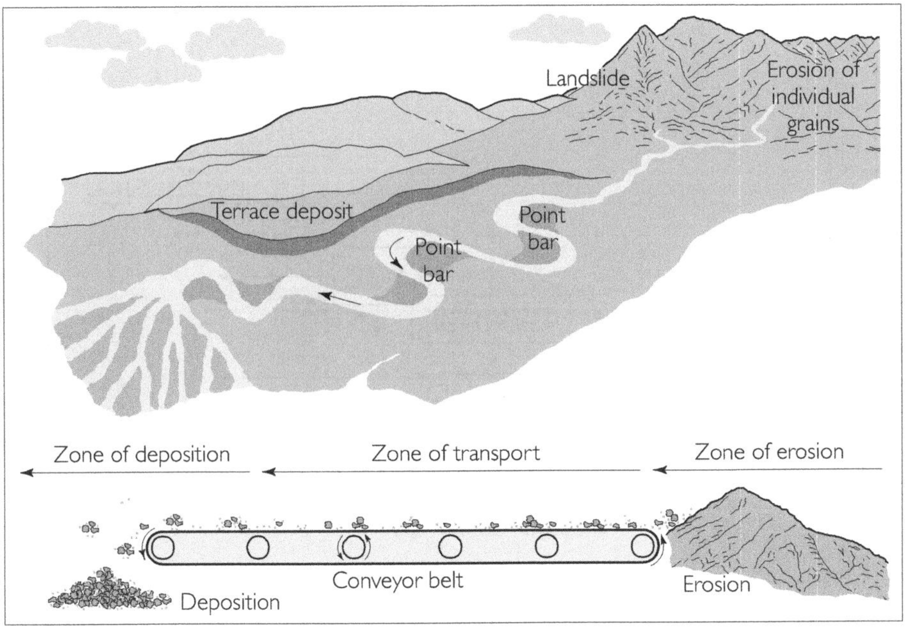

Figure 1.2 *A conveyor belt depiction of the channel system*
Source: Kondolf (1994b).

systems to extend the influence of the river for agriculture, and Biswas (1967) tabulated a chronology of engineering works by the Sumerians, Egyptians and Harappans who, by 2500 BC, had developed a very powerful civilization in the Indus Basin. This involved establishing Water Codes such as that of Hammurabi (1750 BC), and in China, by *c.* 2200 BC, Emperor Yu had begun to control rivers in the interest of land reclamation. Achievements by these hydraulic civilizations were particularly impressive in the absence of any theoretical basis for hydraulic calculation and design (Vallentine, 1967).

Each phase of river use is associated with a particular imperative which, at the time of the hydraulic civilizations, was large-scale water use. However, following the end of the Roman Empire (in the fifth century AD) there was a period of 1000 years when the prevailing imperative was very much local and small-scale river use. Thus, a *second* phase can be thought of in terms of pre-industrial revolution river uses with only small-scale river modifications but with extensive land use changes, especially deforestation, which had significant indirect impacts on river systems (see p. 104). Major uses of rivers in this phase devolved upon agriculture, fishing, drainage, water mills and navigation. Irrigation for agriculture had long been deployed and flood farming widely used, in North America and North Africa for example. However, similar practices were also significant in temperate areas whereby flood or water meadows were constructed, with a network of shallow channels that distributed sediment-laden flood water across the floodplain and may have dated back to the mediaeval period or even earlier (Brown, 1997). Structures in British rivers, in the form of weirs to control water depth or availability for fishing, have been recorded from Anglo-Saxon times and an excavated Saxon fish weir is recorded in the

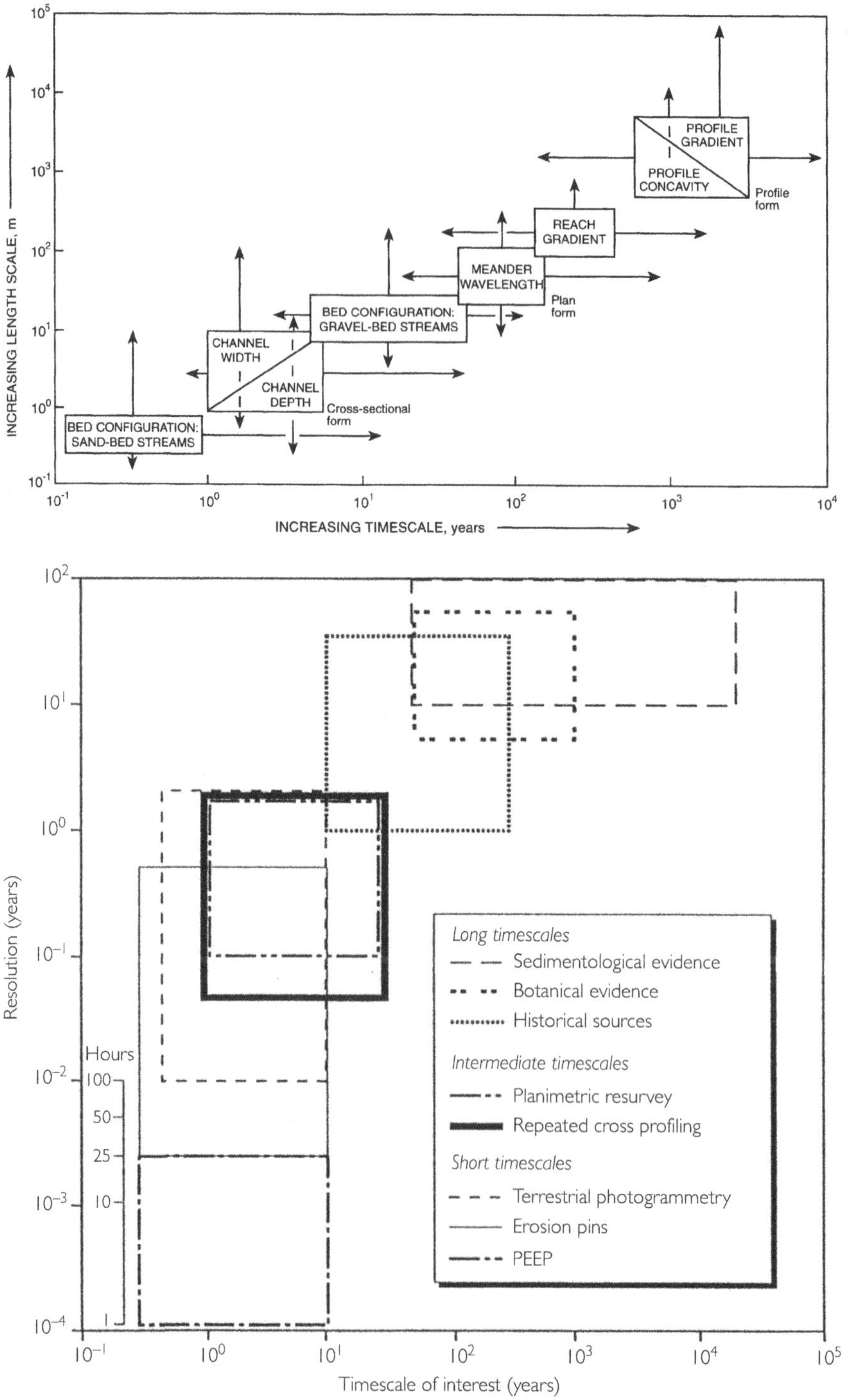

Figure 1.3 *Timescales and river channels. Above relates length to timescale developed from a diagram by Knighton (1998). Below relates timescale of interest to resolution showing how different types of evidence relate to particular timescales (after Lawler, 1993). PEEP is the Photo-Electronic Erosion Pin used to measure bank erosion.*

Table 1.2 Six chronological phases of river use (see Section 1.2) and the management methods used for each (see Section 1.3)

Chronological phase	Characteristic developments (see Section 1.2)	Management methods employed (see Section 1.3)
1. Hydraulic civilizations	River flow regulation Irrigation Land reclamation	Dam construction River diversions Ditch building Land drainage
2. Pre-industrial revolution	Flow regulation Drainage schemes Fish weirs Water mills Navigation Timber transport	Land drainage In-channel structures River diversions Canal construction Dredging Local channelization
3. Industrial revolution	Industrial mills Cooling water Power generation Irrigation Water supply	Dam construction Canal building River diversions Channelization
4. Late nineteenth to mid twentieth century	River flow regulation Conjunctive and multiple use river projects Flood defence	Large dam construction Channelization River diversions Structural revetment River basin planning
5. Second part of twentieth century	River flow regulation Integrated use river projects Flood control Conservation management Re-management of rivers	Large dam construction River basin planning Channelization Structural and bioengineered revetments River diversions Mitigation, enhancement and restoration techniques
6. Late twentieth and early twenty-first centuries	Conservation management Re-management of rivers Sustainable use river projects	Integrated river basin planning Re-regulation of flow Mitigation, enhancement and restoration techniques Hybrid and bioengineered revetments

Trent Valley near Colwick (Brown, 1997: 259). Drainage schemes, implemented to extend areas of agricultural land in relatively wet, lowland drainage basins such as in England, began in Roman times but were greatly accelerated after the pioneering work by the Dutch in the seventeenth century (Darby, 1956), which was especially influential in the English Fenlands (Darby, 1983). Water mills were used from early times and could be employed for irrigation and for grinding cereals and pulses. In AD 1086, the Domesday Survey recorded 5624 grist mills throughout England and in the twelfth and thirteenth centuries water power was applied to fulling stocks by means of undershot wheels (Beckinsale, 1969a) and, as the map of the Domesday mills shows, there was hardly a reach of any lowland river in England not occupied by a mill (Brown, 1997: 263; after Hodgen, 1939). Gradually the undershot wheel gave way to the overshot wheel that consisted of a series of cups filled from above and forced downward by the weight of a relatively small amount of water. Associated with mills were river diversions for mill leats or offtakes designed to provide sufficient hydraulic head to drive the mill wheel before returning the water to the main river channel. Water mills were also used by the paper-making industry: the first English site was at Sele Mill on the Beane, a tributary of the Lea below Hertford. Between 1601 and 1650, 43 paper mills were mapped for England, with 116 in the 1690s (Shorter, 1971). Use of rivers for navigation was reflected in treaties dating back to AD 1177 for the River Po, and to AD 1255 for the Rhine (Beckinsale, 1969a). Along many rivers shoaling and siltation presented navigational problems and in 1575 Acts of Parliament enabled the 'cleansing' of parts of the River Severn by the Commissioners of Sewers (Brown, 1997: 265). A related use of rivers was for the downstream transport of timber from forested areas. The period also saw the beginnings of technologies usually associated with the industrial era: a channelization scheme for the River Adda (Italy) was designed by Bertola in 1400, dredging using endless chain technology was developed by 1561, and a masonry dam 41 m high was constructed at Alicante in Spain in 1594.

A *third* phase of use began with the industrial revolution, and water mills along rivers in the USA and in Europe at the end of the eighteenth century were employed to drive elaborate machinery (Beckinsale, 1969b). The imperative in this phase was related to harnessing new technology and many industrial works were located adjacent to rivers for the water power available and also because they required river water for other purposes such as washing and bleaching. As industry became located along rivers there was a phase of increased transport of coal and other industrial materials along them where feasible, leading also to the construction of canal networks as dedicated watercourses for industrial transportation prior to the onset of rail transport. For instance, by 1825, the Erie Canal made possible the transport of goods from the Atlantic coast of New York state to Lake Erie, using also the Hudson and Mohawk Rivers. In addition to the benefits for textiles and metal (iron) processing and manufacturing, there was the innovation of hydroelectricity generation. The first generation for industrial purposes was probably in 1869 at Lancey in the Dauphine Alps of France. By 1882 the first commercial thermal generating stations were built in London and New York and such stations often utilized river water to meet the demands for cooling. Developments also occurred following the rapid European colonization of the New World. For instance, in 1867, Jack Swilling, a famous frontiersman, organized a company and built the first canal, called the Swilling ditch, that subsequently became the Salt River Valley Canal in Arizona and laid the foundation for the Salt River Project and for the irrigation of 300 000 acres of land. In Australia, the pre-European use of water by Aboriginals had left little tangible impression on the Australian landscape, although three phases of European settlement caused progressively major landscape impacts (Smith, D.I., 1998). In the first phase, up to 1900, rural and urban water developments were instigated, often unplanned, before a subsequent stage of engineering-led development occurred, including the construction of dams. For example, Meckering Weir in Perth's eastern suburbs was built in the 1890s as storage for the supply of water to the gold field towns of the dry interior of Western Australia by diverting to the east a westward flowing river (Smith, D.I., 1998). A consequence of the industrial revolution was that the technology became available for the establishment of water supply systems and these involved river water in a variety of ways by impoundment and direct extraction.

The late nineteenth and the first half of the twentieth centuries brought, *fourthly*, a period when the scale of river use could expand, based upon the precedents of the earlier nineteenth century but harnessing the new technology available. The phase is perhaps best characterized by the imperative 'technology can fix it' (Leopold, 1977). Whereas the water supply of London was provided by a number of water companies in the nineteenth century, eight of these companies and some local water authorities were taken over and their works extended by the Metropolitan Water Board from 1902 (Vallentine, 1967) as a forerunner of modern river management bureaucracies. Nationally, catchment-based management grew following the River Boards Act of 1948 (Pitkethly, 1990). Other cities also developed more integrated water supply systems, involving rivers directly or indirectly according to their environment. Waste water disposal had been developed in the nineteenth century and sewerage systems were initially often combined with stormwater drainage systems, although in the twentieth century the two became increasingly separate. Once separated, the waste water could be returned to rivers from waste water treatment plants whereas stormwater was drained directly into river channels – a potentially growing problem as urban areas increased in size. Increased flooding of urban areas necessitated progress towards more effective flood defence. Embankments or levees are one of the oldest forms of protection, are very extensive along rivers such as the Nile and the Huang He, and became key components in flood control systems to permit industrial development along the lower courses of the Mississippi and Missouri rivers in the USA (Brookes, 1988). As flood control measures became even more necessary because of the increasing flood hazard (White, 1958, 1974) one choice that had to be made was that between big dams and little dams: in some areas using a single flood-control dam whereas in others employing small dams along many headwater river channels (Leopold and Maddock, 1953). The era of substantial dam building commenced in this phase as hydraulic principles were developed and applied to the use of rivers (Vallentine, 1967). Whereas the overall impact of dam construction was relatively slight until 1900, advances in building technology made possible the construction of large dams, culminating most famously in the completion, in 1935, of the Hoover dam on the Colorado River in the USA, as part of an acceleration of dam construction up until 1945 (Figure 1.4).

The second major attribute of this phase is the early inception of multiple purpose and conjunctive use projects as it was realized that larger-scale schemes could be designed to bring multiple benefits at little increased cost. The prime example is the Tennessee Valley Authority (TVA) scheme of river basin management where, in order to achieve the two original primary purposes of flood control and navigation development, the TVA (see Chapter 6, pp. 149–50) developed the Tennessee River and its tributaries into one of the most controlled river systems in the world. Led by the development of hydroelectric power generation facilities at dams, the suite of projects included navigational improvements and flood controls in the Tennessee River, soil erosion control, reforestation, improvement in agricultural land use, and increased and diversified industrial development in the watershed.

By the second half of the twentieth century, a *fifth* phase may be discerned. Flood control projects became more extensive and more pragmatically related to local environment conditions, and agricultural intensification, prompted by the need for self-sufficiency in Europe during the Second World War, resulted in a large number of river corridors being cleared of native vegetation to make way for crop agriculture. Support for these actions often required extensive channelization and, by the late twentieth century, it was estimated that the extent of channelized rivers in England and Wales (Figure 1.5) was equivalent to a drainage density of 0.06 km km^{-2} (Brookes *et al.*, 1983) which compares with a density of 0.003 km km^{-2} in the USA (Leopold, 1977), a figure than is even more dramatic given the far lower population density in most areas of the USA.

The rate of dam building continued to increase (Figure 1.4). The decade of the 1960s saw the greatest number of dams built of any decade in the twentieth century and underlined the pre-eminence of North America in dam-building activity (Beaumont, 1978). Approximately 11% of freshwater runoff in the USA and Canada is now withdrawn for human use, and the Mississippi, St Lawrence, Columbia, Nelson and Colorado basins are described as strongly fragmented because of the regulation of river systems (Karr *et al.*, 2000). Worldwide, there were 427 large dams higher than 15 m in 1900, but this had increased to 5268 in 1950, and to 39 000 in 1986 (International Commission on Large Dams (ICOLD), 1988) so that today few of the world's large rivers remain unregulated.

This phase also saw the further development of integrated river basin management (see Chapter 6) including a definition of the approach as land and water management came together under a unified administration (White, 1957). Several trends were evident including the fact that legislation increased to provide the framework within which the range of river uses could be implemented. For instance, in England, a growing demand for water resources made it increasingly unlikely that localized approaches would succeed while, at the same time, a growing network of gauging stations provided hydrological data from which to plan water resources distribution systems over large areas. Following the Water Resources Act of 1963, 29 regional River Authorities were charged with granting licenses for water abstraction, stipulating 'minimum acceptable flows' to dilute river pollution and creating a National Plan for water resources (Pitkethly, 1990). The approach also began to be exported to less economically developed nations and, in 1970, an expert panel at the United Nations outlined a four-stage framework for river basin planning (United Nations, 1970), later criticized as being unnecessarily technocentric (Saha and Barrow, 1981).

Such criticisms, and concerns over the impact of channelization schemes (e.g., Heuvelmans, 1974) as part of growing environmental awareness (Chapter 7, pp. 180–1)

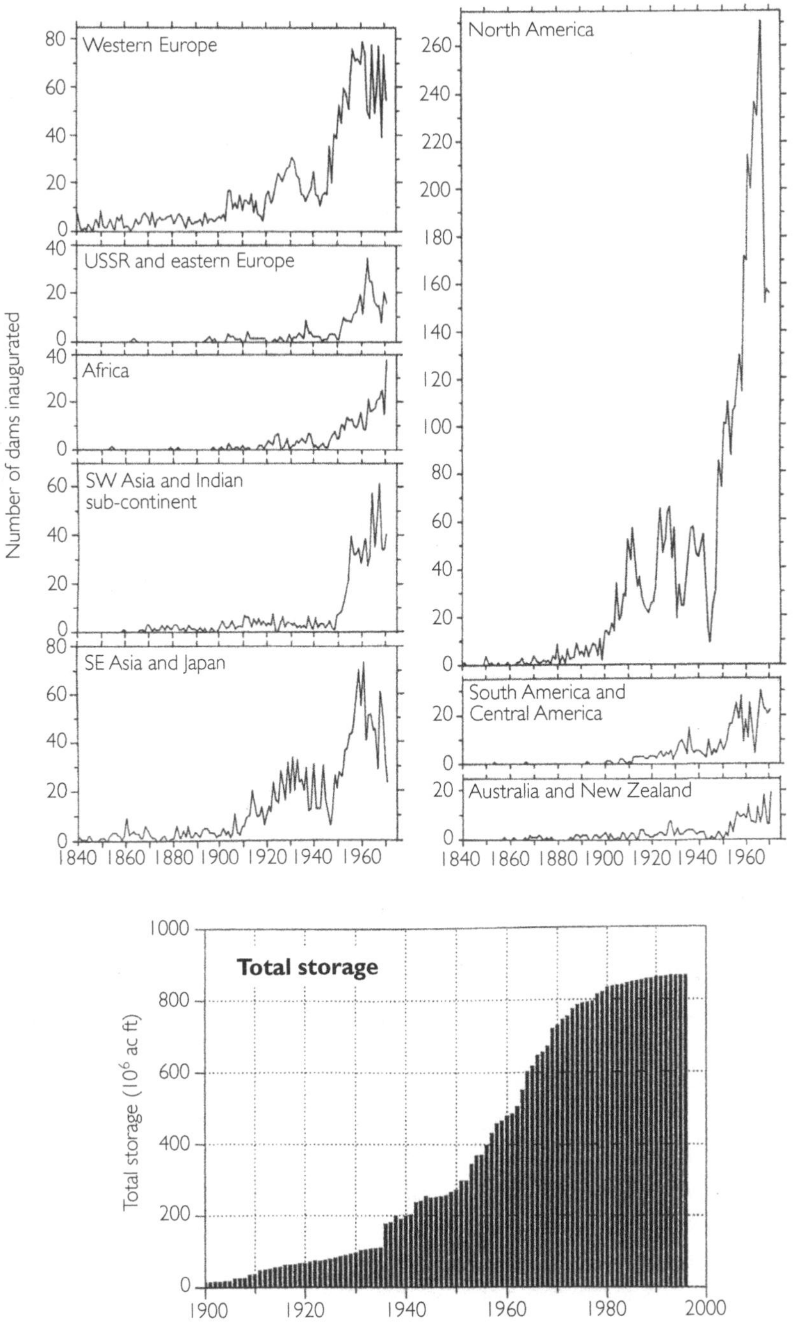

Figure 1.4 *Dam building activity. Above shows dam construction by major regions (after Beaumont, 1978); below shows increasing cumulative storage capacity behind dams in the US (Graf, 2001)*

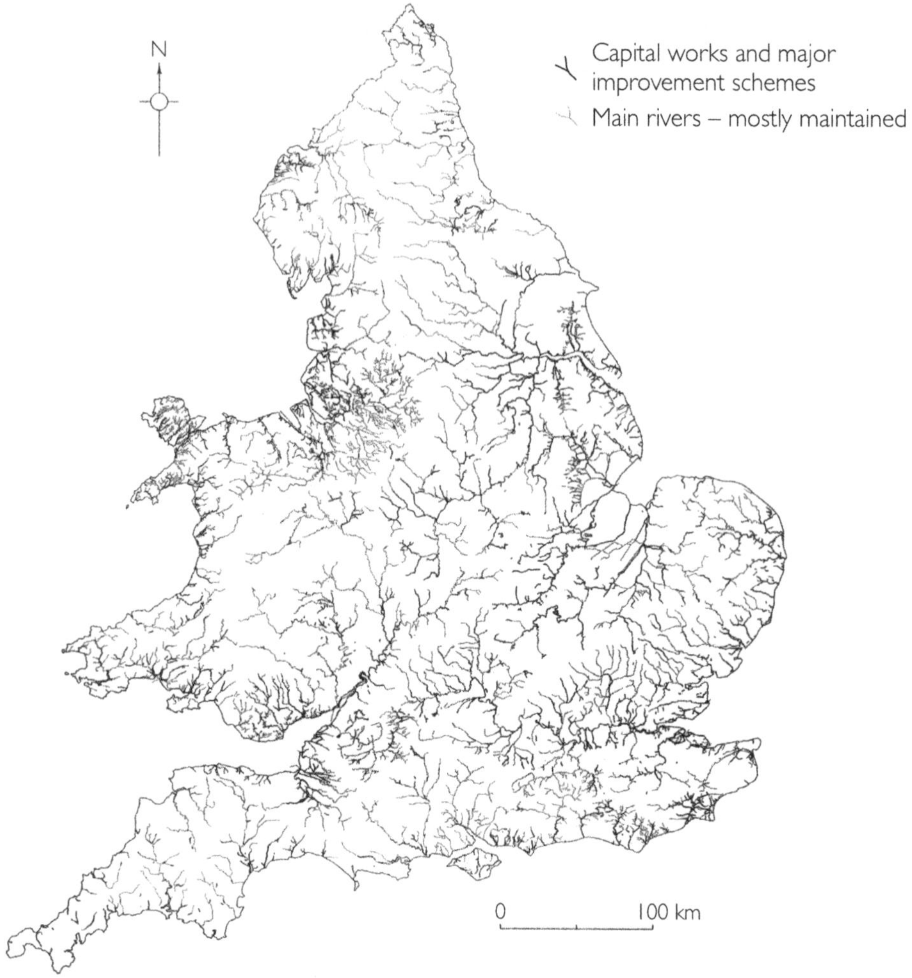

Figure 1.5 *Rivers channelized in England and Wales 1930–80 (after Brookes* et al., *1983)*

also heralded the beginning of a conservation movement in river use in the latter part of the twentieth century. Legislation enacted from the 1970s in different countries began to require environmental awareness prior to river and river channel management projects and the conservation of environmental values (see Chapter 7). Disenchantment with channelization began to affect the way in which channels were managed so that alternatives were sought, resulting in the development of techniques for mitigating the impact of management projects, locally enhancing the channel and finally culminating in river channel restoration (see Table 9.2). In tandem, the major phase of dam building finished at the end of the 1980s as both the environmental and economic rationale for dam building became more difficult to justify, bringing to a close the 'pioneering' period of water resource development (Smith, D. I., 1998).

The very late twentieth century saw the beginning of a *sixth* phase of river use in which the realization of the potential of alternative strategies for managing the legacy of past uses of the river, together with dealing with contemporary hazards and allowing for impacts of global change, meant that the imperative became one of sustainability. While some dam building continues, for example on the Qingjiang River, a tributary of the Yangtse in China where a large dam is planned for completion in 2009 at the same time as the Three Gorges dam, the great burst of dam building is now over and has been succeeded by retirement of dams or at least by changes in their operating rules (re-regulation) (Graf, 1999). The legacy of the dam building phase is the radical alteration of the physical, biological and chemical characteristics of the rivers: 75 000 dams in the USA now fragment the natural flow characteristics and ecology of once-integrated river systems (Graf, 2001) and there are safety concerns over the 32% of all dams in the USA that pose 'significant' or 'high' downstream hazards (Graf, 2001). Likewise, schemes both for water resources development and flood control and hazard avoidance now require integral conservation measures and are often constrained or challenged by previous management operations. Therefore, the development of sustainable approaches (Chapter 6, p. 169) and of restoration-inspired approaches (Chapter 9, p. 237) provides the core for the realization and requirement parts of this book.

1.3 River channel management

Approaches to the management of river channels can be associated with the major uses of rivers (Table 1.2). Whereas early human activities were often aligned to either accommodate or avoid the near-river floodplain and only to modify the channel at flow diversions for irrigation systems, later activities began to involve local channel modifications to facilitate river crossings and bridging points, for example. Channel modifications subsequently began to involve dredging, de-shoaling and snagging removal of woody debris over extensive river reaches, often for the benefit of navigation. Rivers began to be re-aligned, often locally to assist the siting of parallel-running roads and railways and then more extensively for larger-scale river management projects often associated with flood control. By the nineteenth century, channelization of the braided Alsatian section of the River Rhine, initiated in 1817 by J.G. Tulla, introduced the creation of truly 'engineered' river corridors, and the statement by Tulla 'As a rule, no stream or river needs more than one bed' became accepted as guiding policy for engineering design (see Petts, 1989). During the twentieth century, extensive lengths of river channel were channelized as settlements and industries spread increasingly across the hitherto avoided floodplains, and agriculture intensified. A positive feedback cycle of management developed whereby early floodplain settlements demanded flood protection which, when engineered, made the floodplain a more attractive place for building so that the floodplain population increased further, requiring further expenditure on flood protection and so on. This has resulted in the creation of numerous settlements that are inherently vulnerable to flood-related hazards and highly dependent on public funding for their protection.

Unfortunately, the management approach characterized by the imperative 'technology can fix it' (Leopold, 1977) was not wholly benign; some project 'solutions' failed outright as they did not accommodate the natural tendency for the river channel to evolve over time but, more frequently, adverse indirect reactions occurred, involving unforeseen river channel changes prompted by the 'solution'. These reactions frequently involved a progressive impact upstream or downstream of the project which required containment by the authorization of additional management projects. In this way the management approach became self-perpetuating. The essential issue was that engineered river channel form was at variance with the natural tendency of the river system so that the 'solutions' were symptomatic and not sustainable. As a consequence, and also because of increased environmental awareness, there was a movement towards less radical alterations of the river channels, to the use of 'soft' and bioengineered protection wherever possible rather than rock and concrete structures, and towards a philosophy of 'working with the river rather than against it' (Winkley, 1972). This philosophy encouraged approaches that were designed for the particular river and the location that were holistic in character and that had sustainability as characteristics of success. In management terms, this movement was ushered in, first, by techniques to mitigate or enhance previously managed river channels, and then by projects to rehabilitate and restore river channels (see Brookes and Shields, 1996a). As a consequence, some rivers that had previously been straightened and re-sectioned or otherwise modified were returned to nearer their natural condition. Such restoration procedures are a major theme in this book because they require an understanding of what is natural, how much change has to be accommodated and what is sustainable.

In this context, river channel management can be defined as management that is primarily concerned with the condition of the river's bed and banks as they influence the exploitation and delivery of water resources, reducing the distribution and frequency of channel-related hazards, and the achievement of river conservation objectives. Together, these foci require an understanding of the functional dynamics and connectivity of the channel hydrosystem to produce a solution that provides a sustained level of performance over the management period without creating a legacy of environmental degradation.

1.4 A multidisciplinary topic

No single academic discipline focuses solely upon investigation of the river channel, and as rivers and river channels are perceived differently between disciplines according to their paradigms and focal interests, there is a spectrum of views of the river channel. This has the consequence that expert guidance for river channel management derives from many disciplines and, desirably, should be informed by a multidisciplinary approach.

Despite its logical derivation, 'potamology' as a term for the study of rivers has not been widely adopted (Leopold, 1974) possibly because of the pre-existing fragmentation of studies involved with rivers and with river channels. Understanding of rivers is founded in hydrology and a critical step in establishing that science (Dooge, 1974) was appreciation of the hydrological cycle (Nace, 1974). Elements of the cycle were identified by John Ray in the third edition of *Miscellaneous Discourses* published in the early eighteenth century (1713), but it took time for the concept of the hydrological cycle to become widely accepted and, for at least 150 years from 1700 to 1850, it was a handmaiden of natural theology as much as of natural philosophy (Tuan, 1968). Although Palissy (1580) may have had the first accurate insight into the process of runoff (Ward, 1982) and John Dalton, in a paper in 1799, brought rainfall and runoff together quantitatively (Newson, 1997), it was probably Pierre Perrault (1674) who, through his calculations showing that runoff from the

Seine Basin was only one-sixth of the precipitation received, appreciated the functional significance of the drainage basin (Gregory, 1976a). Such fundamental foundations gave rise to the science of hydrology, developed in the early twentieth century and, together with hydraulics, providing the basis for the study of river channels. Initial river channel research in engineering was based upon regime theory (e.g. Blench, 1952, 1957) and the regime equations used by hydraulic and civil engineers to design irrigation channels or canals. Although the regime approach to river channels (e.g. Ackers, 1980) was profitable until more empirical data and studies had been undertaken, it suffered from the assumption that channels were 'in regime' whereas such conditions were rarely fulfilled by natural rivers. From hydraulic and hydrologic foundations, civil and hydraulic engineers (e.g. Shen, 1973) laid the scientific foundations for river channel management.

Complementing the engineer's approach and accelerating after the 1960s was a focus upon rivers by environmental scientists. The need to investigate contemporary environmental processes stimulated sedimentologists (e.g. Allen, 1970) and physical geographers, especially geomorphologists (e.g. Leopold *et al.*, 1964) to study river channel processes including fundamental studies of the hydraulic geometry of river channels (Leopold and Maddock, 1953), of river channel patterns (Leopold and Wolman, 1957) and of the generation of floodplains (Wolman and Leopold, 1957). An approach from a drainage basin perspective (e.g. Gregory and Walling, 1973) was hydrologically informed, initially by Horton (1945) and subsequently gave rise to morphometric approaches such as those of the Columbia School (Strahler, 1992), which led to significant developments in process geomorphology (see Chapter 3, p. 47) and understanding of channel changes and river metamorphosis (Schumm, 1969; Chapter 4, p. 115). Initially, seven separate strands of river channel investigation were identified (Gregory, 1976a) subsequently leading to more applied approaches, including hydrogeomorphology (Gregory, 1979b), and benefiting from knowledge of river channel adjustments to enhance understanding of the fluvial system (Schumm, 1977b). Specifically related to river channel mechanics, especially in the context of gravel bed rivers (Hey *et al.*, 1982), investigations of river channel dynamics were recognized as the means by which contemporary processes could provide a basis for managing channels (Thorne *et al.*, 1997a). In addition, biology and, more specifically, ecology made significant contributions (Chapter 3, p. 48) to the understanding of ecological aspects of river channel behaviour: ecological studies included river channel classifications based upon flora and fauna (e.g. Haslam and Wolseley, 1981), investigations of debris in channels and the associated habitats (e.g. Harmon *et al.*, 1986), and the application of ecosystem perspectives to riparian zones (e.g. Gregory *et al.*, 1991). Although investigations by environmental scientists initially reflected the viewpoints of their parent disciplines, they became progressively less distinct over time and fused to become more inter- and multidisciplinary as reflected in the collaboration of physical and biological scientists (e.g. Boon *et al.*, 1992; Calow and Petts, 1992, 1994) and in the development of integrative approaches such as the fluvial hydrosystem (Petts and Amoros, 1996b).

Disciplines concerned with aspects of environmental management and planning have also shown increasing interest in river channel management. Thus, landscape architecture (e.g. McHarg, 1969) has provided a general planning and philosophical context for management, as well as specific inputs such as plans that integrate recreational opportunities in river corridors with flood control requirements. In addition there have been contributions from social science disciplines, including studies related to the costs and benefits of river uses from economics, to the perception of river landscapes from psychology and human geography, and to the cultural basis of river understanding from sociology. Some contributions have deliberately encompassed the interface between the physical/environmental and the social sciences (e.g. Chorley, 1969; Cosgrove and Petts, 1990). Practitioners have also made very significant contributions founded in long practical experience and knowledge (e.g. Winkley, 1972; Purseglove, 1988; Gardiner, 1991a).

A range of disciplines is therefore relevant to river channel management and chapters in this book refer to research, approaches and policy deriving from across the disciplinary range. As explained in the preface (p. xi), this book stems from a geomorphological viewpoint, necessarily so as the discipline of geomorphology has the greatest direct concern with the river channel and the dynamics of river channel systems, but endeavours to reflect the increasingly multidisciplinary nature of contemporary river channel management in the context of the catchment hydrosystem (Table 1.1).

1.5 Book structure

The structure of this book arises partly as a sequel to other texts concerned with aspects of river channel management. Volumes have been produced providing relevant background to river channel management based on ecology (e.g. Moss, 1988; Jeffries and Mills, 1990; Allan, 1995) or geomorphology (Thorne *et al.*, 1997a) or both disciplines (e.g. Calow and Petts, 1992), but no one book has addressed the topic explicitly. Books concerned with river basin or watershed management often include sections dealing implicitly with river channel management (e.g. Dunne and Leopold, 1978; Naiman, 1992; Newson, 1997). Some volumes have concentrated upon management for particular types of river channel, as in the case of channelized rivers (e.g. Brookes, 1988), mountain rivers (Wohl, 2000), the restoration of urban rivers (Riley, 1998) or more generally upon water resource projects (Canter, 1985). An extensive range of edited volumes has included those focusing upon particular aspects of river channel management including restoration (Gore, 1985b; Brookes and Shields, 1996a) and rehabilitation (De Waal *et al.*, 1998), flow regulation (Petts and Wood, 1988), the ecological dimension (Harper and Ferguson, 1995), applied fluvial geomorphology (Thorne *et al.*, 1997a), floodplain processes (Anderson *et al.*, 1996), or river conservation (Boon *et al.*, 1992, 2000). Other books have used examples to provide a manual (Gardiner, 1991a), while others have initiated an integrated approach either by juxtaposing experience from different countries (Petts and Amoros, 1996b) or by embracing the interface between the environmental and social sciences (Cosgrove and Petts, 1990). As river management has attracted more attention, new volumes have been produced for particular parts of the world including Australasia (Brizga and Finlayson, 2000), approaching management from the viewpoint of experience of channel adjustments and change, and the northern Pacific Coast of the USA, centred on river ecology (Naiman and Bilby, 1998). However, as no one book focuses explicitly on river channel management, this volume is written to fill the gap, attempting a perspective with an awareness of river channel adjustments and their consequences.

Our objective is to reflect the available material relevant to river channel management, not as a manual or do-it-yourself kit, but rather providing the background to the way in which approaches have been, are and will be developed. This approach requires reference to the very significant advances made in a range of disciplines but particularly from fluvial geomorphology, ecology, engineering, hydrology, physical geography and environmental sciences, and extending to social sciences and philosophy. As experience in river channel management has shown, there is no single solution guaranteed to work in any specific area, instead there is a range from which a selection has to be made, influenced by the setting of the particular river in its catchment, its evolutionary history, the cultural context of the area, the community priorities for management, and the public response to prior projects. For any adopted solution to be sustainable, future adjustments, including those arising from global change, should also be considered.

After this introduction the book is constructed in four sections. The first, *Retrospect*, in two chapters explains why river management is needed, what needs to be managed and how management has been undertaken over the historic timescale. It does this by auditing the consequences of twentieth-century, engineering-dominated, approaches (Chapter 2) relative to the contemporary understanding of the dynamics of river channel systems (3). A second section, *Realization*, shows how approaches to management should be conditioned by knowledge of river channel adjustments (Chapter 4), the sensitivity of river channels to change (5), river basin planning (6) and the needs for a sustainable foundation for river channel management (7). In the light of this realization, a third section, *Requirements*, outlines the present state of the art in river channel management. This includes environmental assessment and post-project appraisal (Chapter 8), conservation-based management approaches (9), and techniques for working with the river (10). A fourth and concluding section, *Revision* (Chapter 11), attempts to outline the rudiments of a 'management with nature' approach to river channel management suitable for the early twenty-first century and involving a necessarily pluralist focus to cater for the diversity of riverine environments that have to be managed.

As the pertinent background literature for river channel management is now so vast, we assume that the reader is familiar with the fundamentals of hydrology, fluvial geomorphology and ecology as an essential foundation for river channel management. We therefore introduce background material expressly as relevant to the management perspective.

1.6 Key assumptions

The culmination of centuries of river use and river management means that river channels are now seldom natural but are constrained in many different ways; for the future we need a way of treating our rivers sympathetically that acknowledges their dynamics and their history, and the cultural context for management. Underpinning the approach in this book and elaborated in the conclusions in the Retrospect and Realization sections are eight basic assumptions that form the core of developing this approach.

1 *River channel management requires an understanding of* ***river channel processes****, including the effects of flow and sediment transport, and of the interactions between river channel morphology and river channel processes.* Once the emphasis in environmen-

tal sciences was directed towards a greater emphasis upon processes, the form-process paradigm was clearly embraced (e.g. Leopold *et al.*, 1964; see Chapter 3).

2 *Interaction between river channel form and river channel process involves the ecology and the environs of the river channel.* In research and practical applications since the 1980s it has been increasingly appreciated that it is counter-productive to separate river channel ecology from river channel form and process or to emphasize one particular aspect. The ***fluvial hydrosystems approach*** (Petts and Amoros, 1996b) has been advocated as a way of avoiding this selective emphasis (see Chapter 3).

3 *River channel management should be envisioned in the* ***drainage basin context***. Several spatial scales apply to the drainage basin (Figure 1.1); the relationship between each of these and the river channel must be considered. Thus, the interrelation of the river channel and floodplain, and of the river channel and the river corridor, must be analysed and taken into account both physically (see Chapters 3, 4, 5) and from a management perspective. Schemes of integrated river basin management employed in many countries provide an overall spatial framework and context for management, although the potential for integration has been difficult to match in practice (e.g. Adams, 1985; Mitchell, 1990; Downs *et al.*, 1991; see Chapter 6).

4 *Every river channel should be managed within its* ***dynamic temporal context.*** Knowledge of the sensitivity of the river channel (Newson, 1995) provides information for understanding the short-term significance of channel processes but also is important over longer periods of time (see Chapters 4 and 5). Previously, there was insufficient appreciation of the way in which river channels react to management and how long the reactions take. The temporal setting of any river channel (Figure 1.3) should inform management because, first, understanding of past channel changes can indicate what was 'natural' and therefore guide an appropriate target for management under contemporary catchment conditions; and, second that alternative forms of adjustment occur in time, even in response to the same cause, so that there are possibly ten ways to be wrong in interpreting landforms including river channels (Schumm, 1991; see Chapter 3; Table 4.12).

5 *To be as* ***sustainable*** *as possible, management of river channels should minimize adverse feedback effects.* This arises as a corollary of past temporal change and requires that channel management is conceived against an understanding of the potential for future river channel adjustments, including those due to the management project. It also requires that monitoring and evaluation follow the implementation of a management scheme in order to document and evaluate the success of the scheme (see Chapters 2, 4, 5, 8).

6 *A* ***holistic approach*** *to environment is necessary to provide the context for river channel management.* Management projects should consider all relevant environmental factors in design in a process that has become known as 'design with nature' (e.g. McHarg, 1969, 1992). Such an approach has been embraced by landscape architecture and has been advocated for environmental designs based in the environmental sciences (Gregory, 1985, 1992a, 2000). For river channel management, a holistic approach must encompass knowledge of river behaviour and river channel adjustments (Newson, 1995; see Chapter 7).

7 *The* ***legal and political context*** *of river channels influences available river channel management options.* Management must be undertaken within the prevailing legislative framework, sometimes causing particular difficulties where river channels form regional or national boundaries. Since the late twentieth century, many countries have adopted legislation based on environmental assessment that requires river channel management to respond to conservation objectives, in addition to water resources and hazard avoidance objectives (see Chapters 6, 7).

8 *River channel management has to reflect the needs and requirements of the* ***culture*** *of the region or nation within which the river occurs.* Any approach to river channel management must be seen in the context of when it is written, from which areas and from which cultural perspectives it derives. This means that the solution to a particular management problem may differ substantially from one country to another. For instance, by the end of the twentieth century it had become appreciated that, although structural solutions were necessary in some management scenarios, a range of alternative techniques including nonstructural options were viable in many instances, requiring a choice to be made between several potential management approaches. The choices were usually made to maximize the conservation advantage (or minimize further environmental degradation), and reflect a general public awareness of environmental issues (Chapters 6, 7, 8, 9, 10).

II

Retrospect

These two chapters explain, first, why river channel management is needed, what needs to be managed, and how management has been undertaken over the historic timescale; and second, how research on river channel systems has contributed to an understanding of river channel system structures, their components and connectivity, and of the dynamics of channel behaviour.

2
River Channels Engineered

Synopsis

Consideration of how and why river channels were engineered as part of river channel management proceeds from the purposes of management (2.1), to the development of an approach based on hydraulic design (2.2), followed by the range of engineering methods employed (2.3). The culmination of river channel engineering (2.4) by the late twentieth century has led to a re-appraisal of methods and the evolution of river channel management (2.5) which, while recognizing that hard engineering is still required in some instances, is progressing towards approaches based on principles of sustainable management.

2.1 The purposes of river channel management

Since the earliest civilizations, rivers have been a vital resource in the landscape, explaining why they have assumed a very significant role in relation to economic and social activity. River channel uses (Table 1.2) indicate the plethora of ways in which river channel management was required in order to effect those uses.

Traditional river uses (Table 1.2) required channel modifications and discharge regulation, and can be viewed according to whether actions were necessary, supplementary or optional in relation to the desired river use. Modification of river channels for *flood control* involved alterations of the channel by widening, dredging and straightening the channel, relocating small streams, removing vegetation and/or the construction of levees or flood embankments. Subsequent developments often included more complex channel modifications, such as the development of diversions as an integral part of particular flood protection schemes, and the use of upstream storage reservoirs to control downstream flood impacts. *Irrigation* is even longer established in some areas of the world, with technological developments allowing flooding methods that utilize temporary ditches to guide flood flows to give way to basin, border and furrow methods with more sophisticated ditch and irrigation canals, justifying the development of a specialized branch of hydraulic engineering, namely fluvial hydraulics. *Land reclamation*, often associated with flood control schemes and land drainage in wetlands, low-lying areas and coastal zones, frequently necessitated changes to river channels occasioning development of an associated field of applied science, for example in the Netherlands, where 96% of all streams are degraded to some degree with only 4% remaining as semi-natural (Jasperse, 1998). *Local river use* embraces a diversity of small, site-specific uses of the river channel including water mills or fish weirs, which may require localized channel modification, some subsequently absorbed into other river channel developments. *Navigation* is a long-existing use of the river channel, more extensive when the draught of river craft was small, and becoming mainly restricted to larger river channels as larger craft were used. Whereas early use of river channels for navigation required clearing of the channel of gravel shoals, vegetation 'snags' and debris, and the maintenance of a sufficiently deep channel, later requirements involved weirs and lock systems to provide an assured depth of water for larger vessels, variable sluices to maintain standard high water levels, systematic dredging, river training works and sometimes straightening of the river channel. *River crossings* have been necessary since the time of the earliest land communications, developing from shallow fords or reaches, where ferries could be established, to bridge crossings that required attendant modifications of river channels and so also the development of a branch of hydraulic engineering concerned with bridge structures and design. Crossings may often have been associated with *industrial mills*, and the harnessing of water power was a locating influence in many countries for flour milling, paper making or fulling.

The modern counterparts of such long-standing uses continue to require channel modification or discharge regulation. However, a number of other river uses (Table 1.2) depended upon technological developments after the industrial revolution (Table 2.1), each affecting the river channel in some way. *Cooling water* was extracted from rivers when power stations were established; in 1984 in England and Wales 36% of the water abstracted was for cooling systems at thermal power stations (Department of the Environment, 1985; Petts, 1990). *Irrigation schemes* became more extensive with the advent of mechanical water pumps, and technological developments also enabled substantial advances in *water supply systems*. Prior to the industrial revolution, water supply by direct abstraction from rivers and wells was essentially local and small scale. Subsequent rapid growth of urban populations and a greater dependence upon an assured supply of water in large quantities meant that river sources for water supply became increasingly risky. In addition, the over-exploitation of ground water and the contamination of potable supplies led to the increased use of dams in unpolluted headwaters, with surface and ground water reservoirs regulating river flows, and the river used as a natural aqueduct to convey water for multiple uses. As urban populations spread over once-avoided floodplains and encroached to the river's edge, *river stabilization and training*, including bank protection, became increasingly important in order to prevent, or at least control, erosion of the bed and banks, shoaling or channel migration. *Hydroelectric power* generation developed with the invention of the turbine; between 1890 and 1895 a tunnel 4 km long was constructed from the foot of Niagara Falls with vertical shafts to the turbines, allowing the generation of electricity on a substantial scale (Cosgrove, 1990). Although used for hundreds of years, *dams and reservoirs* became increasingly influential when engineering techniques enabled large dams to be created (Figure 2.1) and large volumes of river water to be impounded (Figure 1.4). Likewise, with the advent of new machinery, *commercial dredging*, previously undertaken on a small scale at particular locations, could become more significant.

Table 2.1 Examples of developments and inventions critically significant for river channel engineering and use

Development	Time and details
Reinforced concrete	Concrete first invented by the Romans, but 'lost' in the Middle Ages, was gradually rediscovered during the eighteenth century. Joseph Monier, a Parisian gardener, made garden pots of concrete reinforced with iron mesh, and received a patent in 1867. The invention of reinforced cement in 1848 led to the invention of reinforced concrete in 1892, consisting of hydraulic concrete into which a metallic framework has been inserted during casting so that blocks are obtained that have a bending and tensile strength superior to ordinary concrete.
Prestressed concrete	Steel wire or cables or ropes placed in empty mould, then stretched and reanchored. After concrete placed and set the anchors are released and as the steel attempts to return to its original length it compresses the concrete.
Pumps	Although the oldest known suction pump for lifting water was in use in Egypt by the third century BC, and the force pump was employed in the Middle Ages in European cities, a great variety of rotary pumps were developed in the nineteenth century.
Dredgers	Date back to 4000 BC but the first steam dredge used in England in 1796, steam-driven bucket and chain dredges used in the early 1800s and diesel engines first installed on dredges in the USA in the early 1920s.
Snagging machines	Early channel clearance in New Zealand involved the complete removal of willows by mechanical methods such as traction engines and bulldozers (Brookes, 1988).
Turbines	In the 1800s, engineers and inventors began developing more efficient enclosed turbines. A French engineer built the first successful enclosed water turbine in 1827, the world's first hydro plant (Appleton, Wisconsin) began producing power in 1882, being succeeded by hydro dams in Italy and Norway. After 1900 the size of dams and power stations increased rapidly and progress in turbine design increased the head at which turbines could operate from 30 m in 1900 to >200 m by 1930s (McCully, 1996).
Gabions	Used extensively in Europe for many years but very limited use in USA before 1960 (Petersen, 1986).
Cutoffs	Resisted on Mississippi until 1932 when cutoff programme initiated (Petersen, 1986).
River training	Training by embankments along the banks in order to confine the flow to a single channel and carry sediment to sea practised in China as early as end of sixteenth century (Przedwojski *et al.*, 1995).
Geotextiles and geomembranes	Permeable synthetic fabric used in conjunction with soil for the function of filtration, drainage, soil reinforcement or erosion protection (Hemphill and Bramley, 1989).

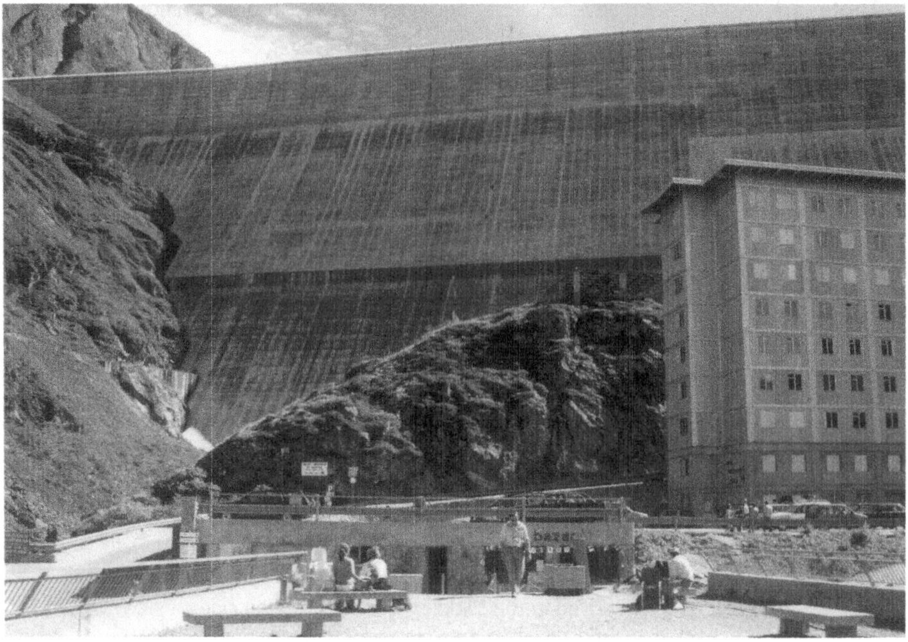

Figure 2.1 *Grande Dixence Dam, Switzerland (photograph G.H. Brown)*

River channel modifications for various river uses have become increasingly inter-related as multipurpose schemes have become more popular. Although many of the uses of rivers and river channels have existed for hundreds of years, developments in science and a series of technology thresholds allowed very significant developments (Table 2.1) in the last two centuries, with the development of specialized subfields such as hydraulic engineering.

2.2 The development of hydraulic design

The purposes of river engineering include municipal and industrial water supply; irrigation water supply; hydroelectric power; flood control; cooling water; recreation; fish and wildlife; and navigation (Petersen, 1986). In summary, three chronological phases of channel management can be distinguished (Petts, 1990): the first concerned with management of perennial water sources for local agriculture and domestic supplies; the second, approximately from 1600 to 1900, embracing river management for navigation and water power and the local regulation of floods for irrigation and drainage; and the third phase, including channelization, in which rivers are completely regulated by large structures often as part of complex basin and interbasin development for power generation, water supply and flood control (see pp. 6–15, section 1.2). These three phases operated sequentially at particular times in different parts of the world according to breakthroughs in technology (Table 2.1) and the establishment of a sound scientific foundation in hydrology and particularly in hydraulics. Basic understanding of the hydrological cycle and formative developments in the field of hydraulics were achieved in the eighteenth century, but it was in the nineteenth century that scientific approaches to river channel management really began to develop (Biswas, 1967), and led to the hydraulic engineering approach to modern river channel management.

Until the nineteenth century, river channel modification was regarded as more an art than a science (e.g., Straub, 1942, quoted in Petersen, 1986), engineers having acquired over 1000 years a large practical knowledge and designed schemes according to the rules derived from their own working experience. In *Natura de fiumi*, Guglielmini (1697) established the principles of river channel regulation and, like Tulla in France, favoured single channels to replace braided rivers. In the nineteenth century the principles and practice of canal and river engineering were established (Stephenson, 1858). Calculations necessary in the emerging discipline of hydraulic engineering (Table 2.2) centred on the discharge entering a particular reach and its interaction with the channel geometry of that reach. Resulting estimates provided information fundamental to modifying the channel bed or banks, the entire cross-section or the planform. The newly found scientific foundation for hydraulics was quickly adopted by engineers as the basis for decision-making on river channel management issues.

The problem of inadequate data and information was dealt with in a number of ways (Table 2.2). Estimation of flow discharge from weather data and upstream drainage basin characteristics depended upon use of the rational formula (Mulvaney, 1850; see Linsley, 1982); subsequently requirements for design discharges, which included not only discharge amount but also some indication of probability and frequency of occurrence, were enhanced by flood frequency analyses (Dalrymple, 1960). For specific reaches or channel cross-sections, an influential approach was based upon the demonstration (Kennedy, 1895) that Punjab irrigation system channels follow the relation:

$$U = 0.84\, D^{0.64}$$

where U is the mean velocity of flow in feet per second and D is the average depth of flow in feet. It was found subsequently that the constants varied from one system to another, and equations developed from canal systems in other areas including India, Egypt, Pakistan and Russia were developed eventually into the regime theory approach (p. 31). For channel design it was appreciated that the necessary conditions to be specified related to:

- flow variables: water discharge (Q), sediment discharge (Q_s), velocity (u), average depth (d) or hydraulic radius (R), water surface slope (S);
- sediment variables: mean diameter (D) standard deviation of size, difference in specific weights of sediment and water;
- fluid variables: mass density, dynamic viscosity, gravitational acceleration;
- channel geometry variables: average width (w) or wetted perimeter (P) and shape factor (n).

Two alternative approaches were used to relate these four groups of variables for the engineering design of river channels: the regime approach and the tractive force approach. Founded upon experience gained from irrigation canal design (Kennedy, 1895), the regime approach defined an alluvial stream reach to be in 'regime' or in equilibrium when the amount of sediment calculated to enter the upstream end of the reach was equal to the amount calculated to exit the downstream end, being equivalent to the

Table 2.2 Some pertinent developments in hydraulics and hydrology that have supported development of river channel management

Development	Implications	Relevant references
Rational formula	Estimation of peak discharge (Q_p) from runoff coefficient *C*, intensity of rainfall in time T_c and area of catchment (*A*) in form $Q_p = CiA$	Concept originated from Mulvaney (1850); see Linsley (1982); Shaw (1994)
Tractive force approach	Design of channels in noncohesive sediments using tractive force theory is dependent upon four principal assumptions namely: 1 The channel side slope at or above the water surface is equal to the angle of internal friction of the sediment. 2 At incipient motion it is assumed that traction is opposed by the component of the submerged weight (*W′*) of the grain acting normal to the bed, multiplied by the tangent of the angle of internal friction. 3 It is assumed that in the middle of the channel motion is initiated only by the shear stress exerted by the flow. 4 On the side slopes, where the angle is greater than zero, there is an additional downslope component of the submerged weight of the grains.	Developed by du Boys (1879); see Brookes (1988)
Regime theory	Developed from design of irrigation canals based on correlations among variables for sand bed canals. Designed channels intended to convey a specific discharge and sediment supply.	Methods developed by Kennedy (1895); Lacey (1930); Blench (1952, 1969); see Chang (1988)
Flood frequency analysis	Analysis of data from hydrological records to establish relationship between discharge and return period or probability of occurrence as a basis for estimating discharges of specific recurrence intervals.	Developed by Dalrymple (1960); see Shaw (1994); Ward (1978)

Sediment transport	Following pioneering studies in the 1930s by Shields and White the importance of shear stress and buoyant particle weight were recognized in determining conditions for incipient particle movement and sediment models were subsequently developed.	Developed by Shields (1936) and White (1940); see Lawson and O'Neill (1975)
Floodplain modelling	Development of floodplain models and integration of floodplain hydraulics with overbank sedimentation, and of hillslope–channel coupling.	See Anderson *et al.*, (1996); Siggers *et al.*, (1999); and Michaelides and Wainwright (2002)

sediment transport capacity of the stream for given characteristics of sediment, flow and fluid (Table 2.2); this approach was formalized 'to develop and apply to any river capable of self-adjustment, the quantitative laws of self-formation of channels formed as a consequence of their flow moving boundary material' (Blench, 1957). Difficulties encountered when utilizing the regime approach were that the equations:

- were developed for irrigation canals whose dynamics of flow and sediment transport differ from alluvial river channels;
- apply to conditions of river channel 'stability' that assume a static channel whereas in nature, stable river channels undergo morphological change (dynamic stability, p. 76);
- are empirical in derivation so that the engineer needed to choose which of the alternative available equations best matched the conditions at the channel reach under analysis.

The second approach, used especially in North America, was based on tractive force theory, and developed from the du Boys (1879) tractive force concept:

$$\tau o = \psi Rs$$

where τo is unit tractive force; ψ is specific weight of water; R is hydraulic radius; and s is bed slope. The design of channels in noncohesive sediments using tractive force theory is dependent upon four principal assumptions (Table 2.2) and also assumes steady, uniform flow. Sinuous channels present particular problems (Hey and Thorne, 1986; Hey, 1997b) that cannot be resolved using the tractive force method. A successful design was perceived to be one where the slope is set so that no sediment accumulates at the design discharge.

These two approaches, founded on an evolving understanding of one-dimensional hydraulics, provided a basis for river channel design that persisted for almost a century: straight channel reaches were designed with a uniform trapezoidal cross-section based on the notion of local channel 'stability', wherein the ideal channel is regarded as a conduit for flow at the maximum permissible velocity that does not cause channel bed or bank erosion. Manuals for river channel design were written for specific design situations, such as a *Manual on river behaviour control and training* (Joglekar, 1956, 1971) written to guide river and canal management in India and set against the background of types of Indian rivers but with other examples. Specifically related to irrigation and hydraulic design were volumes dealing with general principles of channel design (Leliavsky, 1955), with irrigation works (Leliavsky, 1957) and with river and canal hydraulics (Leliavsky, 1965). Such specialized works provided the foundation for operating practice in specific cases. Although knowledge of the hydraulics of rigid boundary open channels had given adequate tools with which to design channels with a fixed boundary and carrying clear water, the design of a channel flowing through alluvial material and carrying varying amounts of sediment in the water remained extremely difficult (Garde and Ranga Raju, 1977: 6). Difficulties relating to natural channels were ascribed to three degrees of freedom, namely slope, width, and depth of flow in the channel under equilibrium conditions. The objectives of hydrologic design had come to include the process of assessing the impact of hydrologic events on water resource systems and choosing values for the key variables of the system so that it will perform adequately (Chow *et al.*, 1988); but it was appreciated that not all problems in river engineering could be solved yet (Jansen *et al.*, 1979: xi). In fluid mechanics, also known as fluvial hydraulics, the further understanding required of the complex nature of water and sediment movement, and of changes in bed configuration, could be obtained and was supplemented by the further use and development of scale and mathematical models.

As regime and tractive force approaches provided the basis for river engineering, five additional developments occurred in hydraulic engineering throughout the twentieth century that influenced river channel design. First, in addition to hydraulic designs for irrigation canals and hydraulic structures, calculations were required to avoid problems in particular local situations, such as scour below dams, scour at river crossings around bridge piers (e.g. Przedwojski *et al.*, 1995: 353–409), and backwater flows into drainage systems. Second, a more dynamic appreciation of the river channel was achieved through greater knowledge of sediment transport processes and equations, with understanding of entrainment thresholds, of bedload and suspended sediment transport and flux, and of the limitations of bedload transport formulae (e.g. Reid *et al.*, 1997). Accumulated empirical data showed how suspended sediment was not always directly related to discharge, so that suspended sediment rating curves over-simplified the real situation, and how bedload transport formulae had often assumed capacity loads, although in many environments the supply of sediment was limited and often considerably less than the potential maxima indicated by the transport equations. Third, one of the primary effects of river engineering methods was to impose static equilibrium on the river channel (see p. 77) where a stable channel was defined (Lane, 1955b) as an unlined earth channel that (a) carries water, (b) has bed and banks not scoured by moving water and (c) has no objectionable deposits of sediment. Lane (1955a) also provided a basis for studying adjusting channels that was subsequently developed (Schumm, 1969, 1977a) as a presage for more understanding of river channel adjustments with, fourthly, degrees of freedom as characteristics of river channels (see p. 91) which can adjust (Hey *et al.*, 1982; Thorne *et al.*, 1987; Billi *et al.*, 1992; Hey, 1997b). Fifth, river channel engineering was undertaken in an increasingly broader context with greater concern for other parts of the basin and for the ecology of the channel and the corridor. The link between hydrology and hydraulics became clearer with the development of hydraulic models that can operate network-wide if needed, and which established the link between flood frequency analysis and flood levels as a method for sizing the channel (Table 2.2). Thus, in Australasia (Brizga and Finlayson, 2000: 5), an inception phase was followed by an engineering phase dominated by engineering works and schemes, in turn succeeded by an environmental phase that embraces concern for ecology and more holistic management.

Therefore these developments show how *river* engineering was emphasized (Brandon, 1987a, b) although the *river channel* was central to the procedures employed.

2.3 Methods of river engineering

Hydraulic principles underpinning channel design enabled the development from the late nineteenth century of hydraulic engineering of river channels, which added scientific rigour and extended the range and spatial extent of methods employed beyond that developed from centuries of local or regional engineering expertise and preferences (Petts, 1989). Such spatial extension had already begun in nineteenth-century Europe in response to the need for more agricultural land, the need to combat malaria by eliminating the mosquito habitat and the problems of increased flooding due to deforestation and land drainage. For these reasons Tulla initiated channelization of the braided Alsatian section of the Rhine in 1817, and possibly the greatest engineering work of the nineteenth century was the regulation of the Tisza in the south and west Carpathians. Beginning in 1845, 12.5×10^6 ha of floodplain marsh were drained and the river course was shortened by 340 km (Szilagy, 1932). By the end of the nineteenth century many small streams in England were straightened or tamed (Lamplugh, 1914) while most large rivers had been channelized to some degree.

Increases in the extent of river channel engineering also related to advances in construction technology (Table 2.1), including dam building technology that was well-established by the end of the eighteenth century (pp. 11–12). Two other factors were of significance. First, the incidence of catastrophic floods throughout Europe from the mid-seventeenth to the mid-nineteenth century, reflecting the oscillation of climate conditions from cold to warm, combined with catchment land use changes, especially deforestation, demanded a political response: the governments of Rome and Florence, for example, appointed several commissions of engineers and mathematicians using the laws of hydraulics to consider how to regulate rivers. Second, river channel engineering methods, developed to address problems in specific environments, produced concentrations of expertise in the application of particular methods (Petts, 1989). In the Netherlands, particularly renowned for the development of dredging technology, the design of flood gates, retaining walls and groynes was a specialism that had been developing since the end of the sixteenth century; in Italy, land reclamation had stimulated advances in technology of river training and regulation that were used in the systematic control of rivers from the mid-fifteenth century onwards. In Switzerland, the nineteenth century was viewed as the century of river training works stimulated by an unusual number of disastrous floods, the need to control malaria and other diseases, the demand for land to meet population growth and the change in political structure (Vischer, 1989). In southern France, the Garonne River saw a progressive increase in the percentage of the disturbed area within the floodplain, sequential periods of river use, dominated by navigation in the eighteenth century and agriculture in the nineteenth, were followed in the twentieth century by the intensification of agriculture coupled with industrialization and urbanization (Decamps *et al.*, 1989).

Engineering methods are broadly of two kinds (Table 2.3): channel modification and discharge regulation. Individual engineering practices in these two categories are outlined below, although a combination of techniques is often employed. It was suggested that schemes for channel modification could be accomplished in a particular order such as working in a downstream direction and letting the river do most of the work (Winkley, 1972):

1 begin any realignment or control at a fixed hard point;
2 realign a river where possible to build in more depth and curvature at bends. A smooth transition from one curve to the next, such that the tangent of each thalweg radius will coincide at the crossing, produces the best alignment and stability;
3 plan areas of deposition according to normal bar spacing for each individual reach of the river;
4 design and locate any structure to take advantage of secondary currents and build in stages or steps that take best advantage of the work the river can do;
5 any future work should be done with the above in mind except where flood protection is in danger.

Channel modification

Channel modification includes several procedures (e.g. Hemphill and Bramley, 1989), implemented separately or in combination, to control the channel morphology and often included channelization (Table 2.3).

Maintenance/clearing and snagging

Maintenance/clearing and snagging includes a number of techniques generally referred to as operational maintenance, and involves an ongoing recurrent commitment such as tree clearance (clearing and snagging), weed control, clearance of trash from urban areas and, potentially, dredging (Brookes, 1997). Where maintenance became highly mechanized, the end product often began to resemble channel modification by re-sectioning (see p. 36). Annual growth of aquatic plants can add significant additional roughness and thus seriously reduce channel capacity during flood events, so that emergent or marginal bank plants are cut manually or by specially equipped boats, chemically controlled by herbicides or grazed by fish such as carp (Brookes, 1988); in some rivers plants are controlled several times each year. Clearing and snagging procedures range from the removal of fallen trees and of debris jams from the channel, cutting overhanging branches of riparian trees that may act to reduce flow capacities or contribute through breakage, to removal of in-channel large woody debris or of all trees growing on the channel banks and floodplain where their growth conflicts with the roughness characterization modelled under the ideal flood flow scenarios. The upper Nepean River was badly choked with debris in 1968 so that a policy of controlled dredging and stream clearing was undertaken (Erskine and Green, 2000).

Table 2.3 Possible channel modifications required for different types of river channel use

River channel use	Channel modification											Flow regulation	
	Maintenance	Dredging	Channel training	Re-sectioning	Embankments	Realignment	Canalization	Bank and bed protection	Bed protection	Flood walls/lined channels	Floodplain modification	Weirs	Dams
Irrigation								**3**				**2**	**1**
Land reclamation			**2**								**1**		
Local river use	**2**			**2**			**2**						
Navigation	**1**	**1**	**1**						**2**				
River crossings					**2**	**3**		**1**	**2**	**3**	**2**		
Industrial mills			**1**	**1**					**1**			**1**	**2**
Cooling water			**2**	**3**									
Hydropower generation												**1**	**1**
Water supply			**3**										**1**
Flood control			**3**				**2**	**3**			**1**		**1**
River stabilization			**1**	**1**					**2**				
Instream mining	**1**	**1**	**1**			**2**						**2**	
Restoration	**1**		**1**	**1**					**2**		**2**		

Notes: Action required: 1, measure necessary to achieve the required use (necessary); 2, measure of secondary importance (supplementary); 3, measure which might be necessary as a result of the action taken to achieve the use (optional).
Source: Developed from Table 1.2 (Chapter 1, p. 9) and adapting a table compiled by Jansen *et al.* (1979: table1.1-1) on river uses and measures to achieve them.

Dredging

Dredging involves sediment removal from the channel bed to create or maintain a given volume of flow capacity or flow depth suitable for intended uses of the river. Dredging is particularly associated with the maintenance of flow depth for the purposes of navigation in larger rivers, for instance, the channel of the Mississippi has been maintained according to decree of Congress since 1896. Another use is for maintenance of flood flow capacity in locations where population is located close to river channels that have become prone to sedimentation. In some rivers, the frequency of dredging increased once appropriate technological developments allowed rapid and effective mechanization (Table 2.1). Dredgers are either mechanical (using diggers or buckets to lift sediment) or hydraulic (removing sediment by using suction pipes or pumps) (Petersen, 1986). In addition to determining where and how often to employ them, it is also necessary to decide how many dredgers to use, how to transport the spoil that is removed and where to dispose of it; it is often transferred only as far as the channel bank where it is used to build embankments as a further flood defence by increasing conveyance capacity still further.

Channel training

Channel training involves temporary or permanent structures designed to modify and regulate the channel bed; the structures direct channel flows to scour a channel for purposes of navigation or water abstraction. Permanent structures include longitudinal and lateral dikes (shaped groynes) to establish the channel boundaries within designed regulation lines, accelerating the formation of a channel of gentler current line and the removal of shoals. If the river is trained on both sides by long dikes, then permanent width is obtained. Temporary structures are positioned as high flows recede and removed after the channel has been scoured (floating panels) or before the next flood season (bandalls). Bandalls consist of bamboo mats supported by stakes placed obliquely to the river flow on the submerged sandbanks at one or both sides of the channel to be eroded. Developed in the Ganges-Brahmaputra river area they are still used where particular conditions apply, although they are very labour intensive and so are not capable of widespread application. Conversely, bottom panels are usually left in place, at an angle of about 45° to the current at the edge of the channel to be eroded. Overflowing water is accelerated causing scouring that becomes deeper and wider further downstream and this may be semi-permanent if the channel is reasonably stable. Such temporary structures can be effective but their efficiency level is low and their success cannot be guaranteed (Jansen *et al.*, 1979), so that mechanical dredging offers a more efficient solution, although at higher cost. The preferred structure tends to vary with location.

Re-sectioning

Re-sectioning involves channel modification by widening and/or deepening for flood control purposes. Widening is frequently used in lowland rivers to provide additional cross-sectional capacity often along with deepening, where gradients permit. Deepening of the channel may be undertaken as part of a system of regional drainage, to lower the water table often in conjunction with channel straightening to 'speed the water away' (Figure 2.2).

Embankments

Embankments, or levees, are a long-used, expedient method of flood control, especially in lowland areas, and effectively enlarge the capacity of the channel. They are earth embankments constructed along a channel to prevent land being flooded by high streamflow or backwater, they aim to give protection against floods with a long return period, and they have been used effectively along the Huang He (see pp. 150–3).

Realignment

Realignment can take several forms including meander cut-offs, general straightening of the channel over a channel reach and river diversions that may result in a longer channel course. Cutting off meanders to improve river alignment for navigation, or as a flood con-

Figure 2.2 *Channel straightening of Yalobusha River, Mississippi (see Downs and Simon, 2001, photograph P.W. Downs)*

trol measure, may be effected by cutting a small channel across the neck and allowing the river to develop and enlarge the channel. Whereas the Mississippi made 13–15 natural cut-offs per century, they were resisted by the Corps of Engineers by construction of revetments to prevent bank erosion and by dikes to prevent overbank flow, until the cut-off programme was initiated in 1932 (Petersen, 1986). Diversion channels are probably one of oldest methods of flood protection (Przedwojski *et al.*, 1995).

Construction is often enabled by a temporary diversion of the river or a cofferdam, a temporary structure of earth, rock or piling enclosing part of the stream bed that is de-watered.

Canalization

Canalization is the transformation from a free-flowing stream to a series of slackwater pools by construction of a series of locks and dams (Petersen, 1986) and occurs where structures intersect the river or where major dam developments or installations are emplaced.

Bank protection

Bank protection includes revetments and lining materials, while training using groynes can also be used for reasons of bank protection. Revetments, which can be permeable or impermeable (Przedwojski *et al.*, 1995), are structures constructed on sloping soil banks to protect and stabilize their surface against erosion by currents, waves, surface water and ground water flows. One of most common types of cover is rip-rap, consisting of a loose armour made up of randomly quarried rock (Przedwojski *et al.*, 1995). Alternatives include container systems or gabions, which are usually permeable containers filled with ballast. The containers may be jute bags, tubes, wire or plastic netting baskets or mattresses, which are thinner, more flexible versions. Bituminous linings include asphaltic concrete, asphalt mastic, grouting mortars, stone asphalt, lean sand asphalt and membranes. Other systems used include sheet piling, soil reinforcement systems, three-dimensional synthetic mats and tyre rolls (Przedwojski *et al.*, 1995).

Groynes are stone, gravel, rock, earth or pile structures built at an angle to a river bank to deflect flowing water away from critical zones, to prevent erosion of the bank, in order to establish a more desirable channel for flood control, navigation and erosion control. They can also be impermeable, made of concrete, boulders or gabions, or permeable, composed of timber or lines of stakes interwoven with willows. In designing groynes it is necessary to select their width, slope, shape and spacing according to the current and the character of the channel. Spur dikes are attached to one of the banks and induce a new pattern of lateral erosion, thus changing the depth and slope of the reach. Dikes have been used extensively on the Lower Mississippi since 1960 to help maintain navigation channels (Brookes, 1988). Revetments can be used to stabilize concave banks of bends, whereas dikes (usually pile structures with stone fill), at an angle to the current, are used to direct flow from one bend into the next, to give sharp bends a larger radius of curvature, to close off secondary channels or to concentrate flow in a limited width (Petersen, 1986). They are also often used on wide braided rivers to establish well defined channels and on meandering rivers to control flow into or out of a bend or through a crossing (Przedwojski *et al.*, 1995).

Bank protection methods work by using structures for regulation of the low water bed, construction of bank revetments, longitudinal dikes, groynes or spur dikes, ramparts, separation dams between river channels and closure dams through river branches.

Design criteria for bank protection relate to the type, materials, shape, dimensions of the bank and the construction process. Structures, where installed, may be closed, where bodies of soil or stone material are protected by revetments, or open, to permit part of the flow to pass through at all water stages. In determining the shape of bank revetments, of longitudinal dikes, of groynes and of closure dams, it is important to determine precisely the toe level, the crest level and the steepness of slopes, and the level and width of berms (Jansen *et al.*, 1979: 357–59).

Bed protection

Bed protection can be achieved by:

- armouring – to protect the channel bed from erosion and scour, especially if appropriate materials occur locally so that they can be augmented as necessary;
- fixed weirs – to control bed slope, but cannot be used if navigation is to be maintained;
- submerged sills – installed perpendicular to the direction of flow and used to control bed slope, bed and water surface elevation, and to prevent stream bed degradation and upstream headcutting (Figure 2.3). The two types are channel stabilizers (with a crest at approximately bed elevation) and drop structures (designed to limit and stabilize channel bed slope by means of a vertical drop) (Petersen, 1986).

Other techniques are explained in Chapter 10 (pp. 303–6).

Flood walls and lined channels

Flood walls and lined channels are expensive protection measures and are usually used only in very constrained environments, such as in urban settings where houses or roads

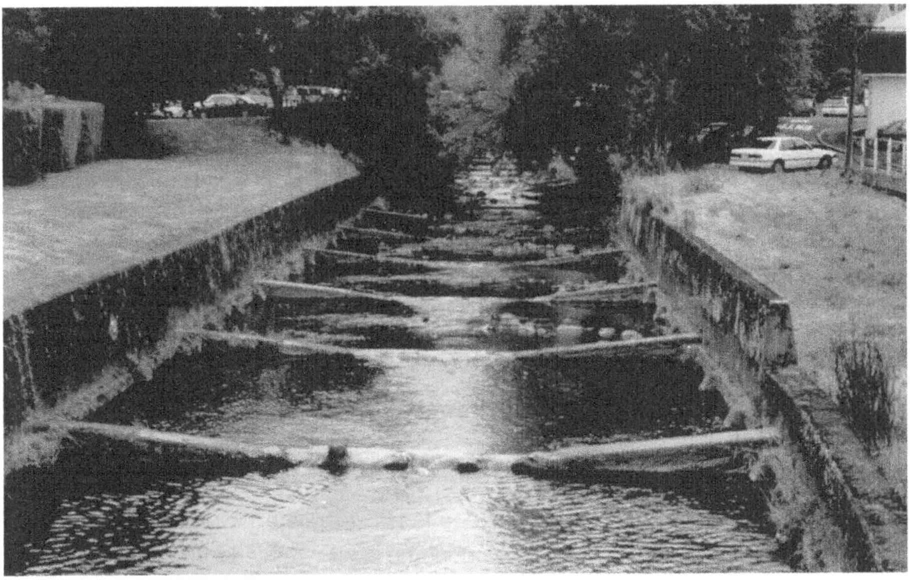

Figure 2. 3 *Bed protection measures using sills, Water of Leith, New Zealand (photograph P.W. Downs)*

Figure 2.4 *Bank protection of the Monks Brook, Chandlers Ford, UK (see Gregory et al., 1992, photograph P.W. Downs)*

abut the channel, where absolute channel rigidity is required or where a vertical or very steep bank face is required in order to save space. They generally consist of sheet steel piling (Figure 2.4) or a rectangular concrete section, but also include trapezoidal channels lined with concrete or paving slabs.

Floodplain modifications

Floodplain modifications include measures that can increase storage and flow capability by (Przedwojski *et al.*, 1995):

- changing floodplain roughness, floodplain level by lowering or raising the floodplain width, or the crest height of levees;

- culverts that effect drainage of the floodplain with sluice gates or valves kept closed during high river levels and opened after flood recession (Przedwojski *et al.*, 1995);
- other floodplain changes including retarded floodplain storage, used to reduce flood levels, for example the River Po system in Italy where summer (or secondary) dikes are overtopped only during high floods.

The Rhine river and floodplain, suffering from long-lasting floods, were regulated in the nineteenth century by replacing the branched channel system in the upstream reach by a single faintly curved channel and by cutting off meander bends in the downstream meandering reach (Jansen *et al.*, 1979).

River discharge regulation

River discharge regulation has been achieved by dams and weirs.

Weirs

Weirs are bed-fixing structures that raise bed levels upstream; they can instigate erosion downstream (see p. 98), which has to be considered in weir design. Fixed weirs are energy-absorbing structures that decrease the capacity of the flow to transport sediments; they are installed in headwater streams and in river reaches without shipping, often built in series with crest height and distance apart determined by an optimization procedure (Przedwojski *et al.*, 1995).

Dams

Compared with the rest of the world, Europe had the greatest rate of dam building up to 1880. Masonry dams up to 50 m high were common in headwater areas by 1850–1900, but big dams of the twentieth century were made possible by technology improvements (Table 2.1). An increase in dam building from 1900 to 1940 was succeeded by a peak from 1950 to 1970; Grande Dixence, a 285 m-high dam in Switzerland, is the world's second highest (Figure 2.1). There are three types of dam configuration (Petts, 1999): large dams in headwaters or in canyons downstream; chains of major dams such as the Colorado with 19 dams (see p. 94) or the Columbia with 23; or a series of run-of-river impoundments by navigation weirs and locks often involving hydroelectric power (HEP) plants, illustrated by the Murray below the Darling confluence, and by the Rhone downstream from Lyon, regulated by 12 low dams. Storage reservoirs for flood control can have either uncontrolled or controlled releases, as control gates determine how much water is released (Przedwojski *et al.*, 1995).

2.4 The culmination of river channel engineering

Implementation of works of the kind reviewed above (2.3) meant that, by the late twentieth century, large stretches of river channel had become extremely modified and constrained in planform and cross-section as a culmination of a series of engineering works designed to enable water resources management and to exercise control over flooding, erosion and sediment deposition. The cumulative effect of successive engineering is demonstrated on the River Waal, the main branch of the Rhine in the Netherlands.

Landowners built the first structures to protect their land from erosion, dike boards and municipalities later continued the work, and subsequently federal authorities undertook the regulation of large river reaches, not just for navigation but also to prevent formation of ice jams which could cause flooding of adjacent land (Jansen *et al.*, 1979). The jamming risk was reduced by eliminating sand banks, islands and obstacles along the banks, while, along the Rhine backwaters, braids and side channels have been greatly reduced. The measures have achieved their original purpose in that bed degradation up to 7 m has occurred, and the area subject to flooding has been reduced by 85–94% (Mays, 2001). The Mississippi River has also been subject to a sequence of engineering measures (Leopold and Maddock, 1954) and these can be traced as:

- first levee at New Orleans in 1717; individual landowners built levees with their own resources;
- Corps of Engineers began work on navigation improvements in 1824;
- Congress authorized survey of Mississippi river in 1850 to determine most practical plan to preclude flooding;
- 1879 establishment of Mississippi River Commission, until 1900 most programmes of the Corps of Engineers dealt with navigation improvements, with special studies and work associated with expansion of roads and railroads, and with mapping;
- first Flood Control Act 1917 authorized work on the Mississippi; Rivers and Harbors Act of 1920; large scale federal flood control activity in 1928 following the disastrous flood of 1927;
- Flood Control Act of 1936 saw federal participation in flood control on a nationwide basis; reinforced by subsequent Acts;
- post-1945, river basin development took shape with all water uses and problems considered.

As a consequence of these events 45% of the river bank of the Mississippi was revetted, the river length shortened by 229 km, and the floodplain extent subject to regular inundation reduced by 90% by the building of levees (Mays, 2001), revealing how different techniques used on the lower Mississippi River could produce widely different river reactions (Winkley, 1982). In regions with a long-established tradition of river management, methods adopted can vary from one area to another. Thus, in the case of the suitability and effectiveness of flood embankments in India, opinions about the methods to use have differed from state to state and from time to time (Sinha and Jain, 1998).

Modification of river channels by engineering works is encapsulated in the term 'channelization' (Figure 2.5) although other terms for the same group of engineering methods include kanalisation in Germany and chenalisation in France (Brookes, 1988). In the light of other definitions Brookes (1988: 5) proposed channelization to be:

> *... the term used to embrace all processes of river channel engineering for the purposes of flood control, drainage improvement, maintenance of navigation, reduction of bank erosion, or relocation for highway construction ... may also be associated with programmes of forest or field drainage.*

Channelization (Figure 2.5) can therefore be envisaged as involving some combination of straightening, widening or deepening of the channel, thus reducing the frequency of inun-

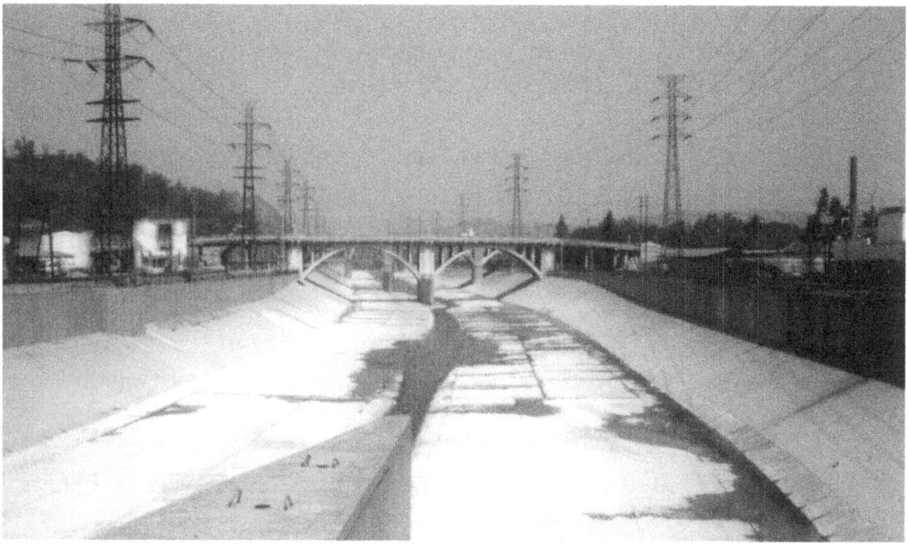

Figure 2. 5 *Channelization of the Los Angeles River (photograph K.J. Gregory)*

dation of the floodplain. It may include wholescale diversion of the channel for flood control purposes; supplemented by works designed to protect river banks from erosion; and including periodic maintenance activity to remove deposited sediment and to cut back vegetation where it threatens to reduce the discharge capacity of the channel cross-section or is able to increase erosion. In many countries, river channelization was pursued particularly vigorously after the Second World War until the 1980s, by which time it is estimated that, for instance, 98% of all rivers in mainland Denmark had been modified (Brookes, 1987a, b) as had up to 96% of river channels in river catchments in lowland Britain. Overall in England and Wales, 8500 km of rivers were channelized (density 0.03 km km^{-2}; Brookes *et al.*, 1983), and in the USA 26 500 km of rivers were channelized by 1977 (0.006 km km^{-2}, Leopold, 1977). In the Hunter Valley of New South Wales from 1955 to 1985 there were 90 major river training work schemes affecting 853 km of rivers, plus 436 minor schemes (Erskine, 1990). In addition, by 1973, at least 15% of stable world annual runoff was contributed by storage reservoirs (Lvovitch, 1973) as evidence of an extensive flow regulation. As such, in the industrialized nations very few rivers were left in their natural state. For instance, in the USA, where flow regulation has been most intensive, less than 2% of rivers are now in their wild and scenic condition (Graf, 2001). It was partly the increasing evidence of environmental damage arising from such extensive channelization programmes that caused river managers to reconsider their options for providing river channel management (see Chapter 7).

2.5 Reactions in the late twentieth century

Population increase following the industrial revolution often prompted concerted local and national efforts to resolve water resources issues and to mitigate riverine flood and erosion problems but, with the knowledge available at the time, few alternatives to an engineering solution could be envisaged. The industrial revolution brought technological advances that unleashed the powerful notion that nature could be conquered and its resources utilized and exploited for the benefit of humanity (Williams, 2001), illustrated by the founding statement of the Institute of Civil Engineers in 1830 which was 'to harness the great sources of power in nature for the use and convenience of man'. As successive generations of river engineers were trained in hydraulics and civil engineering, they tended to view rivers as large-scale plumbing problems, at a time when the prevailing paradigm was to control nature, when river channels were regarded as flumes that needed to be made as efficient as possible, and when developments in technology were available to engineer channels. The emphasis on water resources management and flood control meant that hydraulics and structures could produce solutions that worked in most cases by providing short-term mitigation of the flood or other problem addressed.

Any assessment of the hydraulic engineering approach to river channels must begin by acknowledging that it worked in many cases by relieving the flood or other problem addressed and by providing for water resources management in a manner compatible with the prevailing management paradigm of the period (Williams, 2001). However, by the late twentieth century, what has come to be thought of as the 'hard' or structural engineering approach to river channel management was being reconsidered in a variety of ways (Table 2.4): four aspects in particular led to questioning the established approach. First, the environmental effects of river engineering upon the dynamic hydrosystem of channel processes, the morphology, biology including ecology and aesthetics of river channels were progressively demonstrated by fluvial geomorphologists and ecologists (Chapter 4, p. 99), prompting the realization that alternative approaches should be sought (see Chapter 7). Second, concerns were raised about the apparent longer-term failures of schemes to sustain the design solution without costly additional capital works or maintenance. This encompasses a number of cases where major functional problems arose following river engineering (see Table 2.5), including 48 examples of recorded dam failures since 1860 that have killed more than ten people (McCully, 1996) but, more common were cases where implementation of a structural solution simply moved a problem elsewhere or intensified problems in other locations. For instance, many erosion control measures instigated increased erosion upstream or downstream of the 'solution', and many flood control channels led to greater flood risk downstream and to serious disruption to the regional water table. More problematic still, many flood control channels were found to be inadequate during their first major flood event, sometimes as a result of hydraulic calculations that did not incorporate the effects of sediment and vegetation on flood flows. One answer was a recommendation for further structural measures so that the approach became self-perpetuating, continually justifying finance for further measures. Eventually, softer measures were sought including more comprehensive approaches that tried to tackle the source of an issue rather than the symptoms. Third, a change in economic outlook from 'river engineering at all costs' to management measures that could be justified according to their benefit:cost ratio nullified the use of the most expensive hard engineering techniques for all but the most demanding situations. Fourth, a change in the intellectual climate occurred towards a context in which the idea of human domination over nature gave way to a more

Table 2.4 Perceptions and characterization of the hydraulic engineering approach to river channel management

View expressed	Source
In a book titled *The River Killers*: 'For the growing masses concerned with saving outdoor America. There is only one positive answer: completely abolish The Civil Works Branch of the Army Corps of Engineers'.	Heuvelmans (1974: 183)
Our technology will fix it. Man's engineering capabilities are nearly limitless ... What is needed is a gentler basis for perceiving the effects of our engineering capabilities. This more humble view of our relation to the hydrologic system requires a modicum of reverence for rivers.	Leopold (1977: 430)
The art of the engineer is to establish systems and to build relevant structures which may change the face of the Earth with inadequate data and information. This applies especially to the engineer who has to do with the complexity and diversity of the large natural surface and subsurface water resource systems on Earth.	Volker and Henry (1988: iii)
The UK legacy of damaged rivers, floodplains and wetlands traced back to early marshland drainage but also to the 'dig (and drain) for victory' campaign of the Second World War. By the time grants for new land drainage works gave way to urban flood defence in the mid 1980s, few rivers were left in their natural state. Many other countries have a similar history.	Gardiner (1998)
Developments at the end of the twentieth century were leading from the era of conventional channelization and riprap projects to the restoration era requiring the commitment of top level federal agency managers to change their agency missions.	Riley (1998: 95)
The problems engineers were asked to address were how to design and construct river engineering works to achieve the exploitation of a river's resources. Engineers were not asked to be river or watershed managers, nor were they asked to consider the ecologic, or long term consequences of their actions.	Williams (2001)

Table 2.5 Instances of failure of river management schemes

Location	Type of installation or structure	Problem or failure	Consequences
Vaiont Dam, Italy	Large concrete dam, 265 m high	Dam completed in 1960, in 1963 landslide (c. 270 million m^3) into the reservoir caused water to overtop dam and send flood wave (initially 70 m high) downstream	2600 deaths and much property damage in area adjacent to river channel
I–5 Arroyo Pasajero Tuni Bridge, USA	Bridge over sandy bottom arroyo	Scour of bridge foundations after flood on 10 March 1995	Five people killed
Walnut Street Bridge, USA	Two spans washed away January 1996	Lifted off foundations by high flood waters and ice built up	Bridge spans moved c. 100 m down Susquehanna River

symbiotic notion of human relations with nature, and in which the paradigm of river engineering was succeeded by a paradigm emphasizing sustainable management (Williams, 2001).

Changing the doctrine of river channel management required more than replacing hard engineering by a softer approach, but rather of complementing river engineering (principles of hydraulic design are still vital as *part* of the process) by additional ideas and new techniques for management that function within a catchment-based approach (see Chapter 6). Certainly it has been aided by innovative techniques and new materials, including geotextiles being employed for more environmentally sensitive approaches (Haltiner, 1995) and greater sophistication of hydraulic modelling capabilities (see Table 2.2). However, in addition, it has required adoption of a more overt consideration of sustainability, the essence of which is that we use our environment in ways that do not detract from its ecological integrity or value to future generations. The new approaches have been characterized in several ways including evolution from a product-oriented engineering approach to a *dynamic multi-objective management approach* (Hunt, 2000); a switch from technological river management to *ecological river management* (Nienhuis *et al.*, 1998); or *sustainable river management* involving the use of resources without degrading their quality or reducing their quantity (Smits *et al.*, 2000).

As the focus of new flood control projects in the USA shifted from control of hydraulic structures to more sustainable approaches it was suggested (Galloway, 2000) that perhaps these approaches 'may also be able to convince Mark Twain that while engineers may not ever tame the river, they may well be able to work with it in harmony' (p. 63). However, just as risks and uncertainties were associated with hydraulic engineering solutions, there is no panacea that eliminates the same risks and uncertainties. The primary difference is that approaches to river channel management are now founded on an understanding of river channel systems, as introduced in the next chapter.

3
River Channel Systems

Synopsis

Knowledge of processes in river channel systems informs ways in which river channels are managed, but this chapter is necessarily very selective by assuming familiarity with basic hydrological and ecological processes. A basic understanding of the composition, character and function of river channel systems is provided by outlining the way in which river channel systems are researched (3.1), described and characterized (3.2), leading to a composite classification of channels suitable as a framework for initiating management (Table 3.3). A review of individual components of the river channel system and aspects of their functional connectivity (3.3) extends to an outline of channel dynamics (3.4) and to elements in the hydrosystem that are vital considerations in engineering and management in the drainage basin context, although the way in which channel behaviour is perceived is not always identical with physical reality (3.5).

Methods of river channel management during the majority of the twentieth century (Chapter 2) tended to concentrate upon structural solutions to single issues in particular reaches, may have instigated problems elsewhere and gave insufficient attention to ecology, aesthetics and recreation. Whereas these approaches relied upon knowledge of the hydraulics of particular reaches of channels, management now relies upon more environmentally friendly and sustainable alternatives that require understanding of the composition, character and function of river channel systems, the basis of which is outlined in this chapter.

3.1 Research on river channel systems

Understanding of the character and dynamics of river channel systems has been advanced by a number of disciplines (Chapter 1, p. 16), often reflecting paradigmatic shifts in the approach taken to investigations within drainage basins. In geomorphology and other environmental sciences, the dominant approach until the mid-twentieth century emphasized long-term evolution of landscapes often employing the Davisian cycle of erosion that emphasized stage or time of landscape development and minimized the study of process (Gregory, 2000) that had been advocated by G.K. Gilbert. Greater emphasis upon processes (Leopold *et al.*, 1964) and by investigations of the Columbia School (Strahler, 1992) led to a focus upon the fluvial system (Schumm, 1977b) and geomorphic engineering (Coates, 1976). Process-based approaches to river channels grew also from sedimentology research (e.g., Allen, 1970) undertaken with keen awareness of flow hydraulics and sediment movement; from the uptake of process hydrology reflected in numerous text books (e.g. Linsley *et al.*, 1949; Chow, 1964; Ward, 1967); and the increase in related scientific research journals (e.g. *Journal of Hydrology* 1963–; *Water Resources Research*, 1965–; and *Hydrological Processes*, 1987–). Such developments provided a solid foundation for understanding river systems as reflected in textbooks for fluvial geomorphology research (Morisawa, 1968; Richards, 1982; Knighton, 1984, 1998; Morisawa, 1985), with emphasis upon the drainage basin in relation to form and process (Gregory and Walling, 1973), relevance to environmental planning (Dunne and Leopold, 1978), to land use and water resources (e.g. Newson, 1992, 1997) and to engineering (Thorne *et al.*, 1997a).

In the second half of the twentieth century research interests developed according to discipline, so that approaches to rivers tended to be developed somewhat independently from hydrology, geomorphology and ecology. Thus *The ecology of running waters* (Hynes, 1970) had an impact in ecology reminiscent of that by Leopold, Wolman and Miller (Leopold *et al.*, 1964) in geomorphology, and the ecological strand was strongly presented in *Freshwater Ecology: Principles and Applications* (Jeffries and Mills, 1990) with greater emphasis upon techniques (Haslam and Wolseley, 1981; Hauer and Lamberti, 1996). Subsequent attempts to link ecology and hydrology more closely were demonstrated by Oglesby *et al.* (1972), Whitton (1975) and Gordon *et al.* (1992).

Such developments complemented long-standing engineering approaches (Chapter 1, pp. 11–15; Chapter 2, p. 27), and generated a revised focus for river studies in the latter part of the twentieth century. Four significant trends are evident. First, scientific approaches from disciplines of biology/ecology, engineering and the environmental sciences including physical geography, geomorphology and environmental geology became progressively less discrete and more multidisciplinary (e.g. Boon *et al.*, 1992) and more internationally founded collaborative approaches (e.g. Petts and Amoros, 1996a) together with new multidisciplinary journals (e.g. *Regulated Rivers; Research and Management* 1984–2001; *River Research and Applications* 2001–; *Aquatic Conservation: Marine and Freshwater Ecosystems*, 1990–) became more influential. The links between hydrology and ecology could then be viewed from management, hydrological, historical and science-linked perspectives (Petts *et al.* 1995). Multidisciplinary approaches were often catalysed by conferences (e.g. Harper and Ferguson, 1995) or by the enterprise of a key individual (e.g. Shen, 1973). Second, greater attention was devoted to temporal changes of river channels (e.g. Gregory, 1977b; Richards, 1987), to such changes over palaeohydrological timescales (e.g. Gregory, 1998) generating a growing appreciation of the value of an understanding of river channel history. Third, experience of particular areas influenced river channel management from a fluvial geomorphology or ecological perspective, as exemplified in the UK (Carling and Petts, 1992) and Europe (De Waal *et al.*, 1998), in Australia (Brizga and Finlayson, 2000) and in New Zealand (Mosley and Jowett, 1999). Fourth, there was concentration upon some particular management themes such as river regulation (Petts, 1984) and channelization (Brookes, 1988).

Growth in knowledge of river channel dynamics came to provide a sound contribution to river channel management. Whereas in 1970 river management remained an 'art' rather than a science (Petts *et al.*, 1995), the developing vision expressed by Mellquist (1992) was that we need politicians who *dare*, bureaucrats who *want*, and scientists/engineers who *can*. Although management had been undertaken without science by using trial and error and folklore, it was increasingly appreciated that, to be effective, river channel management required improved application of environmental sciences; putting principles into practice can be the ultimate arbiter of sound science (Calow and Petts, 1992). By the end of the twentieth century, previously distinct approaches were beginning to fuse, responding to concerns for environment, global change and sustainability, and providing a holistic foundation for the better understanding of drainage basin dynamics and for the integrated role of river channels in that system. In the light of these developments this chapter focuses on major characteristics of the river system (3.2) as a basis for understanding its functional connectivity (3.3) and channel dynamics (3.4), with applications qualified according to perceptions of river processes (3.5).

3.2 River channel system structures

Knowledge of the variety of river channels and their arrangement within the drainage basin, that is the network structure, is necessary to understand river channel functions as a basis for river channel management. Proceeding downstream in a river channel system changes occur in water, sediment and solute discharge, and in the morphological and ecological characteristics of the channel. Generally, as the drainage area and the length of river channels increase so do the discharge and the size of the river channel, in both cross-section and planform, and the floodplain also grows in size. The ecology of the channel changes downstream because the net primary productivity of aquatic plants varies according to light availability, temperature, nutrient concentrations, inorganic carbon concentration and water movement, and these conditions alter progressively, so that regularity occurs in the downstream lateral distribution and production of plants. In addition to this in-stream (autochthonous) biological production source, there are sources related to inputs from terrestrial vegetation (an allochthonous source) and to the transport of organic matter from upstream. These latter two conditions also change progressively downstream. Therefore species (particularly invertebrates) that have clearly defined niche habitat requirements related to physical conditions (determined by hydraulic stresses) and vegetation type, also alter systematically downstream. When considering any river channel reach it is necessary to locate that reach within its structure and basin context to appreciate how it is affected by reaches upstream and how it may affect reaches downstream.

It has been concluded that streams are best viewed as continua (Cushing *et al.*, 1980, 1983), that the view of river channels as a continuum is important, especially for biological classification (Vannote *et al.*, 1980), and is capable of further elaboration (e.g. Ward *et al.*, 1999b) by relating to the connectivity of river ecosystems and ecotones. Based on the energy equilibrium concept (see p. 75) often used in fluvial geomorphology, the river continuum concept contends that understanding of biological strategies and dynamics of river systems requires consideration of the gradient of physical factors provided by the drainage network (Vannote *et al.*, 1980). The structural and functional characteristics of stream communities are seen as adapted to conform to the most probable position or mean state of the physical system, so that producer and consumer communities characteristic of a given river reach become established in harmony with the dynamic physical conditions of the channel.

The continuum concept (e.g. Petts, 1984; Petts and Calow, 1996) does not negate attempts to classify channels into sections of particular types, and indeed the proposers of the concept (Vannote *et al.*, 1980) suggested grouping lotic communities into headwater streams, middle order streams and large rivers on the basis of invertebrate functional feeding groups. Existence of a continuum means that, in view of the lack of natural discontinuities, any classification is bound to be arbitrary (Wright *et al.*, 1984) but classification of river channels and categorization of drainage basin structure is necessary to:

- establish the range of riverine environments that exist, using their classification as the basis for extending understanding from one area to another; including predicting a river's behaviour from its appearance (Rosgen, 1996);
- develop specific hydraulic and sediment relationships for a given stream type and its state;
- characterize river habitats for fish, for leisure and other uses;
- enhance awareness of the importance of historical evolution of an area – an approach developed since an early genetic classification for the Colorado (Powell, 1875);
- reflect the management problems presented, which may include the maintenance of physical conditions or habitat requirements or the restoration of previous characteristics – a diagnostic procedure (Montgomery and Macdonald, 2002) can be developed similar to that of medical practice;
- provide a constant frame of reference for communicating details of stream morphology and condition among a variety of disciplines and interested parties.

Categorization of river channels is useful because, as suggested for rangeland environments, stream types can limit the effectiveness of management when they have different potential conditions and recovery rates (Myers and Swanson, 1996), and different stream types in urban areas are likely to exhibit varying degrees and types of instability (Bledsoe and Watson, 2001).

In seeking a basis for the differentiation of stream channel types it is necessary to remember the distinction between characterization of the network of river channels within the drainage basin that makes no prior assumption about the channel types included (Mosley, 1992); classification requiring a definition of channel types, grouping entities by similarity (Newson *et al.*, 1998a); taxonomy, which is an objective procedure allocating cases on the basis of measured attributes; and typology, which is conceptual and depends upon the subjective judgements of class definitions and boundaries (Newson *et al.*, 1998a).

Each of these four approaches has been used in the research literature since 1980, but classification, taxonomy and typology have been emphasized. In developing a very comprehensive review of characterization and classification approaches (Mosley, 1987) six major approaches to classification of channels in the drainage basin structure can be discerned (Table 3.1). Some disciplines have emphasized particular attributes of the basin. Thus, channel types and morphological classification can be from an expressly geomorphological point of view (Thorne, 1997) with little attention to the ecology; whereas conversely, ecological classifications often pay less attention to the geomorphological characteristics. Characterization (see Church, 1992) is exemplified by the system for classifying river channel processes especially in Canada (Kellerhals *et al.*, 1976). Classification approaches became more wide-ranging as they were expanded to include fish, invertebrates or use for recreation (Mosley, 1987).

A comprehensive framework for river channel classification should: encompass the character of the river channel that was, is and could be developed in any particular location, reflect diverse approaches (Table 3.1), and acknowledge existence of a continuum. Any framework therefore needs to achieve four objectives:

1 *Comprehensively embrace all aspects of river channel character.* including morphology, channel pattern and floodplain, ecology of flora, fauna and their habitats. This is not easily achieved because there has been a tendency for geomorphological and ecological/biological classifications to be constructed separately (Table 3.1), although some are more comprehensive (e.g. Stanford and Ward, 1992; Bisson and Montgomery, 1996; Ward *et al.*, 1999b).

2 *Be basin-wide, embracing all scales*, relating the detailed field-surveyed channel reach or habitat to the basin or catchment as a whole. Integrated treatment has been achieved in various ways, by including catchment, sector, reach and section scales as appropriate for river channel typology (Newson *et al.*, 1998b), by a hierarchical subdivision of basins into watersheds, valley segments and channel reaches (Bisson and Montgomery, 1996) and by distinguishing channel, floodplain, terrace and hillslope as they relate to pioneer riparian forest, climax riparian forest and upland forest (Stanford, 1996). The idealized fluvial system can be characterized (Schumm, 1977b) as composed of (Figure 3.1A) production, transfer and deposition zones. The uppermost, the area from which water and sediment are derived, is the sediment source area where some sediment storage can occur (zone 1). The transfer zone (zone 2) is predominantly one of transport where, for a stable channel, input of sediment can equal output. Zone 3 is the sediment sink area of deposition, which can occur on an alluvial fan, alluvial plain, delta or in deeper waters. Zone 1 is argued to be of greatest interest to watershed scientists and to hydrologists as well as to geomorphologists concerned with the evolution and growth of drainage systems; zone 2 is of major concern to the hydraulic and river control engineer, to geomorphologists involved primarily with river channel morphology, and to stratigraphers and sedimentologists in view of palaeochannels and valley fill deposits (Schumm, 1977b). Zone 3 is of prime concern to the geologist and the coastal engineer, in view of the internal structure, stratigraphy and morphology of the deposits. Similar broad classifications offered by Palmer (1976), based upon Bauer's geohydraulic zones, and by Warner (1987), developing Pickup's classification (Pickup, 1984, 1986) for New Guinea and applied to Australia, can be reconciled with the Schumm zones. To ensure that the classification framework relates to the most detailed subdivision scale it is necessary to link to the field survey schemes developed (e.g. Rosgen,

Table 3.1 Some approaches to the classification of river channels

Approach	Author	Designated purpose	Major characteristics used	Number of zones
General/whole river	Nevins (1965)	Management-related	Lithology and position in long profile	4 (mountain or torrent, shingle, silt, tidal phases)
	Mosley (1987)	Various purposes	Nature of source; physiography; chemical characteristics	Varying numbers of rivers according to criterion used
	Mosley (1985, 1987)	Characterization of scenic beauty	Water colour	8 classes of water colour
	Strahler, 1952	Characterization	Stream order	Numerical ordering developed from Horton (1945) method. Major world rivers seldom greater than 12th order
	Gregory and Walling (1973) and others	Stream network dynamics	Frequency of flow in channels	3 types of channel (perennial, intermittent, ephemeral)
Basin wide	Schumm (1977b)	Geomorphology palaeohydrology	Dominant process	3 zones (production, transfer, deposition) of idealized fluvial system
	Palmer (1976)	Management	Developed from Bauer geohydraulic zones	4 zones (boulder, floodway, pastoral, estuarine)
	Warner (1987)	Typology based on channel bed sediments and channel stability	Based upon approach developed in New Guinea by Pickup (1984, 1986)	5 types (source, armoured, gravel–sand transition, sand, backwater (sink))
Basin-wide: zonation geomorphology	Schumm (1963)	Tentative classification of alluvial channels	Sediment transported	3 types (bedload, mixed load, suspended load)
	Brice (1975)	Channel patterns	Degree of sinuosity, braiding and anabranching and the character of planform	27 types are basis for complex classification
Basin-wide: ecological	Ricker (1934) and Mosley (1987)	Ecological stream classification	Summer flow and channel width + bed sediment, water temperature, current velocity, water quality, vegetation and invertebrates to identify subgroups	4 classes (spring creeks, drainage creeks, trout streams, warm rivers) with up to 6 sub-groups

Approach	Author	Designated purpose	Major characteristics used	Number of zones
	Huet (1949, 1954)	Fish habitats	Size of stream, velocity, slope and morphology	4 zones (Trout, Grayling, Barbel, Bream)
	Pennack (1971)	Classification of lotic habitats	13 criteria (channel width, flow, water quality, vegetation characteristics)	6 habitats (Dace, Trout feeder, Trout, Bass or pickerel, catfish or carp, tidal stream)
	Frissell *et al.* 1986	Hierarchical		
	Ward and Stanford (1987)	Landscape ecology		
	Naiman *et al.* (1992)	Vertebrate and invertebrate biotic zonation		
'Objective' multivariate analysis	Wright *et al.* (1984)	Classification of running water sites in Great Britain	Macro-invertebrate species lists and 28 characteristics of the river environment	Two-way indicator species analysis classified 268 sites into 16 groups and multiple discriminant analysis explored relationship between site groupings and environmental data
	Cushing *et al.* (1980)	Stream site classification	15 physical and chemical variables on 34 rivers	5 groups identified by cluster analysis
Dominantly field-based	Kellerhals *et al.* (1976)	Characterization	River processes	3 groups: channel pattern (7 types), islands (4), bars (7)
	Rosgen (1994)	Morphological description	Dominant bed material, channel characteristics	41 stream types
	Bisson and Montgomery (1996)	Method of characterizing valley segments, stream reaches and channel units	Channel materials, flow types	3 valley segment types, 8 stream reach types
	Thorne (1998)	Stream reconnaissance	Valley floor, channel morphology and bed material characteristics	Many
	Raven *et al.* (1998)	River habitat survey for conservation and restoration of wildlife habitats	Data collected from 500-m lengths of channel at 5600 sites in UK	Database resource for interrogation and relation to management problems

1994, 1996; Thorne, 1997, 1998) and to incremental analysis (Mosley, 1987) as used for analysis of aquatic habitats.

3 *Relate to river channel processes, reflecting seasonal and short-term dynamics*: requiring consideration of variations at a point, within the channel and at the scale of the channel unit throughout the channel network. Whereas classifications have often been constructed using static, morphological or ecological attributes, habitats and channel morphology are affected by dynamic river channel processes, including the magnitude, duration, timing and frequency of flood events, and by seasonal variations in hydrological regime.

4 *Allow for channel change*, and channel sensitivity (see Chapters 4, 5 and 8).

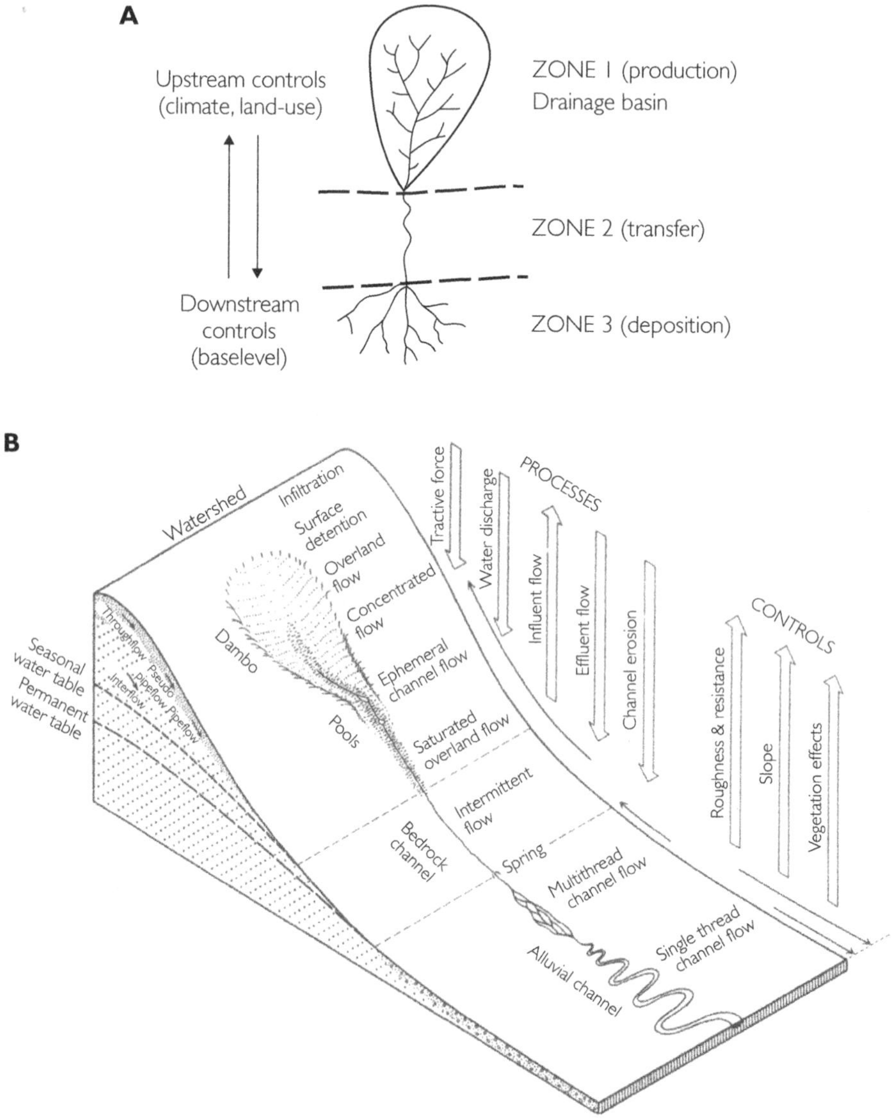

Integration of the various approaches used is necessary to relate to field-based classifications (Table 3.1) and has been attempted in several ways focusing upon the dynamic aspects of the network (Figure 3.1B) combined with the character of the channel and its planform (Gregory, 1979a). Rosgen (1994, 1996) devised a hierarchically assessed, field-founded approach that reflects the evolutionary history of the area and provides an organized procedure for later determining the agents of formation and the processes utilized by different stream types for maintaining their form and function (Figure 3.1C), and so approaching the specific objectives stated above. There is still insufficient agreement on scale terminology, as many methods (e.g. Table 3.1) approach classification from one particular discipline. Hierarchical principles and a truly geomorphological channel classification based on reaches incorporating mesoscale habitat typologies is a way (see Table 3.2) of identifying physical biotopes (Newson and Newson, 2000).

Approaches to classifying the composition of the river channel network within the drainage basin (Table 3.1) provide a series of inter-related classifications that can be superimposed one upon another. In the light of the various classifications previously proposed and the four objectives suggested above, an integrated framework is required and several have been suggested. A hierarchy that is primarily morphological divided streams into seven major types according to degree of entrenchment, gradient, width/depth ratio and sinuosity; with six sub-categories within each major category depending on the dominant type of bed/bank materials (Rosgen, 1996). A field-based stream reconnaissance approach (e.g. Thorne, 1998) provides a systematic method for recording information so that it can be interpreted to describe channel process and stability problems. The River Habitat

C

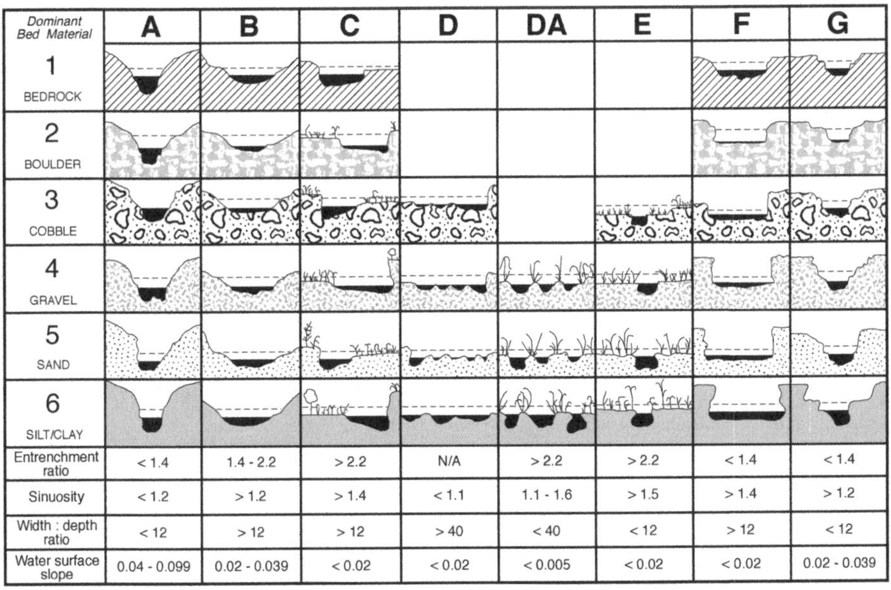

	A	B	C	D	DA	E	F	G
Entrenchment ratio	< 1.4	1.4 - 2.2	> 2.2	N/A	> 2.2	> 2.2	< 1.4	< 1.4
Sinuosity	< 1.2	> 1.2	> 1.4	< 1.1	1.1 - 1.6	> 1.5	> 1.4	> 1.2
Width : depth ratio	< 12	> 12	> 12	> 40	< 40	< 12	> 12	< 12
Water surface slope	0.04 - 0.099	0.02 - 0.039	< 0.02	< 0.02	< 0.005	< 0.02	< 0.02	0.02 - 0.039

***Figure 3.1** Approaches to channel classification. **A**. Basin classification (Schumm, 1977b); **B** basin dynamics (Gregory, 1979b); **C** morphological stream types (Rosgen, 1994)*

Table 3.2 Flow types used for field identification of physical biotopes

Flow type	Associated biotope(s)	Description
Free fall	*Waterfall*	Water falls vertically and without obstruction from a distinct feature, generally more than 1 m high and often across the full channel width.
Chute	*Spill* – chute flow over areas of exposed bedrock. *Cascade* – chute flow over individual boulders	Fast, smooth boundary turbulent flow over boulders or bedrock. Flow is in contact with the substrate, and exhibits upstream convergence and downstream divergence.
Broken standing waves	*Cascade* – at the downstream side of the boulder flow diverges or 'breaks'	White-water 'tumbling' waves with the crest facing in an upstream direction. Associated with 'surging' flow.
Unbroken standing waves	*Riffle*	Undular standing waves in which the crest faces upstream without 'breaking'.
Rippled	*Run*	Surface turbulence does not produce waves, but symmetrical ripples which move in a downstream direction.
Upwelling	*Boil*	Secondary flow cells visible at the water surface by vertical 'boils' or circular horizontal eddies.
Smooth boundary turbulent	*Glide*	Flow in which relative roughness is sufficiently low that very little surface turbulence occurs. Very small turbulent flow cells are visible, reflections are distorted and surface 'foam' moves in a downstream direction. A stick placed vertically in the flow creates an upstream-facing 'V'.
Scarcely perceptible flow	*Pool* – occupies the full channel width. *Marginal deadwater* – does not occupy the full channel width	Surface foam appears to be stationary and reflections are not distorted. A stick placed on the water's surface will remain still.

Source: After Newson and Newson, 2000.

Survey (RHS) shows how a scheme can be nationally based on habitat quality data from a sample of sites taken as representing the 85 000 km of rivers classified for water quality in the UK (Raven *et al.*, 1998). The RHS database contains some morphological information and, with additional data, might be extended to make the resulting typology dynamic and therefore capable of predicting channel stability as an important component of the information needed for sustainable river management (Newson *et al.*, 1998b).

The need to set streams into a hierarchical system within the watershed context was advocated by Frissell *et al.* (1986) to give a perspective that should allow more systematic interpretation and description of watershed–stream relationships. Valley segments, stream reaches and channel geomorphic units were recognized as three hierarchically nested subdivisions of the drainage network by Bisson and Montgomery (1996) who developed a classification scheme for characterizing aquatic habitats at intermediate landscape scales, by combining approaches that had focused on valley segments and stream reaches (Montgomery and Buffington, 1998) and on channel geomorphic units (Hawkins *et al.*, 1993). Such progress accommodates the requirement for field-identified stream reaches to be incorporated within an overall drainage basin scheme. In the light of previous classifications an original framework is proffered here (Table 3.3) involving seven nested scales, and this could be used in relation to a range of river channel management problems. However, although such a general framework (Table 3.3) is necessary to provide a hierarchical context for a particular stream or river, the individual distinctiveness of that river can provide the challenge for management. For this reason it has been argued that the unique nature of each lotic ecosystem should be considered within the river discontinuum that recognizes the general trends along the longitudinal profile but also creates a framework for studying the importance of the character of each stream (Poole, 2002) thus emphasizing the concept that 'every stream is likely to be individual' (Hynes, 1975) and reflecting the primacy of place (Phillips, 2001).

3.3 Components and connectivity

The framework (Table 3.3) provides a hierarchical structure for analysing individual catchment hydrosystem components requiring focus upon their characteristics, and upon the linkages between them, leading to dynamics of channel behaviour (3.4). Only aspects particularly pertinent to river channel management are outlined here.

Components of the catchment hydrosystem

These range from those at a point in a channel, through the channel cross-section and channel pattern, to the floodplain and the network.

A point in a channel

At a point in a channel conditions relate to changes in:

- water: depth, volume, velocity, temperature, turbidity, chemistry (including dissolved oxygen, nutrients especially nitrogen and phosphorus compounds, organic and inorganic chemicals natural and synthetic, heavy metals and toxic substances, pH);
- sediment: bed material, suspended sediment.

Table 3.3 A channel classification framework compiled as a basis for river channel management

• **Drainage basin**/watershed/catchment				
• **zones:**	production transfer deposition	Schumm (1977b)	Boulder + floodway Pastoral Estuarine	Palmer (1976)
• valley **segments:**	bedrock alluvial colluvial	Montgomery and Buffington (1993)	**valley floor** **floodplain** river **corridor**	
• stream **reaches:**	cascade step-pool plane-bed pool-riffle regime braided	Bisson and Montgomery (1996)		
• **channel unit:** habitats: (**section** of Newson *et al.* (1998a)	fast water: turbulent nonturbulent slow water: scour pools dammed pools	Hawkins *et al.* (1993)		
• **within channel:**	aquatic habitats aquatic communities sedimentary assemblage			
• river **environment at a point** – incremental analysis				

Note: Where sections of classifications previously proposed are referred to they are cited by the appropriate reference.

It is such variations 'at a point' that are monitored at river gauging stations with values usually integrated for the entire gauged cross-section. Variations in these process variables, together with morphological characteristics of the channel and riparian vegetation, determine shade, cover and concealment available, providing aquatic habitats which are useful in river management (Harper and Everard, 1998). Habitat-scale investigations have been particularly prominent in headwater streams, especially regulated ones, and also in downstream reaches affected by management procedures. River managers can use instream flow incremental methodology (IFIM) and relatively simple two-dimensional hydraulic models to assess instream habitat impacts of flow regulation and abstraction (Newson and Newson, 2000). Incremental analysis arose from detailed studies of the intimate relationship between stream biota and the physical characteristics of river channels as biologists considered the influence of microhabitat characteristics on stream communities because freshwater biota respond to conditions at a point (Mosley, 1987). A top-down approach has been used by freshwater ecologists proceeding from biota in the river via multivariate analysis to river typologies (e.g. Holmes *et al.*, 1998), to functional habitats (Harper and Ferguson, 1995) or mesohabitats (Pardo and Armitage, 1997), thus determining which physical characteristics control the spatial distribution of habitats. Alternatively, in a bottom-up approach, habitat hydraulics (or ecohydraulics) (Norwegian Insititute of Technology, 1994; Leclerc et *al.*, 1996) are used to predict biotic patterns of the lotic/benthic environment from a knowledge of flow processes. Although not easily combined, these two approaches together can give a more complete understanding of temporal variations in channel environment characteristics (pp. 201–12, Section 8.2), of habitat patch dynamics (Townsend, 1989) and hydraulic stream ecology (Statzner et *al.*, 1988). Combining channel geomorphology and habitat approaches (Frissell et *al.*, 1986) gives the spatial and temporal variability of habitat described as 'patch dynamics' (Bovee, 1996) elaborated on (pp. 74–5).

'Point' data is also often collected at a scale greater than a point, including the column to obtain depth-integrated or depth-differentiated samples or using a grid which may be as little as 1 m or less to obtain small-scale spatial patterns. The range of terms applied to such scales includes hydraulic or physical biotopes, or mesohabitats, which are visually distinct units of habitat within the stream, recognizable from the bank with apparent physical uniformity (Pardo and Armitage, 1997). Flow types are one way of characterizing the physical habitats of the in-channel environment; eight flow types (Table 3.2) can be used to identify physical biotopes analogous to terms used for physical habitat descriptors (Newson and Newson, 2000).

Sediment movement is also an integral varying part of instream habitat conditions. A hierarchy of channel morphological units (Church, 1992), reminiscent of physical biotopes, reflects the hydraulics of flow and sediment movement in the channel (Carling, 1996). Sediment can be visualized hierarchically from individual particles to sedimentary structures, facies models and thence sediment architecture and stratigraphy (Gregory and Maizels, 1991). Sedimentary structures include aggregates of particles as bedforms, with a sequence of bedforms in alluvial channels dependent upon the type of flow regime (e.g. Carling, 1996). Deposits associated with particular fluvial sedimentary environments can include (Lewin, 1996) lag deposits, active channel bars, abandoned channel fills, channel margin deposits (levee, crevasse, crevasse splay), floodplains and backswamps and nonfluvial materials (aeolian, glacial, colluvial). Flow and sediment types combine to constitute habitats identified according to either substrate particle size, aquatic plants, riparian terrestrial vegetation or exposed substrata (Harper et *al.*, 1995) but for management purposes need to be related to the river channel and other scales (Newson and Newson, 2000).

The channel cross-section

The *river channel* is the primary element in the catchment hydrosystem (Table 1.1) and, geomorphologically, the fundamental fluvial landform. For management purposes, it is necessary to know how the river channel is delimited, defined, controlled, what types of cross-section and reach exist, and how these are differentiated by channel vegetation and debris. Definition is not easy because channels can be compound in cross-section, they may not be clearly differentiated from the floodplain and they may alter because of short-term storm events. Definition of the river channel cross-section can be based upon:

- morphology – guided by minimum width–depth ratio, bench index, definition of the active channel (Osterkamp *et al.*, 1983) or overtopping level (Wharton, 1995);
- sedimentological evidence including the height of point and alluvial bars;
- ecological/biotic evidence using trees, grass, shrubs, indicator species or lichen limits (Gregory, 1976b);
- evidence from recent flood events by referring to trash lines.

Defining river channel cross-section end points to a consistent level is necessary so that channels can be compared spatially, throughout one drainage basin or between drainage basins. Channel capacity is usually defined as the cross-sectional area of the river channel as far as the sharp morphological break in slope at the contact with the floodplain. Considerable variety of river channel form, including bedrock channels, is now appreciated so that the channel capacity simply determined in many hydraulic analyses by connecting four points is an over-simplification. Therefore the several possible definitions of capacity need to be borne in mind (Figure 3.2) particularly when relating channel capacity (m^2) to controlling variables including discharge (e.g. Wharton, 1995). Incised channels, which at some point in their history have undergone, or are currently undergoing, bed-level lowering (Simon and Darby, 1999), present a particular problem. Relationships between the dimensions of the cross-section, such as width, depth and slope as dependent variables, with independent variables such as discharge, give a set of hydraulic geometry relationships providing a basis for comparisons of rivers (Park, 1977b). The morphological character of river channel form in a specific basin can be provided by the regression relation between drainage area as the independent variable and channel capacity as the dependent variable; data from 60 different basins (Gregory and Maizels, 1991) showed how channel capacity varies within particular basins and from one to another (Figure 3.3). A decision about channel size appropriate for a particular location needs to consider not only position in the basin but also characteristics of that basin – there can be up to ten times difference in channel size (Figure 3.3), with ephemeral channels in arid basins larger than equivalent ones in humid temperate areas (Chin and Gregory, 2001).

At any specific location along a river, river channel capacity reflects the interaction of water and sediment discharge with local factors including sediment in the bed and banks, vegetation and slope. The frequency of discharge at the bankfull stage, or the channel capacity discharge, was thought to be a 'dominant discharge' corresponding broadly to the mean annual flood with a recurrence interval of 1.58 years on the annual flood frequency series (Gregory and Madew, 1982). However, subsequent research showed that the frequency of occurrence of the bankfull discharge was a range of flows with recurrence intervals in the range 1–10 years (Williams, 1978). Early research concluded that dominant discharge with a particular recurrence interval accounted for the cross-sectional area of the channel. It was later realized that a range of flows controls channel landform (see Knighton, 1998), that the significant controlling discharges can vary along the course of any one river and, in some areas, short-term sequences of events of differing magnitude (e.g. flood- or drought-dominated, see p. 108) may affect channel morphology. River discharges can be estimated from channel dimensions, and such relations have been used in the USA (e.g. Osterkamp and Hedman, 1982) and in the UK

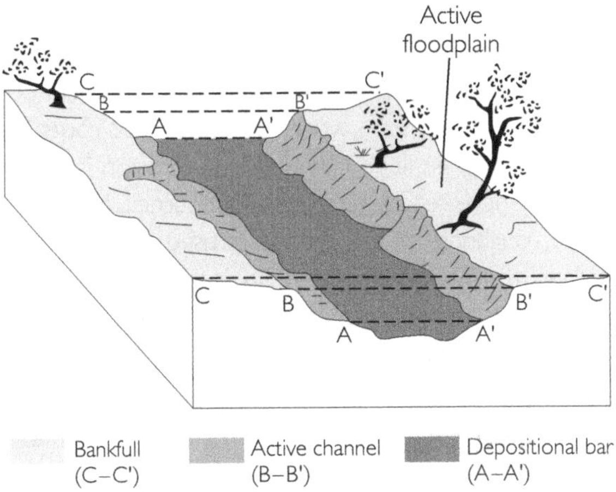

Figure 3.2 *Definitions of channel capacity (Osterkamp and Hedman, 1982)*

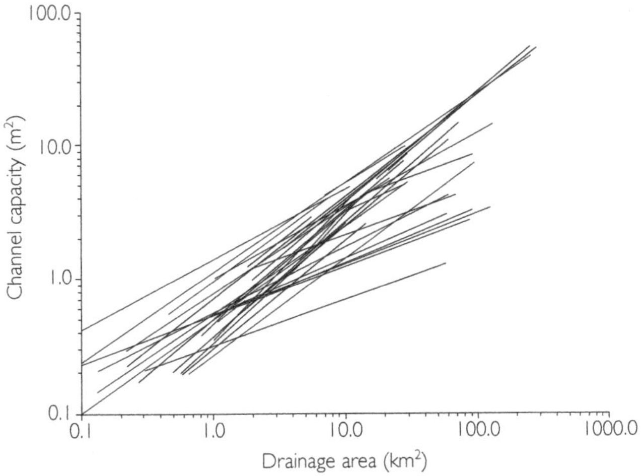

Figure 3.3 *Relation between channel capacity and drainage area (from Gregory in Gregory and Maizels, 1991)*

described as the channel geometry technique (Wharton *et al.*, 1989; Wharton, 1992, 1995), and provide a useful means of first estimation of discharge for management of ungauged river channels.

Sediment discharge, as a fundamental influence on the channel, can be the basis for a classification of alluvial channels with suspended load, bed load and mixed load channels each having characteristic values of width–depth ratio, sinuosity and slope (Schumm, 1963). Channel morphology, especially channel shape, is affected by sediment in the bed and banks because channels in cohesive sediments tend to be relatively deeper with lower width–depth ratios than those for an equivalent controlling discharge in less cohesive sediments. Local slope influences channel capacity, form and shape because, through the continuity equation, higher channel slope is associated with higher velocities, which in turn require a smaller-capacity channel to convey equivalent discharges. Riparian vegetation influences the channel form in a manner akin to that of bank sediment because it increases the cohesiveness of the channel banks – where bound by tree roots width–depth ratios tend to be lower, whereas in grass-lined channels bank erosion is more readily achieved and the channels can shift more readily (Gregory and Gurnell, 1988). Vegetation and sediment influence on the roughness of the channel is used in flow equations such as the Manning equation (Manning, 1891) to estimate velocities and discharges at ungauged sites. In addition to providing flow resistance, vegetation also influences bank strength, bar sedimentation, the formation of log jams and the incidence of concave bank bench deposition (Hickin and Nanson, 1984). Because of the numerous ways in which the distribution, character and presence of riparian vegetation has been modified along world rivers, it is inevitable that vegetation influences have changed over time.

Controls upon channel capacity need to be considered for river channel management because it is imperative that the inter-relations between discharge, local slope, vegetation and sediment are analysed together with knowledge of potential for instability through channel bank and bed erosion. Channels can be categorized according to their intrinsic characteristics (Schumm, 1963) and in a range of ways shown in Table 3.1; in an idealized river system, four channel types (Bathurst, 1993) can be distinguished according to bed material size and channel slope, which both decrease downstream, namely steep pool/fall, boulder bed, gravel bed and sand bed channels, each characterized according to ranges of channel slope, bed material size and resistance coefficients (Table 3.4).

There can be significant spatial variations in morphology that reflect the nature and rate of sediment transported according to the slope and character of *the reach*, with several types of reach recognized (Tables 3.3, 3.4; also see Church, 1992). Plane bed channels may be formed on bedrock or have a surface layer of coarse material, and typically occur at slopes of 0.01–0.03 (Montgomery and Buffington, 1997). However, along many bedrock or mountain channels there is a regular alternation of steps, composed of stones, boulders and cobbles, bedrock and woody debris, and plunge pools developed at the foot of each step, together comprising the characteristic bedforms that dominate the morphology of steep mountain streams (Chin, 1989). Such step–pool sequences usually occur at reach

Table 3.4 Channel types and characteristic values

Type of channel	**Approximate range of**		
	Channel slope (%)	**Bed material D_{50} (mm)**	**Manning *n***
Steep pool/fall	> 5.0	variable	0.1–5.0
Boulder-bed	0.5–5.0	>100	0.03–0.2
Gravel/cobble-bed	0.05–0.5	10–100	0.02–0.2
Sand-bed	<0.1	<2	0.01–0.04

Note: D_{50} is the bed material particle size for which 50% of the material is finer (after Bathurst, 1993).

gradients of 0.03–0.10 (Montgomery and Buffington, 1997). Like plane bed channels, step–pool sequences may indicate a sediment-supply-limited system (Whittaker, 1987). The spacing of the steps and pools is related to channel size with the average spacing ranging from one to four channel widths (Chin, 1989, 1999). Along alluvial channels there can be variations in bedform according to sediment calibre, and there is also an alternation of shallow (riffles) and deeper parts of the reach (pools) that occur characteristically along both straight and meandering channels, especially along gravel-bed streams. Along mixed sediment rivers, where gravel can be available to armour the surface, pools and riffles are the natural bedforms where slope is between 0.005 and 0.200 (Brookes and Sear, 1996). The pools are associated characteristically with meander bends whereas the riffle typically occurs in the straighter section between two bends. The spacing of the pool–riffle sequence is related also to channel size, and has been found to occur with a spacing some five to seven times the channel width (Leopold *et al.*, 1964); although some later results show that the spacing in some areas can be less than this (Gregory *et al.*, 1994). Although there is a range of possible explanations for the development and maintenance of the pool–riffle sequence (Clifford, 1993), the sequence is a means of self-adjustment in channels, closely related to the development of the planform through the flow patterns in meandering channels (Thompson, 1986); they can be characterized as response reaches because of the way in which they respond to changes in water and sediment discharges (Montgomery and Buffington, 1997). The pool–riffle sequence is an integral element in river channel process mechanics so that, when channel adjustment occurs, there is likely to be an associated adjustment in the pool–riffle spacing; after a major channelization scheme in a small, relatively low-power river, riffles could be expected to be established within 16 years (Gregory *et al.*, 1994). Furthermore the re-establishment of the pool–riffle sequence has to be considered as an integral part of restoration management (e.g. Nielsen, 1996; see Chapter 10, p. 294).

Along the reach the distribution of vegetation is determined by hydrochemical and hydrodynamic processes, particularly the need for plants to adapt to the hydraulic stress of flowing water. Inchannel plants are often flexible, with long, thin leaves offering minimal resistance to flow, effective anchoring systems to colonize loose sediment, and rapid growth to offset the potential high rate of damage caused by abrasion and movement of their substrate. In lower velocity, but turbid, waters, aquatic plants often develop floating leaves for a more efficient use of solar energy, and in backwater and cut-off channels free-floating plants develop. Aquatic plants on the bankside may have different types of leaves for particular water levels and many riparian plants deal with hydraulic stress through adventitious roots and an ability to survive periods of inundation, taking nutrients from alluvium deposited during floods (Large and Petts, 1994).

Reaches in woodland or forest areas may have coarse woody debris (CWD) accumulating in the river channel, often as debris dams, which can be classified as *active*, forming a complete barrier to water and sediment movement and create a step or fall in the long profile; *complete*, forming a barrier across the channel without interrupting the slope; and *partial* accumulation that does not cross the channel or has been partly destroyed (Gregory *et al.*, 1985). Such debris was of greater significance before deforestation occurred in many areas of the world and before management of river channels removed the woody debris (Gregory, 2003a). When the influence of vegetation and woody debris on stream channel morphology was first demonstrated in the Sleepers River basin of northern Vermont (Zimmerman *et al.*, 1967), thresholds were identified at drainage areas of 2 and 12 km^2. In basins up to 2 km^2 trees and woody debris were the dominant influence upon channel morphology; from 2 to 12 km^2 channel width increased downstream although varying locally according to vegetation and tree influences; for drainage areas greater than 12 km^2 the effect of vegetation became less dominant. This shows how the influence of CWD decreases markedly once the river channel width exceeds the height of the local trees (Gregory, 1992b), although downstream of such a threshold major debris accumulations can still occur, albeit with less frequent spacing. Interactions between woodland and river channel form and process is illustrated in Figure 3.4; the research accomplished since 1967 was revealed by the bibliography of more than 1000 references compiled for a major international conference in 2000 (Gregory *et al.*, 2000).

Channel features affected by debris accumulations and debris dams (Gregory and Davis, 1992) include the pool–riffle sequence, channel roughness, bank stability and the precise location of channel change. These effects arise from the way in which debris dams affect water and sediment routing along the channel and especially sediment storage, the distribution of erosion (e.g. Davis and Gregory, 1994) and the dissipation of potential energy. Such influences upon the morphology and hydrological and hydraulic processes mean that the ecology of the channel is affected, including the effects of debris dams upon fish populations, the diversity of instream habitats and the variety of instream species. Studies of the spatial distribution, or loading, of debris within channel systems indicate that dam densities can be up to 40 per 100 m of channel (Likens and Bilby, 1982) and can involve a loading value of up to 225 kg m^{-2} (cited in Harmon *et al.*, 1986). In a study of spatial variations in the 110 km^2 Lymington River basin, UK, the average density of dams is 1.15 per 100 m with downstream variations affected by land use of the riparian areas, and the cumulative effects of management over many years (Gregory *et al.*, 1993). The amount of wood in any stream channel is affected by the forest type, successional stage, disturbance history, decomposition rate and channel size (Sedell *et al.*, 1988) so that the time duration or permanence of the debris accumulations is an important consideration. Along the Highland Water, UK, about one-third of the dams changed in some way during a year (Gregory *et al.*, 1985) and in California there can be 65% redistribution within six years (Lienkaemper and Swanson, 1987). Contemporary river channel management often involves selective removal or even restoration of wood in channels with management alternatives proposed according to the characteristics of the channel and the amount of wood present, Gregory and Davis, 1992; table 1).

Channel pattern

Channel pattern or planform, literally the plan of the river channel from the air, like channel capacity is not always easy to define because it can be stage-dependent; a channel that is braided at low flows may become single thread at higher flows. Although patterns were initially subdivided into straight (relatively uncommon), meandering and braided (Leopold and Wolman, 1957), and subsequently into single-thread and multithread channels

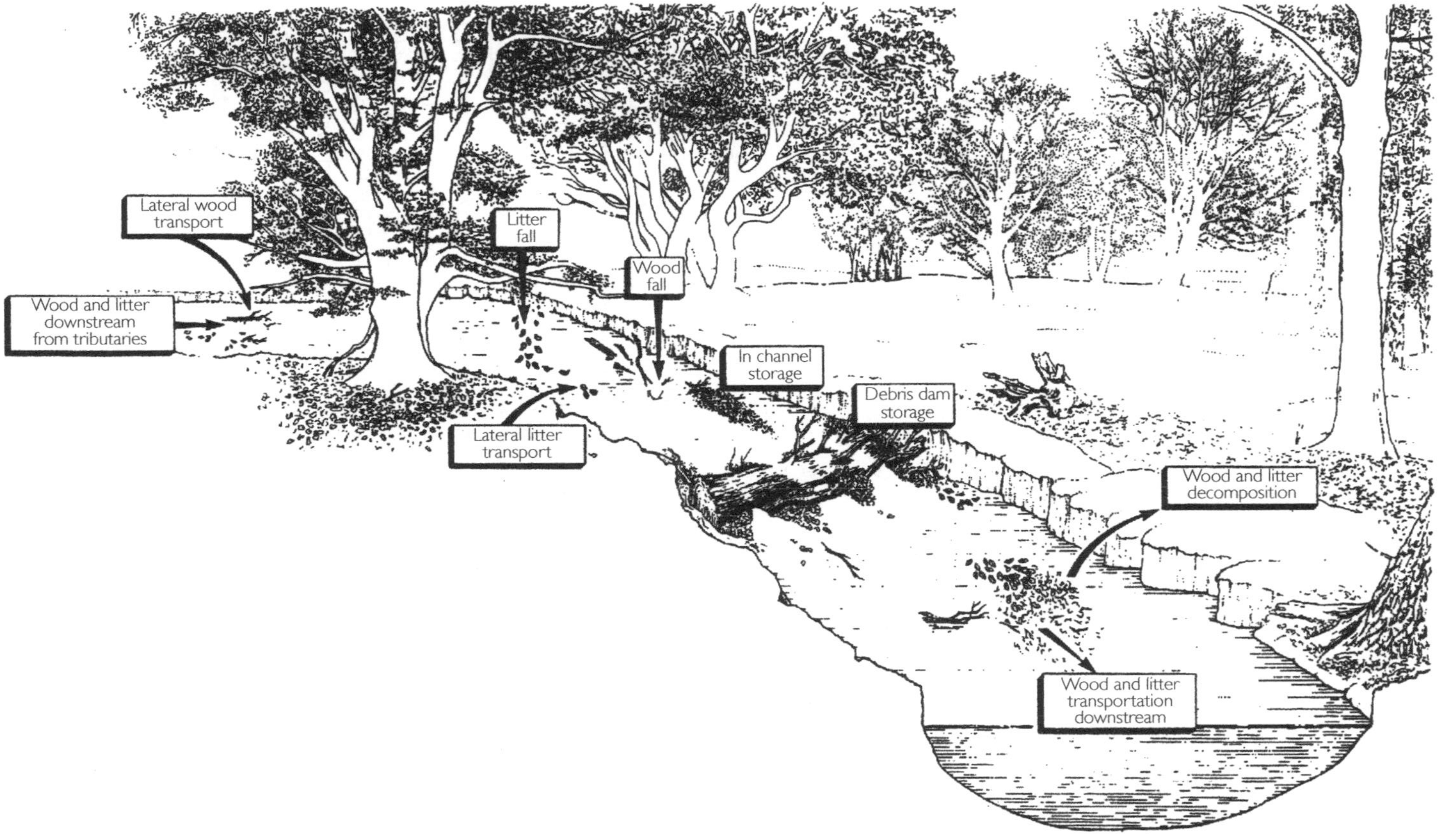

Figure 3.4 *Interaction between river channel form and process in woodland areas (Gregory, 1992b)*

(Chitlale, 1973), a much broader range of patterns is now recognized. Although there is a continuum, whereby one category of pattern grades into another, classifications have been based upon planform (sinuosity, degree of braiding or relative stability), sedimentological characteristics (islands, bar structures, type of sediment load) and channel processes (lateral activity) from which indications of stability can be derived (Church, 1992). A classification based upon any one factor is unsatisfactory and Rust (1978) distinguished straight, meandering, braided and anastomosing according to degree of sinuosity and of division (Figure 3.5A), whereas Schumm (1981, 1985b) suggested 14 patterns (Figure 3.5B) related to sediment characteristics and relative stability. Such an approach may be extended to a classification that reflects relative stability and types of hazard (pp. 122–3) associated with each pattern (Shen *et al.*, 1981). An anabranching river consists of multiple channels separated by stable islands, which are large relative to the size of the channels. Islands in anabranching rivers usually persist for decades or centuries, unlike the more transient bars of braided systems, and are at the same level as the surrounding floodplain; and the channels in anabranching systems behave as independent channels. Characteristics of anabranching systems can be assessed in relation to the continuum of river channel patterns by using three variables – flow strength, bank erodibility and relative sediment supply (Knighton and Nanson, 1993). A further variety of river channel pattern is a wandering gravel river (Church, 1983), which occurs where there has been multiple flow dissection of the active floodplain rather than multiple bars featuring in a single channel as is the case in a braided channel system (Gregory, 1987a).

Size of channel planform, measured by values such as meander wavelength, is related to the controlling discharge, so that the size of meander bends or of braided channels increases downstream and any relationship between pattern and drainage area uses basin size as a surrogate for discharge. Discharge is not the only controlling variable, sediment characteristics and slope also being important, not only determining the incidence of particular kinds of channel patterns but also governing the transition from one type of pattern to another. Thresholds have been specified, either graphically or in equation form, to describe the limiting situation between types of channel pattern. One of the first was the use of bankfull discharge (Q_b) and slope (S) which enabled Leopold and Wolman (1957) to separate meandering and braided planforms as: $S = 0.013Q_b^{-0.44}$; Lane (1957) separated meandering and transitional channels by $S_b = 0.007Q_m^{0.25}$ and transitional from braided threshold by $S_b = 0.0041Q_m^{-0.25}$ where Q_m is the mean annual discharge (see also p. 220). The range of channel planforms affords a variety of diverse habitats, each often associated with particular patterns of vegetation community, affected by discharge frequency and sediment transport characteristics. The role of vegetation in driving riverine landscapes may have been underestimated in highly developed temperate regions such as Europe (Gurnell and Petts, 2002).

Floodplain

Floodplain definition is probably more difficult than river channel cross-section or a channel pattern; not only is it more extensive but also there are several possible definitions of the floodplain and the river corridor, embracing the river and river channel together with their associated wildlife and the adjacent riparian ecosystem (Gurnell *et al.*, 1994), to be considered (Table 1.1). Understanding of floodplain morphology and process was initially domi-

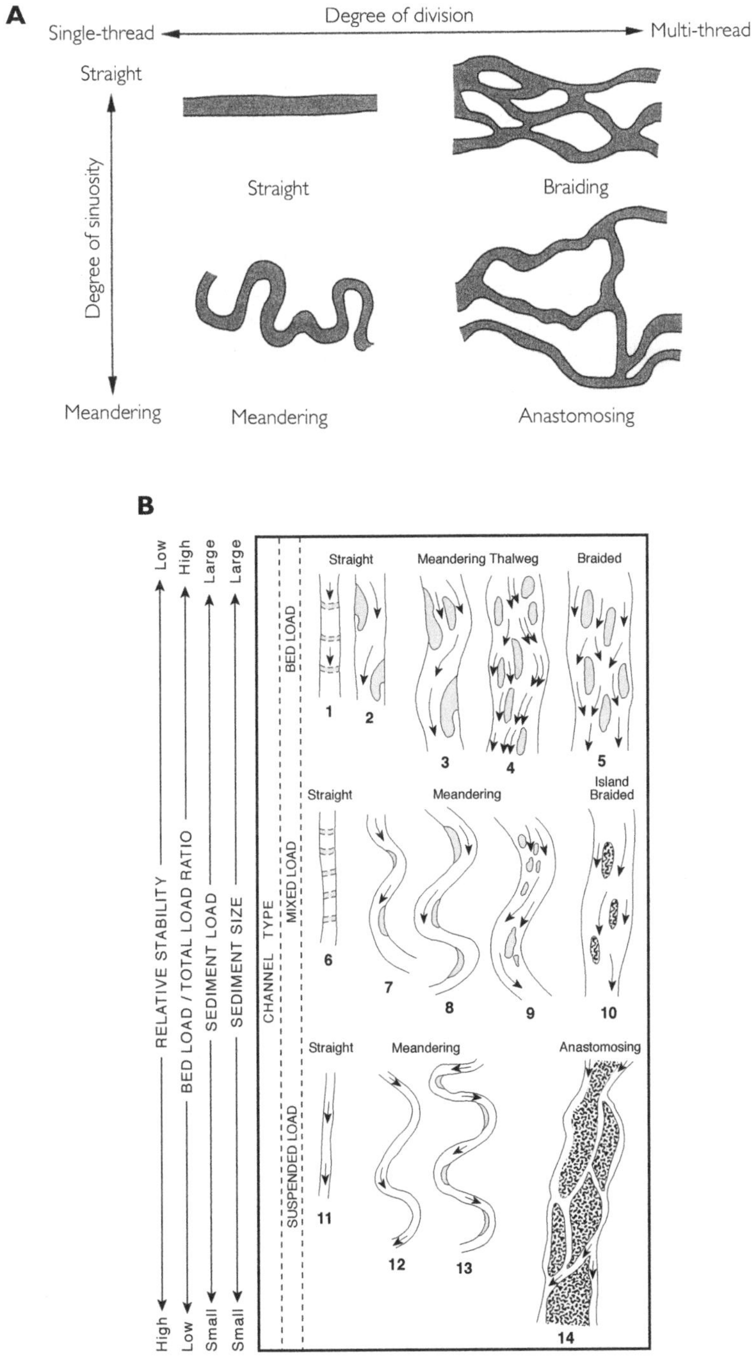

Figure 3.5 *Channel patterns.* ***A*** *Classification of channels according to sinuosity and division (adapted by Lane, 2000 from Rust, 1978).* ***B*** *Classification of channel pattern developed by Knighton (1998) based upon Schumm (1981, 1985b)*

nated by experience from particular rivers, including the Mississippi, Watts Branch, Klaralven and Brahmaputra, each the subject of early influential research (Nanson and Croke, 1992). Physical processes of river deposition operating to varying degrees in a range of sedimentary subenvironments produce substantial variety in floodplain forms (Lewin, 1996) and floodplain assemblage (Figure 3.6A). Research from a great variety of areas, increasing the knowledge of the range of processes involved in floodplain formation (Anderson *et al.*, 1996), has meant that the variety of floodplain types is so diverse that almost every study might require a new model (Nanson and Croke, 1992). A distinction is made between *hydraulic floodplain* recognized by hydrologists and engineers as the surface next to the channel that is inundated at least once during a given return period, and the *genetic floodplain* defined as 'the largely horizontally bedded alluvial landform adjacent to a channel, separated from the channel by banks and built of sediment transported by the present flow regime' (Nanson and Croke, 1992: 460). This landform is contemporaneous with present hydroclimatic conditions, although there may be polyphase floodplains, the product of secular climate or other environmental change, and persistent passive disequilibrium floodplains (Ferguson, 1981) that relate closely to the controls of past hydrological and sedimentological regimes. Other definitions of floodplains have included the active floodplain (Leopold *et al.*, 1964), which is the area subject to annual flooding, or the topographic floodplain which may include part of the relict floodplain surface and is comparable with the genetic floodplain.

Floodplains are formed by channel and overbank accretion and the three main processes are (Nanson and Croke, 1992):

- lateral point bar accretion;
- overbank vertical accretion;
- braid-channel accretion;

together with three less common processes, namely:

- oblique accretion (including inner accretionary bank deposits);
- counterpoint accretion;
- abandoned channel accretion which includes cut-offs and crevasse splays.

An energy-based systematic ordering of floodplains (Nanson and Croke, 1992) recognized three floodplain classes (based on stream power and sediment characteristics) and a combination of 13 floodplain orders and suborders (based on primary and secondary factors, largely geomorphic processes):

A high energy noncohesive floodplains – disequilibrium landforms that erode either completely or partially as a result of infrequent extreme events;

B medium energy noncohesive floodplains – in dynamic equilibrium with the annual to decadal flow regime of the channel and not usually affected by extreme events. Preferred mechanism of floodplain construction is by lateral point bar accretion or braid channel accretion;

C Low energy cohesive floodplains. Usually associated with laterally stable single-thread or anastomosing channels. Formed primarily by vertical accretion of fine-grained deposits and by infrequent channel avulsion.

This classification (Figure 3.6B) is useful for contemporary river channel management but also links with past development to sedimentary architecture models (Miall, 1985).

The morphological variety of floodplains and diversity within them is echoed by their habitats and ecology. Ecologically a floodplain comprises a mosaic of inter-related and interconnected patches, and the functional relationships between the patches and their biological communities depend upon hydrological connectivity with the river and with the alluvial aquifer (Petts, 1998). Spatial patterns of vegetation communities are largely dictated by the floodplain elevation relative to the river, are influenced by the hydrodynamic and hydrochemical gradients but complicated by successional factors. Significant differences occur between terrestrial successions, developing on infilling channel courses (e.g. cut-off meanders, abandoned braids) where the amount of organic material may reach 80%, and those developing on elevated banks, islands and levees where there may be less than 5% organic material. The timing and duration of flooding further differentiates vegetation assemblages. As the frequency and intensity of flooding decreases away from the channel, more woody species generally prevail. The floodplain is increasingly seen as an integral part of the river corridor, which is a linear feature of the landscape, structured along ribbons of alluvium extending from headwaters to the sea (Ward *et al.*, 2002), and more generally as part of the riverine landscape that benefits from a multidisciplinary landscape ecology approach (Tockner *et al.*, 2002).

River channel network

The river channel network, especially its dynamic properties, has frequently been considered at the basin scale in terms of flood routing and hydrological basin modelling, separate from the channel at a cross-section (the hydraulic geometry), and from the river channel pattern along a specific reach, including the relationship with the floodplain. A drainage network is dynamic, embracing perennial, intermittent and ephemeral channels and, in a humid area drainage basin, the range of density of the network composed of these three types of channel varies according to storm events and to prevailing antecedent conditions, from 1.0 to 5.0 km km^{-2}, whereas in arid regions, although the density of perennial channels is usually very low, the density of all channels, including the ephemeral ones, is comparatively high (up to 10 km km^{-2}) because low vegetation density affords little resistance to the production of ephemeral channels during the infrequent storms with high rainfall intensities. Such variations in drainage density between different areas, and within a single basin according to prevailing hydrological conditions, have to be related to the channel and floodplain dynamics elsewhere in the drainage basin (Gregory, 1976a). In a humid basin discharge increases downstream because there are no effluent or transmission losses below the channel bed. However, in arid areas or on permeable rocks such as limestone, where the saturation level is not at the channel bed, the discharge may decrease downstream as a consequence of effluent seepage or transmission losses into the bed. In March 1978 a flood lasting several days with a peak of 3500 $m^3 s^{-1}$ along the Salt River in central Arizona had transmission losses of 17% (Aldridge and Eychaner, 1984). Such transmission loss varies over time as the saturation level fluctuates, emphasizing the need for a three-dimensional view of the dynamics of the river channel network within the drainage basin.

Each component of the basin responds to dynamic conditions in a particular way.

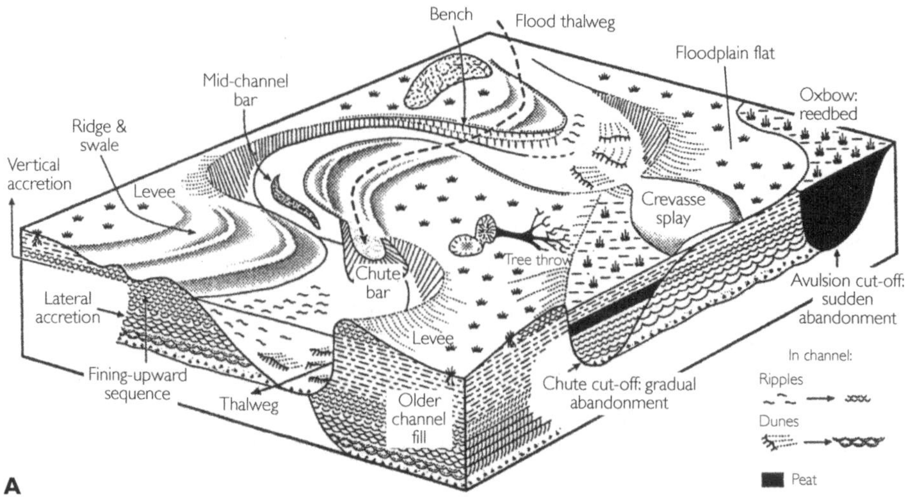

A

Confined coarse-textured floodplain
$\omega = >1000\ \mathrm{Wm^{-2}}$

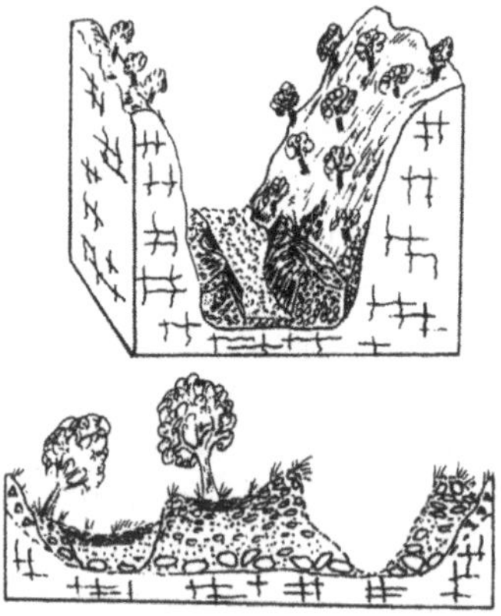

Confined vertical-accretion sandy floodplain
$\omega = 300\text{–}1000\ \mathrm{Wm^{-2}}$

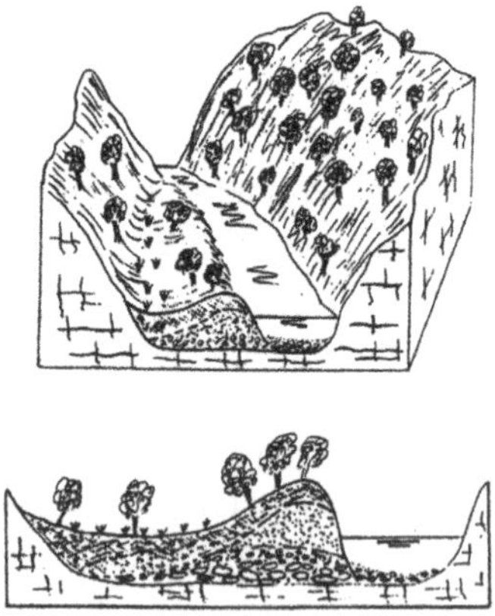

Cut and fill floodplain
$\omega = \sim 300\ \mathrm{Wm^{-2}}$

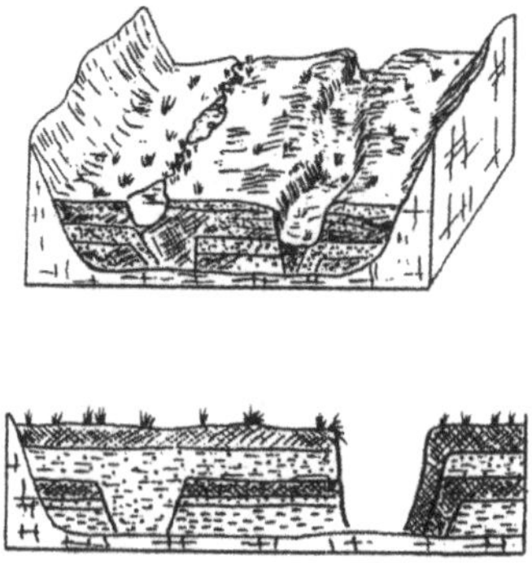

High energy non cohesive floodplains

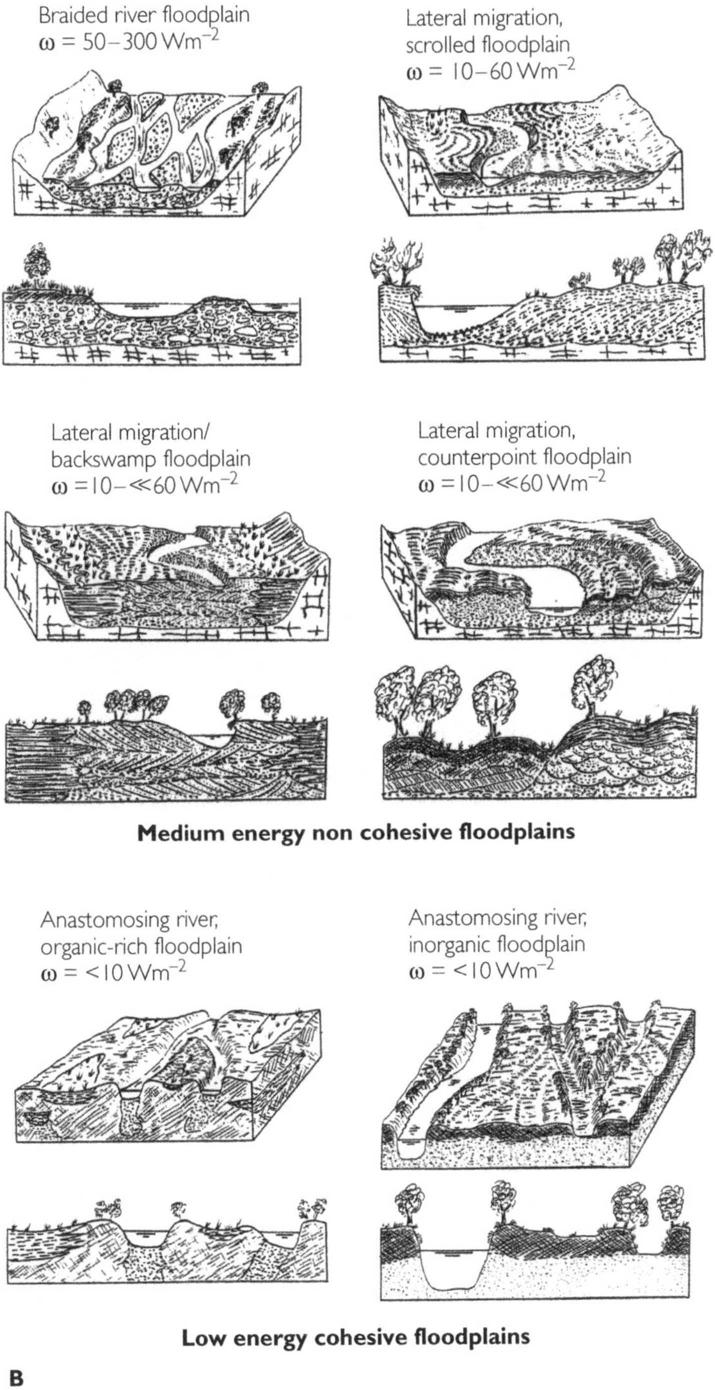

Figure 3.6 *Floodplain development.* ***A*** *A floodplain landform assemblage (after Brown, 1996; adapted from Allen, 1965).* ***B*** *Types of floodplain (after Nanson and Croke, 1992)*

During different storm events, just as the river channel cross-section and the associated floodplain have different degrees of inundation to specific levels (Figure 3.2), so these levels relate to the different stages of river channel pattern in a reach, and to the different stages of extension of the drainage network (Gregory, 1977a). Progressing downstream in a drainage network processes can be 'stepped' as a result of sudden increases in discharge or of surges of different calibre sediment input from a tributary channel.

Connectivity: links between the components

Rivers are transfer systems and also storage systems (Petts, 1998) so that linkages between the separate components have been stressed, and understanding of them is necessary for integrated basin management (Chapter 6, p. 160) and pertinent to river channel management (see Chapter 10, p. 288). River margins are locations where the dynamic interface between terrestrial and aquatic environmental and biological processes and biological diversity tends to be maximized (Gregory *et al.*, 1991) so that connectivity themes include the river continuum concept and serial discontinuity; sediment supply, slugs and storage; nutrient spiralling; the flood pulse concept; ecological patches; and tectonic influences.

The river continuum concept

The river continuum concept and, for lotic ecosystems in regulated rivers, the serial discontinuity concept (Ward and Stanford, 1983), highlighted connectivity (p. 50) and a holistic view of river systems, although the implicit assumption of a stable, single thread channel from headwaters to the sea (Ward *et al.*, 1999a) may have retarded appreciation of the way in which the components of fluvial hydrosystems interact, and underplayed the lateral, or transverse, dimension of river ecosystems. Immediately after a flood-inducing storm event the drainage network is at its maximum extent, including ephemeral as well as intermittent and perennial channels; as the flood wave is generated and moves downstream it may exceed the capacity of the reach, the planform and the channel cross-section so that flood water extends over the hydraulic floodplain. Concepts were therefore necessary to incorporate lateral, vertical and temporal dimensions (Ward, 1989; Sedell *et al.*, 1989; Stanford and Ward, 1993; Ward and Stanford, 1995) so that the basin system at any point can be regarded (Stanford, 1996) as four-dimensional involving:

- the longitudinal river continuum from headwaters to the ocean;
- the lateral aquatic (river channel) to the terrestrial (valley uplands), emphasizing the coupling between components such as hillslope to channel, or channel to floodplain;
- elevational movement along flow pathways – including the flux of water and sediment and solutes from one component to another;
- time emphasizing the discontinuities that can occur within catchment processes. Sediment released upstream from gully erosion can accumulate in floodplains as a sink so that much later the sediment in storage is released to continue through the continuum to the sea. Downstream alluviation and headwater incision can be contemporaneous, and in west central Wisconsin the incision of tributary channels created enlarged channels with capacities to contain floods (Faulkner, 1998) so that it was

possible to compare incised channels in 1939, 1958 and 1965. Four types of incised channels, rills, gullies, entrenched streams and composite incised channels (Schumm, 1999) can have spatial counterparts downstream (Strahler, 1956) that may or may not be contemporaneous.

Like most conceptual models, the river continuum concept is a simplification of reality because it requires modification when applied to rivers beyond small temperate streams (Statzner and Higler, 1985; Junk *et al.*, 1989), considers only certain aspects of the river ecosystem environment (i.e. permanent and lotic, Junk *et al.*, 1989); does not accommodate the impact of significant river–floodplain interactions (Sedell *et al.*, 1989); or flow regulation (Bayley and Li, 1996). The term 'river continuum' has entered general use when describing streams despite the fact that drainage networks often possess significant 'steps' in their hydraulic geometry (Richards, 1973) and habitats according to incoming tributaries (Bruns *et al.*, 1984; Bravard and Gilvear, 1996). Also, the downstream component of the geomorphological template may be far more complex than assumed by the concept, especially related to channel slope (e.g. where bedrock outcrops cause 'discontinuity' in the graded profile).

The serial discontinuity concept

Ecologically, the continuum model assumes continuous transport of nutrients whereas, in reality, downstream transport may be far more 'jerky' with storage alternating with periodic transport. The *serial discontinuity concept* (Ward and Stanford, 1983) emphasizes the potential for the continuum to break down or shift position when the upstream–downstream connectivity is broken (e.g. by large dams). However, the continuum concept has elucidated the general ecological progression of a river from its headwaters downstream and demonstrates from first principles the unified, interdependent nature of a healthy catchment ecosystem and the role of natural hydraulic stresses in determining the biological response.

Sediment supply, slugs and storage

Sediment discharge is as important as river discharge in shaping the river channel system and, in addition to sediment load and calibre, sediment supply is often important. Theoretical sediment transport equations assume unlimited sediment supply but the river transfer system is a storage system in which floodplains can act as sinks for suspended sediment and sediment delivery to particular reaches may be supply-limited. Sediment may remain in floodplain storage for hundreds or thousands of years before removal, posing problems where radioactive materials or other contaminants are associated with the sediment (Marriott, 1998). Sediment storage in the landscape system is illustrated by residence times of up to 10^2 years in channel storage, up to 10^3 years for colluvial sites and up to 10^6 years in alluvial storage on floodplains in the Severn Basin UK (Brown, 1987). On more recent timescales a fluvial audit can demonstrate areas of sediment supply and sinks illustrated by a basin within the Forest of Bowland (Newson *et al.*, 1997). Analogous to the flood pulse are sediment waves or sediment pulses termed *sediment slugs*, defined as bodies of clastic material associated with disequilibrium conditions in fluvial systems over time periods above the event scale (Nicholas *et al.*, 1995). These can range in magnitude from unit bars, at the scale where perturbations in sediment transport are generated by local riverbank erosion and/or bed erosion, to larger scale features resulting from the occurrence of rare, high-magnitude geomorphic events and impacts on water and sediment production of tectonics, glaciation, climate change and anthropogenic effects.

Nutrient spiralling

As nutrients move through the drainage basin, nutrient spiralling (Newbold *et al.*, 1981) involves the cycling of carbon, phosphorus and nitrogen (Newbold, 1996). As a nutrient atom passes downstream it may be used repeatedly, the amount of utilization depending upon the 'tightness' of the spirals or the downstream displacement from one cycle to the next. These are affected by the speed of cycling, the retentiveness of the ecosystem and the degree to which the downstream transport is retarded relative to the water. This resource spiralling concept (Elwood *et al.*, 1983) has analogies with notions of a 'jerky conveyor belt' of downstream sediment transport.

The flood pulse concept

Water and sediment transfer through the basin effect connectivity between the components; the dominant, or channel-forming, discharges that are significant in relation to the channel cross-section and ecology have a significance in relation to a particular state of channel planform and extent of drainage network. In addition to the water balance equation, the type of hydrological regime and the generation of the annual hydrograph, it has been increasingly appreciated how individual storm events or periods of years with above or below average flows can influence morphological and ecological characteristics. The flood pulse concept (Junk *et al.*, 1989; Bayley, 1991) emphasizes the fact that components of the hydrosystem are affected by the frequency and spatial extent of flood pulses. The extent to which the natural river ecosystem benefits from the annual flood has been characterized as the 'flood pulse advantage' (Bayley, 1991), the pulsing of river discharge determining the degree of connectivity and the exchange processes of matter and organisms across river–floodplain gradients. The key physical component in the flood pulse advantage is the 'moving littoral' (Junk *et al.*, 1989), the migration of the aquatic environment laterally onto the floodplain. A slow flood allows time for *in situ* processes to occur, such as release into the water of mineralized products and dry-phase processed organic material. The structure and composition of biological communities is controlled primarily by the seasonal advance and retreat of flood waters across adjacent floodplains, such annual flood pulses enhancing the diversity and productivity of riverine communities. Primary productivity may be enhanced so much that fish populations are not limited by food supply but by factors such as dissolved oxygen levels, availability of spawning levels and predation (Petts and Maddock, 1996). Although developed for large tropical rivers the flood pulse concept can be extended to temperate areas (Tockner *et al.*, 2000) by adding flow pulses (Tockner *et al.*, 2000), because expansion–contraction cycles occurring well below bankfull can influence landscape heterogeneity and biodiversity patterns (p. 284).

Ecological patches
Connectivity is also emphasized through links of freshwater ecology with physical processes, for example on the floodplain (Petts and Maddock, 1996) where the dynamics in terms of structural and functional connectivity of the mosaic of ecological patches, can build on the ecological theory of patch dynamics (Pickett and White, 1985). The natural heterogeneity of patches, in terms of their type, size and age, is a critical component of a healthy river ecosystem. A patchwork of geomorphological units nested at different spatial scales (Bravard and Gilvear, 1996) is analogous and can provide special habitats. The driving forces are summarized as flow variability, elevation relative to river level, sediment permeability and frequency of destructive events, and these physical factors drive a variety of processes fundamental to the floodplain ecology. For instance, geomorphological processes of erosion, deposition and flood disturbance are critical in driving a series of reversible processes (Amoros *et al.*, 1987a,b) that result in an ever-changing succession of 'wetter' and 'drier' habitats; hydrological processes often stimulate biological responses in fish whose life-cycles are adapted to periods of seasonal flooding and in flora that are physiologically adapted to periods of reduced oxygen availability. While the patches are diverse structurally, and can be subdivided geomorphologically into a variety of 'functional sets' centred around the active channel, major aggradational or erosional morphologies and abandoned channels, the truly 'special' characteristic of river ecosystems is the role played by periodic flooding. Flood disturbance provides the basis for the characteristic hydrodynamic and hydrochemical condition in each set and is important in periodically re-setting the physical template on which the species develop. Therefore, in addition to the spatially diverse environment provided by natural river ecosystems, there is a marked temporal component that consistently redefines functional connectivities across the floodplain and rejuvenates ecological successions.

Tectonic influences
Tectonic influences, which can be particularly significant in areas such as Japan and New Zealand, may affect substantial parts of drainage basins (Schumm *et al.*, 2000). Along the Indus, tectonic deformation of the valley floor affects variations in channel planform, channel dimensions and the frequency of flooding, although near Mohengro Daro, a region of forward tilting, the dynamic behaviour of the river and channel avulsion are preferred explanations for the abandonment of Mohengro Daro rather than previous ideas on improbable tectonic and geomorphic conditions (Harvey and Schumm, 1999).

3.4 Dynamics of channel behaviour

Increasing emphasis upon connectivity in the river system encouraged multidisciplinary approaches, to focus on unifying themes for characterizing catchment and channel behaviour, including energy, power and equilibrium.

Energy has been used as a linking variable with underlying principles (Rodriguez-Iturbe and Rinaldo, 1998), which may be specified such that energy expenditure:

- is minimum in any link of the network for the transportation of a given discharge;
- is equal per unit of channel anywhere in the network;
- is minimum in the network as a whole.

An energetic approach (Gregory, 1987b) can employ *stream power* (ω). First introduced in relation to the rate of sediment transport, it was defined (Bagnold, 1960) as the product of fluid density (ρ), slope (s), acceleration due to gravity (g) and discharge (Q) in the form:

$$\omega = \rho Q g s$$

Definition of stream power was clarified by Rhoads (1987) and developed as a unifying theme for urban fluvial geomorphology (Rhoads, 1995). Examples of the analysis of power in the hydrosystem (Table 3.5) show its potential significance for analysis (and see p. 134). The purpose of characterizing the 'natural' system and its dynamics is to provide the basis for management and a further key variable is sensitivity, analogous to disturbance regimes in ecology (Ward, 1998), as developed in Chapter 5.

Most natural stream channels appear as stable and in equilibrium with the prevailing flow conditions so that there have been numerous attempts in ecology, geomorphology and hydrology to specify types of equilibrium condition that exist between the components of the fluvial hydrosystem. Input of mass and energy to a specific channel reach equal to the output could be the basis for specifying an equilibrium condition, indicating a target condition to be maintained or engineered in river channel management. Such an equilibrium condition was subsequently referred to in the engineering literature as 'in regime'. The term grade was originally developed by G.K. Gilbert (Gilbert, 1877) to denote the condition of a channel with a gradient such that a perfect balance occurred among corrasion, resistance and transportation (Graf, 1988b), but has not proved to be easy to apply (Dury, 1966) and has been found to be useful only if substantially modified and developed (Knox, 1975). Rivers are elements of the landscape that change progressively, so that dynamic stability has been defined as 'in a dynamically stable river, reductions in sinuosity through natural cut-offs of meanders are just balanced by the growth of other meanders, so that the overall sinuosity of the reach is steady' (Thorne, 1992: 95–96).

Although it is tempting to assume that there is one simple form of equilibrium that applies to river channels, this has proved to be elusive and several equilibrium concepts are illustrated in Table 3.6. Simple relationships established between the size of the channel cross-section and the discharge of a particular frequency (p. 59) sometimes prompted erroneous notions that channel morphology was controlled by a discharge of a particular magnitude. However, as it was subsequently appreciated that the morphology of a channel is related to a range of discharges and is also influenced by the sediment transport and by the local controlling conditions, it was contended that channel morphology, whether it be channel cross-section or channel pattern, is in some state of equilibrium with the processes operating in the channel. However, any equilibrium is an apparent or quasi equilibrium (Leopold and Maddock, 1953), perhaps more appropriately termed approximate equilibrium (Graf, 1988a), because hydrological events controlling the channel geometry and pattern may occur just once each year or even less frequently so that for the rest of the time the morphology is in a state of apparent equilibrium. The timescale for equilibrium relationships varies according to the components of the hydrosystem and may differ for channel cross-section, planform and network, for example. In addition there may not have been sufficient time available (disequilibrium) a system may be adjusting to a new

Table 3.5 Examples of the use of power in hydrosystems analysis

Used as	Application	Source
Critical power	Defined as the power just sufficient to transport the sediment through the reach	Bull (1979, 1988)
Flood power	Related to palaeofloods	Baker and Costa (1987)
Threshold power	For catastrophic modification of the channel or fluvial landscape, has been suggested as a unit stream power of 300 W m^{-2}	Magilligan (1992)
Stream power used in relation to		
Sediment transport	Stream power used instead of stream discharge, velocity or bed shear stress to relate to sediment motion and transport, especially that of bed load	Allen (1977)
River channel patterns	Patterns classified according to amount and size of bedload and stream power; channel sinuosity has been related to stream power.	Schumm (1977b, 1981)
Floodplains	Types of floodplain differentiated according to stream power values, including high energy (ω>300 W m^{-2}), medium energy (10<ω>300 W m^{-2}), and low energy (ω<10 W m^{-2})	Nanson and Croke (1992)
Channel morphology	British rivers demonstrated a thousand-fold range in the values of specific power with a clear distinction between values of 100 and 1000 W m^{-2} in the high-runoff steep-slope areas of the west, and between 1 and 10 W m^{-2} in the low-slope low-runoff areas of the south and east	Ferguson (1981)
Downstream changes	Downstream changes analysed and shown to be nonlinear along Wisconsin's Blue River	Lecce (1997)

value (unstable equilibrium) or there may be no tendency towards equilibrium (nonequilibrium) with no possibility of identifying an average or characteristic condition (Table 3.6). Whereas linear analysis has long been used for relationships between channel form and process, nonlinear analysis may be more appropriate than hitherto realized (Phillips, 1992), nonlinear frameworks can explain some phenomena not otherwise explained, so that environmental management goals targeted towards a single equilibrium may be inappropriate in landscapes characterized by disequilibrium, nonequilbrium, unstable or multiple equilibria (Phillips, 2003). Furthermore, there are differences between areas so that whereas equilibrium relationships are of greater utility in humid regions, where there is continuous operation of the fluvial system and constant feedback processes operating

Table 3.6 Equilibrium concepts

Term used	Context or meaning	Example reference
Equilibrium	Widely used in relation to aspects of environment to connote stability whereby the system does not undergo any further change, the sum of all the forces acting is equal to zero, input of mass and energy is equal to the outputs from the subsystem, and there is no net transfer of mass or energy between phases (e.g. gas and liquid). Types include: • static equilibrium – no change over time • stable equilibrium – the tendency of the variable to return to its original value • unstable equilibrium – the tendency of the variable to respond to system disturbance by adjustment to a new value • metastable equilibrium – a combination of stable and unstable except that variable settles on new value only after crossing threshold value or returns to original value.	Chorley and Kennedy (1971)
Steady state	A situation whereby input, output and properties of system remain constant over time.	Richards (1982)
Dynamic equilibrium	Used to signify a situation fluctuating about some apparent average state where that average state is itself also changing over time. Used for grade, which is the condition of a channel that has a gradient such that a perfect balance occurs between erosion, resistance, transport and deposition resulting in a channel profile. Most natural stream channels are in dynamic equilibrium with the prevailing flow conditions. Engineers developed 'in regime' to signify where this equilibrium condition obtained. From regime theory of alluvial canals channel form is relatively constant despite variations in flow over periods of 2–10 years.	Blench (1957)
Quasi-equilibrium	An apparent balance between opposing forces and resistances. A tendency towards natural equilibrium state but not realized. Approximate equilibrium would be less confusing.	Leopold and Maddock (1953) Graf (1988a)
Disequilibrium	Tendency towards equilibrium but has not been sufficient time to reach the equilibrium state.	Renwick (1992)
	Alternating flood and drought dominated regimes in New South Wales give instability and incomplete adjustment before the next regime.	Warner (1994)
Nonequilibrium	No net tendency towards equilibrium so that no possibility of identifying an average or characteristic condition.	Renwick (1992)
	Equilibrium conditions in middle reaches of Australian dryland rivers but nonequilibrium conditions in aggradational lower reaches.	Tooth and Nanson (2000)
Multiple equilibria	When several possible equilibrium states may arise from particular environmental conditions.	Phillips (2003)

among the morphology, energy and mass in the system, even an approximate equilibrium may be unlikely in the case of the discontinuous operation of dryland river systems (Graf, 1988b). However, even in humid regions human impact can be sufficient to disrupt any equilibrium conditions that did exist, thus frustrating the goal to maintain, or to restore equilibrium (see p. 109).

To extend data collected from specific areas to more general situations, modelling strategies are required in order to allow spatial or temporal estimates. Quantitative models are essentially of two types: those derived for the basin as catchment models that attempt a solution for the basin as a whole, and those that relate to one specific component of the hydrosystem, including the regime theory models developed for channel patterns or for channel geometry. Many catchment-based models were developed in hydrology, including lumped catchment models in which the physical structure and/or processes involved are simplified; variable source area models; and distributed models (Anderson and Burt, 1985). More hydraulically based modelling, approached by channel routing, is a mathematical way of predicting the changing magnitude, speed and shape of a flood wave as it propagates through the hydrosystem. Similar models have been developed for modelling drainage basin sediment dynamics (Nicholas *et al.*, 1995) but a recurrent problem with sediment and with hydrological modelling has been the difficulty of adequately specifying all components of the system and of modelling the short-term dynamics of that system.

Interconnectivity relates to dynamics and to energy transfer (Table 3.7) in the catchment hydrosystem so that it can be the basis for modelling approaches of components of the hydrosystem, including drainage network inputs (Rodriguez-Iturbe and Rinaldo, 1998), incorporation of dynamic network elements (Wharton, 1994) and sophisticated models developed for floodplain processes (e.g. Anderson and Bates, 1994). Ecological modelling was developed somewhat separately from hydrological, hydraulic and geomorphological modelling, but the need remains to provide comprehensive models embracing all aspects of the hydrosystem. A cellular automaton approach, which divides a catchment into uniform grid cells, has been utilized in ecological modelling and can simulate the dynamic response of a basin to flood events (Coulthard *et al.*, 1999). Considerable recent progress, aided by developments in topographic specification, including that by GIS, has enabled refinement of three types of catchment-scale model, namely: integrated component process models, watershed analysis and conceptual models (see Downs and Priestnall, 2003).

3.5 Prospect and perception of river channel systems

The dynamics of the river channel system arise because climatic inputs, especially through significant storm events, generate discharge hydrographs through the system and in the process mobilize sediment including bedload. The characteristics of the basin, or the channel network, effectively act as the transfer function converting climatic inputs into water, sediment and solute outputs, prompting research that continues to focus on a number of issues providing a prospect shown in Table 3.7. Alongside geomorphological and hydrological research, ecologists and biologists have investigated the associated flora and fauna, although only recently has causal understanding of the intimate links between the physical

Table 3.7 Interconnectivity issues

A. Dynamics:

- Do events of a particular kind or size determine the size, shape and character of a river channel system? Is there a rational basis for analysing inputs to relate to flood frequency and the probability of recurrence of events of a certain size?
- What are the critical values/thresholds beyond which the channel cross-section, planform, floodplain or network respond by change, often involving erosion or deposition?
- Although flows through the system may not change the river channel system, in other cases significant changes do occur; how far do large storm events alter the river channel morphology?
- If morphological changes occur after a large event then how long is necessary for recovery of the channel system?
- What differences occur from one climatic region to another? Do all river channel systems behave in basically the same ways or are there differences particularly between arid and humid systems?

B. Ongoing issues

- Can information on water and sediment processes be obtained for a specified reach from hydrological and sediment models? How do inputs relate to outputs?
- Does the process domain for each reach adequately reflect the reasonable long-term behaviour/experience?
- How does the scale of the river channel system interrelate? (see Table 3.3)
- How does sediment storage affect linkages in the hydrosystem?

habitats and the resulting flora and fauna begun to be fully appreciated, linked by the key concepts of hydraulic stress and disturbance (Petts and Bravard, 1996).

However well the dynamics of the catchment hydrosystem are understood and explained, the perception by individuals may affect all forms of local decision-making. Studies in Australia have shown how there may be a difference between the perception of river channel dynamics and change, which conditions environmental decision making and the record of what actually happens. Thus, basin managers tended to adopt views provided by oral tradition that were not confirmed by the behaviour of the Nogoa River, Queensland and the Avon River, Gippsland, Victoria (Finlayson and Brizga, 2000). In the Herbert River, Queensland the widespread perception of an aggrading river since European settlement is not supported by historical accounts, by gauging station data since 1940 nor by cross-sections compared since 1968 (Ladson and Tilleard, 1999). Along the Missouri river in Montana, landowners believe that the operation of the Fort Peck dam has initiated bank erosion, thus threatening agricultural development, whereas geomorphological evaluation indicates that bed degradation and bank erosion have declined since construction of the dam, with the channel now approaching dynamic equilibrium (Darby and Thorne, 2000) so that public sensitivity is greater than the morphological sensitivity of the river. Such nuances, together with appreciation of the way in which channel change is superimposed upon river channel systems in the context of timescales (Table 3.8) are also relevant to change considered in the next chapter.

Table 3.8 Timescales for river channel systems. Geomorphological components that can affect river ecology over particular timescales

Geomorphological component	Typical timescale of change
Cross-sectional shape (e.g. bank collapse)	Hours, days, months, years (e.g.
Cross-sectional size	during extreme flood events)
Pool and riffle sequence	
Point bars	
Islands	
Substrate including sedimentation	
Pattern (e.g. lateral migration)	Up to 100–500 years
Pattern (e.g. braided to meandering)	Up to 250–15 000 years +

Source: Modified from Brookes (1994); Brookes and Shields (1996b).

Components of the basin can be in equilibrium in various ways (Table 3.6), according to several timescales (see Figure 1.3), and collectively form a necessary basis for understanding the present system that is susceptible to change (Chapter 4). This chapter has demonstrated that:

- *catchment hydrosystems are made up of distinct components which form a continuum*;
- *physical and ecological fluvial system processes are a necessary basis for understanding river channel systems* requiring appreciation of the continuum provided by the fluvial and ecological characteristics of types of channel composing the stream and river network;
- *understanding fluvial systems requires contributions from a number of disciplines* ranging from the patch dynamics, biotopes or hydraulic flow types of ecologists to analysis of river channel patterns by geomorphologists;
- *channel capacity, shape* and types of *channel patterns* at a location in a catchment *reflect the combined effect of discharge and sediment delivered from upstream, the local characteristics of sediment, slope and vegetation, and more recent flow history*;
- *a range of floodplain features relate to their processes of formation, and zones include the river corridor and its associated ecological patterns*;
- *drainage networks*, like channel capacity and pattern *are dynamic*, being made up of perennial, intermittent and ephemeral channels;
- *connectivity of fluvial systems* involves longitudinal, lateral and basin connections and is exemplified by concepts of continuum, as well as sediment storage and slugs, nutrient spiralling, flood pulse, ecological patches and tectonic influences;
- *stream power and equilibrium are especially helpful themes* in linking elements of the fluvial system and assisting analysis of *connectivity dynamics* within the overall catchment context.

III

Realization

These four chapters summarize the way in which the foundation for river channel management at the end of the twentieth century was enhanced by greater understanding of river channel adjustments (Chapter 4), by appreciation of the basis of river channel sensitivity (Chapter 5), and was related to the context of river basin planning (Chapter 6), thus providing themes for a sustainable approach (Chapter 7) to underpin requirements for contemporary river channel management (Section IV).

4
River Channel Adjustments

Synopsis

Four chapters in this Realization section show how research achievements on changing channels created a foundation for river channel management at the end of the twentieth century founded on an understanding that channels will continue to adjust within the lifetime of a management project. Investigations of river channel adjustments have to be seen in relation to river channel stability (4.1) and often occur in response to factors that apply at specific locations such as dams, or along reaches as in the case of channelization, or have occurred as a consequence of inchannel or land use related human impacts over major parts of the basin (4.2). Historical records and palaeohydrological data can inform understanding of channel change, which should also be related to how climate change has influenced channel adjustments (4.3). The world pattern of river channel change (4.4) has to be envisioned against the known complexity of river channel adjustments and several conceptual approaches have been devised to summarize change on a longer timescale (4.5), providing implications for river channel management (4.6).

4.1 River channel stability

Knowledge of the contemporary mechanics of the river channel system (Chapter 3) may appear to provide a sufficient basis for river channel management but because the system evolves, understanding derived from longer time periods should be considered (Baker, 2003; Gregory, 2003b). Perception of how river channels change over time, as well as of channel processes, can affect decision-making and management. River channel change, reviewed in this chapter, is taken to be any alteration of the morphology of the channel from the prevailing condition. Change inherent in the river system involving channel pattern changes by migration, cut-offs, crevassing and avulsion is described as autogenic change (Lewin, 1977) illustrated for the River Po, Italy (Figure 4.1). Thus, the river channel could maintain the same size, shape and pattern but move its position by translation, under equilibrium conditions (Table 3.6). Under some conditions autogenic change may be continuous and progressive. Measured and estimated rates of change (Table 4.1) relate to the size and environmental characteristics of the river basin, but change of channels, patterns and floodplains apply to timescales differentially so that whereas a floodplain may not be

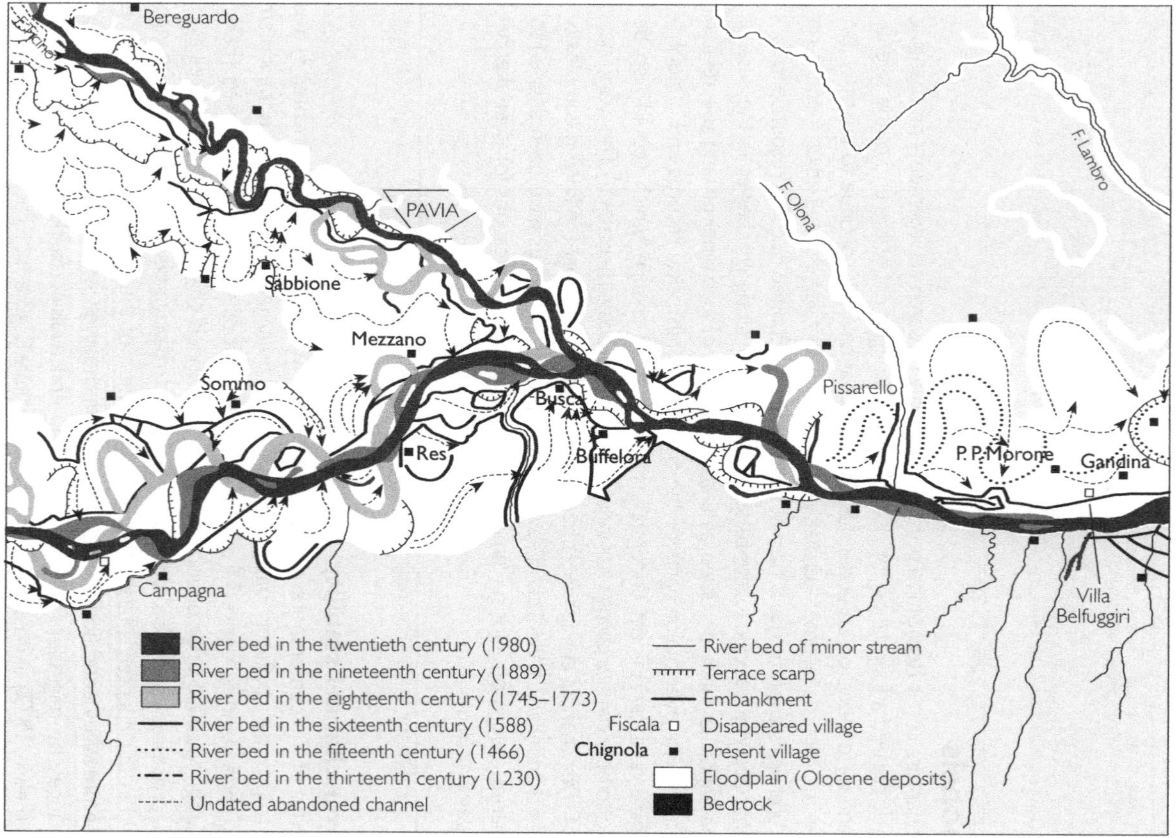

Figure 4.1 *Changes of a section of the River Po, Italy since the twelfth century. The changes, both autogenic and allogenic, were reconstructed from records and historic maps (after Braga and Gervasconi, 1989)*

Table 4.1 Some rates of autogenic river channel change

River channel	Parameter measured	Rates	Source
Cross-section	Bank erosion	793 m yr^{-1} maximum Brahmaputra	Coleman (1969)
Planform	Meander migration	One channel width in several hundred to 1000 years (estimated)	Keller and MacDonald (1995)
	Lateral migration	0.61–305 m yr^{-1} Mississippi	Kolb (1963, in Hooke, 1980)
		0–several hundred metres per year	Wolman and Leopold (1957)
		Russian rivers 10–15 m yr^{-1} unexceptional	Kondrat'yev and Popov (1967)
		6.6 m yr^{-1} downvalley Des Moines River	Handy (1972)
		0.475 m yr^{-1} Beaton River, BC, Canada	Hickin and Nanson (1975)
Floodplain	Accretion rate	0.2–10 mm yr^{-1}	Gomez *et al.* (1998)
		2.4–16 mm yr^{-1} Axe, UK	Macklin (1985)

reworked by channel migration for some hundreds or thousands of years (Nanson and Croke, 1992), channel morphology can change in much shorter time periods reflecting the significance of different timescales (Figure 4.2).

Three time scales (Table 4.2) of steady, graded and cyclic time (Schumm and Lichty, 1965), provide a temporal context showing how certain variables that are dependent at one timescale can be independent or irrelevant at others. This approach has been used extensively in interpreting river environments (Kennedy, 1997), and is paralleled by approaches in biogeography and ecology (Udvardy, 1981; Gregory, 2000). Different types of equilibrium (Table 3.6) relate to these three timescales, which themselves provide a means of linking the short-term concerns of the engineer or manager, possibly focusing on a design period of the order of 50 years at a steady timescale, with the longer, graded timescale investigations of the geomorphologist (Figure 4.2).

Realization, the theme for this section, reflects an appreciation that, in addition to changing autogenically (Table 4.1), river channels can also alter as a result of allogenic changes that occur in response to climate fluctuation or human influence (Lewin, 1977: 175). Allogenic influences can result from a change in climate, from climate forcing or as a result of human activity either on the catchment area or directly on the river channel, often as a

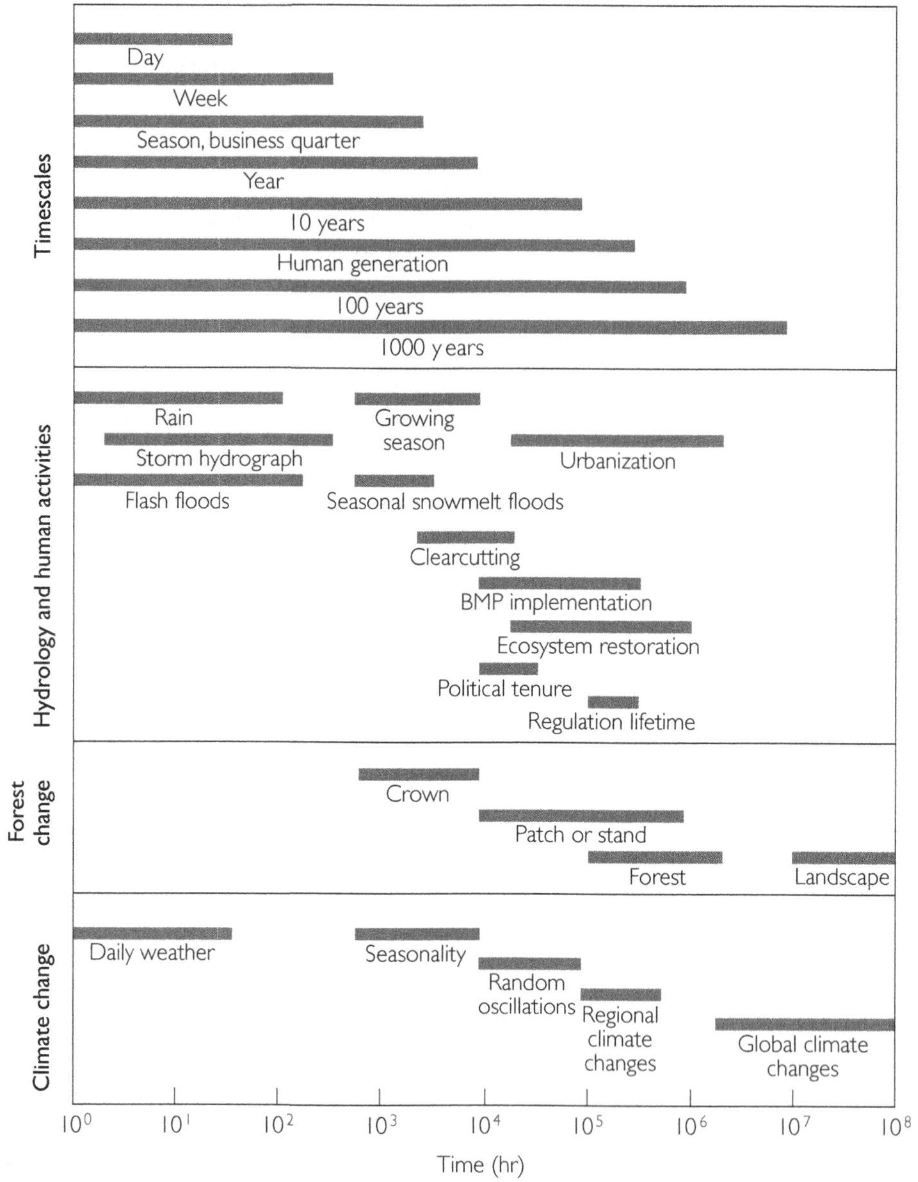

Figure 4.2 *Temporal scales for drainage basins. A framework for considering natural and human processes (from National Research Council, 1999). BMP is Basin Management Plan*

consequence of management of the river and the channel. It is increasingly appreciated that autogenic change is not simple and that it is inextricably linked to forms of allogenic change. Furthermore, it is difficult to differentiate the several causes of human-inspired allogenic change: human impact over a drainage basin can be expressed through a series of vegetation and land use changes and also be affected by changes in climate, so that collectively several changes may fuse together in what has been described as a cultural blur

Table 4.2 Timescales

Drainage basin variables	**Status of variables during designated time spans**[a]		
	Steady	**Graded**	**Cyclic**
Time	Not relevant	Not relevant	Independent
Initial relief	Not relevant	Not relevant	Independent
Geology	Independent	Independent	Independent
Climate	Independent	Independent	Independent
Relief or volume of system above base level	Independent	Independent	Dependent
Vegetation (type and density)	Independent	Independent	Dependent
Hydrology (runoff and sediment yield per unit area within the system)	Independent	Independent	Dependent
Drainage network morphology	Independent	Dependent	Dependent
Hillslope morphology	Independent	Dependent	Dependent
Hydrology (discharge of water and sediment from the system)	Dependent	Dependent	Dependent

Note

[a] Steady state time, typically of the order of a year or less, when a true steady state situation may exist; graded time, may be hundreds of years, during which a graded condition or dynamic equilibrium exists; cyclic or geological time, encompasses millions of years as required to complete an erosion cycle.

Source: After Schumm and Lichty (1965).

(Newson and Lewin, 1991). A frequently cited illustration is channel change of the Cimarron River in south west Kansas where the highly sinuous, relatively narrow and deep River channel was destroyed by a major flood in 1914 (Schumm and Lichty, 1963). Subsequently, from 1914 to 1931, the river widened from an average of 15 m to 366 m, and the entire floodplain was reworked in a period when precipitation was below average. Channel changes were attributed to climatic fluctuations, although an alternative explanation is that agricultural activities that removed natural vegetation may have resulted in increased flood peaks and sediment loads.

Reasons for allogenic river channel changes (Table 4.3) may therefore be difficult to isolate because more than one may affect a basin at the same time, a potential cause may not reach a threshold level sufficient to instigate channel change, there may be a time lag between the initiation of the potential cause and channel change occurring and the duration of change can vary significantly. However, as many river systems are presently in a state of flux as a result of substantial anthropogenic change (James, 1999), separation of the causes remains a necessary challenge. In meandering systems the range of behaviour can vary from the highly stable and ordered to the unstable and chaotic so that research is needed to establish where and how nonlinear and chaotic behaviour occur and hence

Table 4.3 Allogenic changes over the basin that may cause river channel changes

Category	Causes cited in research investigations
Climate	• Vegetation and land use changes
	• Greenhouse effect including thermal pollution, increased CO_2, CO, NO_2
	• Cloud seeding
Drainage basin: spatial	• Deforestation
	• Grazing
	• Fire, burning
	• Agriculture, ploughing
	• Land use, conservation measures
	• Afforestation
	• Building construction
	• Urbanization
Drainage basin: Network	• Drainage schemes
	• Agricultural drains
	• Irrigation networks
	• Ditches
	• Stormwater drains
Channel: Reach	• Desnagging and clearing
	• Clearance of riparian vegetation/tree clearance
	• Sediment removal, mining gravel extraction
	• Sediment addition, mining spoil
	• Boat waves, bank erosion
	• Invasion by exotic vegetation species
	• Afforestation, conservation
	• Restoration and allied techniques
Channel: Cross-section/point	• Channelization
	• Bank protection and stabilization
	• Resectioning, dredging
	• Channel straightening, cut-offs
	• Embankments
	• Diversion of flow, leats, hydroelectric power (HEP)
	• Dam construction
	• Weirs
	• Abstraction of flow
	• Return flow, drains, outfalls
	• Bridge crossings
	• Culverts under roads and crossings

Source: Developed from a table by Gregory and Walling (1973).

how they interrelate with longer-term changes (Hooke, 2003). As autogenic and allogenic change in river channels may be occurring concurrently, it is necessary to appreciate the types of change that can occur, what channel adjustments can take place and how they are identified as a basis for considering the causes responsible.

Any channel change that takes place can be considered in terms of 'degrees of freedom' or permissible 'modes of self-adjustment' (Maddock, 1970) possessed by the river. Since being first suggested (Lane, 1955a, 1957), the approach was developed (Hey, 1982) to specify seven, and later nine, degrees of freedom (Hey, 1997b) listed as:

1 average bankfull channel width (*W*)
2 depth (*d*)
3 maximum depth (d_m)
4 height (Δ) and
5 wavelength (λ) of bedforms
6 slope (*S*)
7 velocity (*V*)
8 sinuosity (*p*)
9 meander arc length (*z*).

It should be possible to define the bankfull hydraulic geometry by simultaneous solution of process equations – one associated with each of the degrees of freedom (Hey, 1982). In the river channel context, it is necessary to visualize how changes that occur within the channel (Table 3.3), the channel cross-section, the channel planform including the floodplain and the channel network are interrelated (Gregory, 1976a, 1987a), expressing each change in terms of size, shape or composition (Table 4.4). Thus, a channel cross-section can adjust by becoming smaller or larger (size); it could maintain its cross-sectional area but

Table 4.4 A holistic context for changes of the river channel

Change of	Inchannel characteristics	River channel cross-section	River channel pattern	Drainage network	Drainage basin
Size	Increase or decrease in bedform size	Increase or decrease of river channel capacity	Increase or decrease of meander wavelength	Increase or decrease of network extent and density	Increase or decrease owing to capture or diversion
Shape	Alteration of shape of channel bars	Adjustment of width/depth ratio	Change from regular to irregular meanders	Drainage pattern changed in shape	Diversion changes basin shape
Composition	Change in sediments in channel	Change in channel sediments	Metamorphosis from single thread to multithread	Incision or gullying of channels	Changes in runoff-producing areas

Source: Developed from Gregory (1987a).

change in shape by becoming more slot-like (shape); or it could change by alteration of the sedimentary materials and deposits that occur within it (composition). Such a perspective on channel change can provide the basis for a holistic view of the catchment hydrosystem as it is affected by temporal change, both autogenic and allogenic, and with which changes in aquatic ecology will be associated.

To date, research on river channel change has placed greatest emphasis upon *what* aspects of river channels can change (Table 4.4) and upon *why*, with less attention given to *where, how,* and *when* changes occur (Gregory, 1987a). These questions are all equally important but, only when answers were produced to what and why, was there a sufficient research basis to inform river channel management. This chapter therefore emphasizes *what* and *why* changes occur with some reference to how much change takes place, although this, and questions of where, how and when it occurs, are elaborated further in Chapter 5.

4.2 Human impacts

The direct effects of human activities, such as inchannel structures or channel modifications (pp. 34–9) leading to river channel changes, have been recognized for many years, but it was not at first realized how this may lead also to channel changes downstream and upstream from engineering schemes, and to secondary impacts along the reach. It also became appreciated that land use changes across the catchment could affect river channel adjustments, although it could be difficult to disentangle such effects from those of climate change. Since 1960 greater understanding has been achieved about *why* changes occur and *what* channel alterations can be instigated.

Point and reach effects

As human impacts on river channels increased throughout the twentieth century, floodplains became more intensively occupied and manipulation of the river channel more adept (Chapter 2 p. 25). River channel adjustments as a result of direct impacts upon the channel cross-section at a point and along reaches include the consequences of river channel engineering, embracing effects of dam and reservoir construction, channelization, embankments, bridge crossings and mining of channel sediments. The potential hydrological effects associated with each of these channel adjustments (Table 4.5) provide the *why* for channel change, thus suggesting *what* effects there can be on system processes and on river channels.

Dam construction

Dam construction can produce major modifications of river channels as sediment trapped above the dam results in clearer water immediately downstream of the dam, potentially inducing scour, so that avoidance of undermining of the dam was for many years an important engineering input for dam design (Komura and Simons, 1967). It was subsequently appreciated that, for considerable distances downstream, the flood frequency, sediment transport, channel morphology and river ecology were also substantially changed, so that there could be a complex sequence of downstream channel responses. In view of the worldwide extent of dams and of large dam technology,

Table 4.5 Point and reach effects on processes and channel morphology

***Why* change occurs**	***What* changes may occur**				
	Process change (Hydrology, H; sediment, S)	**Channel adjustment** (Clearly evident, *)	**Spatial effects**		
			Downstream extensive (#)	**Upstream extensive (#)**	**Complete network (Σ)**
Desnagging and clearing	S+	*			
Clearance of vegetation/ trees	S+	*			
Sediment, gravel extraction	S+	*		#	
Sediment addition	S+	*	#	#	
Boat waves, bank erosion	S+	*			
Invasion by exotic vegetation	S–	*			
Afforestation, conservation	S–	*	#		
Restoration and allied techniques	H– S–	*	#		
Channelization	H+ S+–	*	#	#	
Bank protection	H+ S–	*			
Embankments and levee construction	H+		#		
Resectioning, dredging	H+– S–	*	#		
Channel straightening, cut-offs	H+	*	#		
Diversion of flow, leats, HEP	H–	*	#		
Dam construction	H– S–+	*	#	#	Σ
Weirs	H– S-+	*			
Abstraction of flow	H–	*	#		Σ
Return flow, drains, outfalls	H+	*	#		
Bridge crossings	H+	*	#		
Culverts under crossings	H+	*			

peaking in the 1960s (Chapter 1, p. 13), the significance of such changes was spatially extensive. With 75 000 large dams in the continental USA (Graf, 1999), it has been suggested that the interspacing of dammed, drowned, preserved and restored reaches has fragmented every large river by disconnecting the once integrated free-flowing systems (Graf, 1999), so creating an unprecedented disruption to hydrological and sedimentological connectivity. Similar impacts can affect the sediment concentrations downstream from weirs, and channel-forming discharges can be modified downstream of abstraction points if substantial flows are extracted for irrigation.

Although the river channel is not deliberately modified, disruption in the fluvial system caused by the interposition of a large dam and its impounded reservoir instigate a fundamental change of the hydrology and sediment transport capabilities of the channel downstream, and of the energy conditions of the channel leading into the reservoir upstream. This example of a press disturbance (Underwood, 1994), whereby the hydrological and sediment regimes are permanently changed, means that the river channel behaves as if an abrupt climate change has occurred, producing a condition where flows are usually smaller and sediment transport greatly reduced. Geomorphological and ecological changes in the downstream river channel arise from changes in these driving variables. The complex changes vary between dams according to the magnitude of the regulation of flows and sediment, the resistance of boundary materials downstream, the nature of the flood events following regulation, the location of the dam within the catchment, the number and location of incoming (unregulated) tributaries, the floodplain vegetation response to the altered hydrology and the ecological characteristics of the original channel (Petts, 1979, 1982, 1984). The impact of river regulation upon river channels can be seen (Petts, 1984) against a background of the hydrology of dammed rivers, the quality of reservoir releases and seston transport under the headings of channel morphology, vegetation reaction and structure, macroinvertebrate response to upstream impoundment, and fish and fisheries. Three orders of impact (Petts, 1984: 24–5) have been identified:

- first-order impacts occur simultaneously with dam closure, and affect the transfer of energy and material into, and within, the downstream river, and therefore include impacts on processes, flow regime, sediment load, water quality and plankton;
- second-order impacts are changes in channel structure and primary production that result from the modification of first-order impacts, according to local conditions, and depend upon the characteristics of the river prior to dam closure. These impacts may require a time period of between 1 and 100 years or more to achieve a new 'equilibrium' state, and they therefore influence the channel form, substrate composition, macrophytes and periphyton;
- third-order impacts on fish and invertebrates may occur after a considerable time lag, and reflect all the first- and second-order changes. The fish population is also influenced by the changes of the invertebrate community that provide the major food supply for many species. There may be several phases of population adjustment and readjustment, particularly in response to the second-order factors.

In addition to the potential range of impacts of river regulation on channels, including reductions in capacity to as little as 10% of the original (Table 4.6), the spatial pattern of changes downstream varies according to the incoming tributary influences and the time elapsing since dam closure, summarized diagrammatically (Figure 4.3A). As with other causes of channel adjustment, the resulting changes occur as a 'complex response' (see p. 115) and are predictable only in general terms. However, Petts (1984) was able to suggest how first-order impacts related to the range of biotic and abiotic (instream and floodplain) effects (Figure 4.3B).

Of the many major rivers affected by flow regulation, the Colorado, with a length of 2320 km and draining 8% of the USA, is virtually totally regulated. Legislation from 1922 to 1948 apportioned the flow between seven basin states, Mexico was assured of 10% of the river's annual flow, and construction of works was enabled to regulate, control and

Table 4.6 Impacts of river regulation

Impact	Consequence	Results
Creation of dam	Introduction of barrier – to downstream transfers – to upstream migration of fish	Sediment accretion in reservoir
Creation of lake	Water balance changed	Average runoff reduced and evaporation increased
	Loss of river, wetland and terrestrial ecosystems	
	Barrier to faunal migration (e.g. of fish or caribou)	
	Limnological changes	
Below dam or reservoir	Changes of flows	Reduced seasonal variability and reduced flood magnitudes
	– sediment loads	Bedload transport reduced, suspended load reduced (Willliams and Wolman, 1984)
	– water quality	Significant changes
	– water temperatures	Reduced seasonal and diurnal variability
	Loss of floodplain–channel interactions	Floodplain less frequently inundated with implications for riparian plant communities
	Change of channel morphology	Scour downstream of dam because of clear water erosion; downstream changes in channel capacity, width, depth, width/depth ratio, sinuosity, meander wavelength (Gregory, 1987a)
	Change of instream hydraulic characteristics	
	Change of autogenic:allogenic energy sources	
	Change of channel ecology	Reduction of abundance and diversity of macroinvertebrates; fish species change

Note: Including material from Petts (1999). See also Figure 4.3A. Some impacts may be anticipated e.g. fish ladders.

A

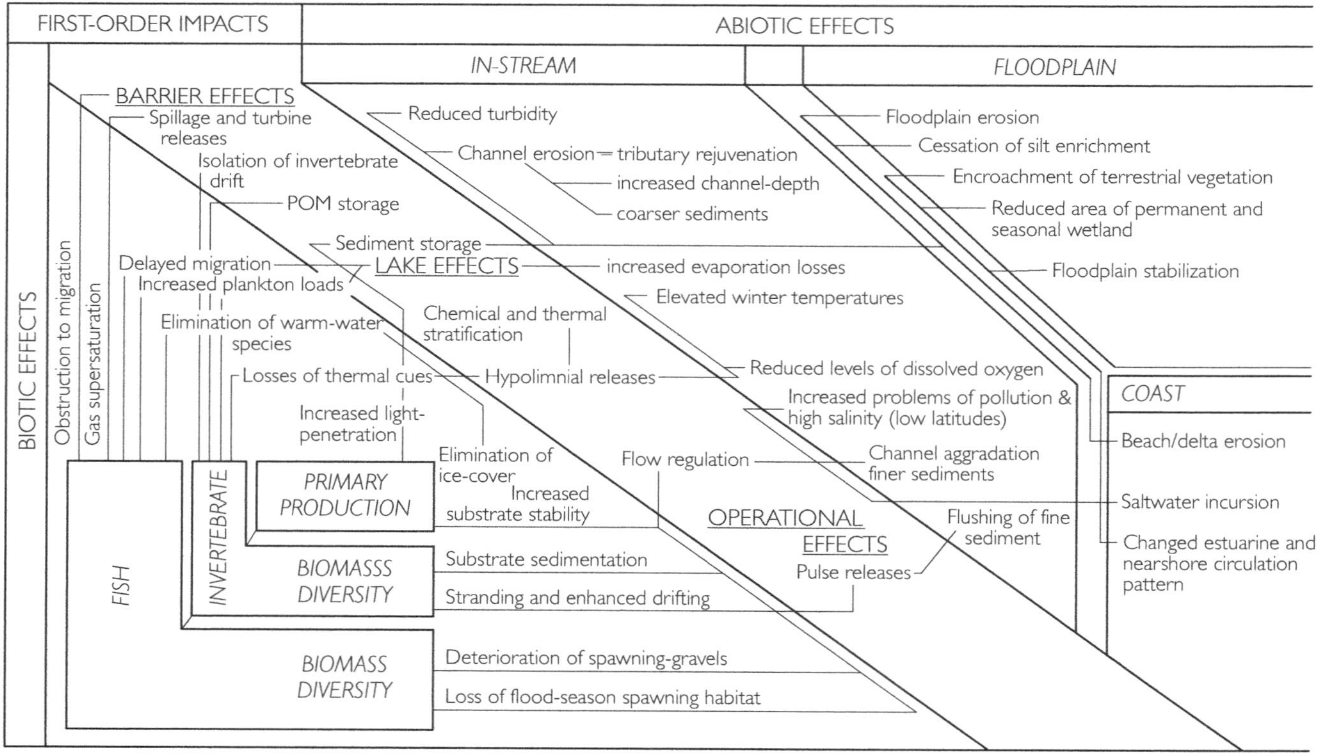
FIRST-ORDER IMPACTS
ABIOTIC EFFECTS
IN-STREAM
FLOODPLAIN
COAST
BIOTIC EFFECTS
BARRIER EFFECTS
Obstruction to migration
Gas supersaturation
Spillage and turbine releases
Isolation of invertebrate drift
POM storage
Delayed migration
Increased plankton loads
Elimination of warm-water species
Losses of thermal cues
Increased light-penetration
Sediment storage
LAKE EFFECTS
Chemical and thermal stratification
Hypolimnial releases
Elimination of ice-cover
Increased substrate stability
Reduced turbidity
Channel erosion
tributary rejuvenation
increased channel-depth
coarser sediments
increased evaporation losses
Elevated winter temperatures
Reduced levels of dissolved oxygen
Increased problems of pollution & high salinity (low latitudes)
Floodplain erosion
Cessation of silt enrichment
Encroachment of terrestrial vegetation
Reduced area of permanent and seasonal wetland
Floodplain stabilization
Beach/delta erosion
Saltwater incursion
Changed estuarine and nearshore circulation pattern
Flow regulation
Channel aggradation finer sediments
OPERATIONAL EFFECTS
Flushing of fine sediment
Pulse releases
FISH
INVERTEBRATE
PRIMARY PRODUCTION
BIOMASSS DIVERSITY
BIOMASS DIVERSITY
Substrate sedimentation
Stranding and enhanced drifting
Deterioration of spawning-gravels
Loss of flood-season spawning habitat

B

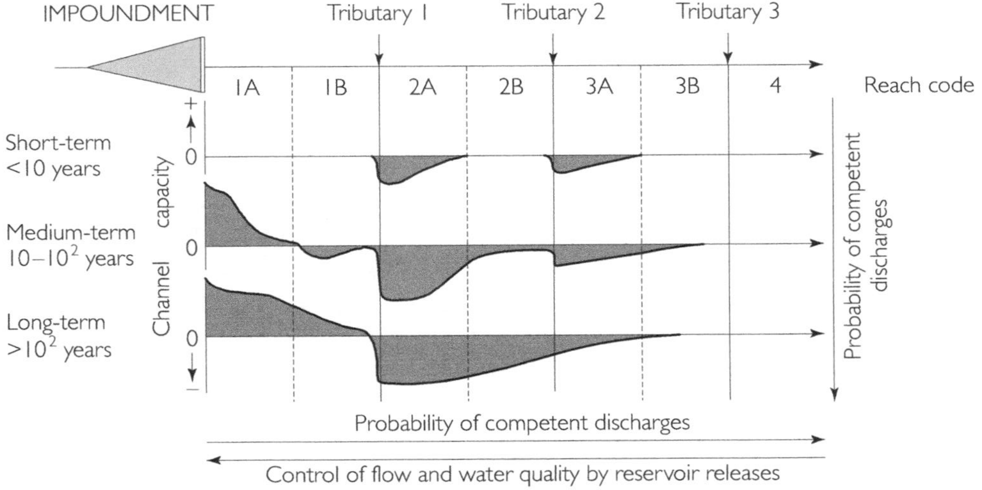

Figure 4.3 ***A*** *Responses to river impoundment (after Petts, 1979, 1984)* ***B*** *Hypothetical model of river responses to impoundment (after Petts, 1984)*

utilize the river. The original river has been transformed into a series of desert lakes as a consequence of dam construction, beginning with the Hoover Dam in 1936. The impact of this pattern of regulation has been very substantial (Graf, 1985) and, in addition to the reduction in channel capacity achieved by the growth of sedimentary bars, there has been increased stability of the channel rapids, and the once barren flood zone, through the Grand Canyon for example, has been transformed into a narrow riparian woodland, of native willows and the exotic salt cedar or *Tamarix chinensis*, with associated fauna. A total of 25% of the river has been converted to standing water (lentic) habitats by impoundment, the backwaters and marshes have been isolated from the river, and the thermal regime and trophic structure of the remaining reaches has been changed by releases from the reservoirs (Petts, 1984). Habitat changes have had major impacts on the biota, many endemic fish have been pushed near to extinction and exotic species have been introduced to exploit the new (especially cold water) habitats created by regulation. Exotic fish have been particularly successful in reservoirs and heavily regulated reaches (Stanford and Ward, 1986).

Analysis of changes at 21 reservoir sites in central and southwestern USA (Williams and Wolman, 1984) showed that on most of the reaches surveyed the channel bed had degraded immediately downstream from the dam, channel width in some cases showed no appreciable change but in others increased by as much as 100% or decreased by as much as 90% and, at many cross-sections, changes in bed elevation and in channel width proceeded irregularly with time. There are many other examples of dramatic changes in river morphology downstream of dams: in France only about 1% of the mean discharge of the Durrance reaches the Rhone as a result of the construction of a sequence of large dams and the diversion of water through 20 power stations, so that the channels are now 35–53% of their former widths in the lower valley and, above the gorge, are reduced to 29–41% (Warner, 2000). On Stony Creek, California, below Black Butte Dam, closed in 1963, the formerly braided channel has metamorphosed into an incised, single-thread, actively migrating meandering channel that has eroded an equivalent of approximately 20% of its average annual bedload from the downstream floodplain (Kondolf and Swanson, 1993). Channel changes can also be associated with weirs, for example on the lower Murray in Australia where the construction of locks 2 and 3 (1925–28) meant that cross-sections have subsequently become wider and shallower in pool 3 but narrower and deeper in pool 2 (Thoms and Walker, 1993).

Channelization

Channelization in various forms constitutes a direct modification to the river channel, and is now very extensive (Brookes, 1985, 1988), involving actions such as channel realignment (straighteneing), resectioning (deepening and widening of cross-sections), construction of embankments, bed and bank protection, dredging and operational maintenance (see Chapter 2). Channelization by realignment and/or resectioning is basically a pulse disturbance (Underwood, 1994) that does not involve any fundamental change to the upstream hydrology or sediment transport processes. Despite the prospect that channelization could be a panacea for local river channel management problems, it did not prove to be the expected technological fix because feedback effects arose within, downstream and upstream from channelization schemes. Concern about its effects by individual citizens and environmental organizations in the USA, the publication of books such as *The River Killers* (Heuvelmans, 1974) suggesting that the US Army Corps of Engineers had 'systematically ruined the nation's rivers' through channelization, together with individual law suits, drew attention to the major environmental consequences. Research investigations since 1970 indicated there to be geomorphological, ecological and biological responses of rivers to channelization (some examples are given in Table 4.7), both in the channelized reaches and also downstream and upstream (e.g. Brookes, 1988). They occur because the channel form created by channelization is not necessarily associated with the prevailing hydrological and

Table 4.7 Impacts of channelization (including material from Shields, 1996)

Type of channelization	Potential responses	Examples
Channel enlargement	Deposition of sediments in enlarged channel	Griggs and Paris (1982)
	Capacity increase up to 1000%	Brookes (1996b)
	Bed lowering	
Straightening	Knickpoint or knickzone formation and migration leading to channel deepening, widening and slope flattening	Harvey and Watson (1986) Shields and Hoover (1991)
	Erosional problems on river and tributaries	
Levees	Vary. Peak discharges may increase owing to elimination of overbank storage. Area between levees may erode or deposit. Meander length and amplitude may increase if banks not stabilized	Khowai and Manu rivers, Bangladesh; Neill and Yaremko (1988)
Bypass channels and diversions	Sedimentation in either the main channel or the diversion channel	Shields and Abt (1989)
Ecological impacts	Reduction of habitat diversity. Species eliminated or reduced. Fish species reduced in number and biomass reduced	Brookes *et al.* (1983)

sediment transport processes and contrasts starkly with upstream and downstream reaches: the consequence of a sharp transition in hydrology and sediment transport capabilities between reaches is that, in order to reduce the contrasts, the channel may attempt to recover its pre-channelization configuration if it possesses sufficient energy and an upstream source of sediment. The severity of the impacts varies over time; sometimes initial dramatic effects are followed by a more stable period, whereas in other cases changes increase and are compounded, but the process of recovery may involve the continued process of sediment exchange between the channel perimeter and the sediment transported from upstream until a new 'equilibrium' morphology is reached.

Within channelized reaches, impacts depend upon the extent of modifications to the channel. If the channel is completely concreted and made impervious then it is effectively converted to a flume that accelerates the progress of flood waves and reduces the locally available sediment supply. If the channel banks are protected but the channel bed is still erodible then, as flood waves progress through the reach more rapidly, scour can occur and a knickpoint or erosional step may develop and progress upstream. Where channels are straightened and bed and banks not protected, then channel bank and bed erosion can be substantial, including a 440% increase in channel capacity over 33 years in the cross-sectional area of the Willow Drainage Ditch, Iowa, USA following straightening (Daniels, 1960), and cross-sections witnessed to increase to as much as ten times the original capacities (Brookes and Gregory, 1988), and the pool–riffle sequence being modified or destroyed. The ecological consequences of channelization are that habitat diversity and niche potential are reduced and the quality and function of the species are changed (Brookes, 1988) so that biomass per unit area of macroinvertebrates is lowered substantially, densities of species can be drastically reduced and fish populations decline, with some species eliminated. Compilations of results from different areas (e.g. Brookes and Gregory, 1988) show how ecological recovery following initial construction can take place in as little as 2 years, although it is more usual for periods of 5 to 15 years to be necessary.

Effects are also very evident downstream of channelized reaches because a successful scheme transmits discharges more rapidly, so that solving a problem in one area can create other problems downstream. These include increased peak discharge, increased flood frequency, adjustments of channel capacity with bank erosion and knickpoint recession, and alterations of water quality (Brookes, 1988). A study of 57 sites of channelization in England and Wales demonstrated the diversity of channel enlargement downstream of channelization and showed how it decreased with distance below the channelization scheme (Brookes, 1985). The hydrological and morphological changes force ecological changes as well, with fish populations, the plant ecology and the aquatic fauna significantly modified (Brookes, 1996a). Changes can also occur upstream of channelized reaches because the net effect of channelization is to increase channel slope so that deposition may occur downstream and incision may occur upstream (e.g. Parker and Andres, 1976).

When a channel is straightened, the gradient is increased and this locally steepened section potentially acts as a knickpoint or knickzone within the channel system. The East and West Prairie rivers in Alberta, Canada, (Parker and Andres, 1976) demonstrated how this knickpoint can initiate erosion of the channel bed, causing the steepened zone to migrate upstream. Indeed, it is the process of knickpoint migration upstream of the engineered reach that often drives within-reach channel changes, providing an example of 'complex response' (Schumm, 1973) whereby, in time, both erosion and deposition can occur in the same channel reach.

The impact of channelization can be a series of supplementary management requirements. Under the best case, in some lowland channels with low stream power, any increase in stream power resulting from channelization may be insufficient to overcome the resistance to erosion offered by the materials and vegetation of the river bank. In these cases, erosion cannot occur and the apparently nonsustainable conditions can prevail with little further human intervention. However, to prevent erosion in channelized reaches where the increase in stream power is sufficient to erode the channel banks, revetments or other bank protection works are required using, for instance, rip-rap, sheet piling, gabion baskets or concrete walling to markedly increase the bank resistance to erosion beyond the resistance provided by the bank's natural materials. Erosion of the channel bed as the knickpoint migrates upstream often requires the provision of hard points in the form of grade control structures to prevent knickpoint migration. Conversely, where the overall stream power of the reach is lowered, periodic maintenance is often required to excavate the sediment deposited as the river attempts to regain its equilibrium conditions through reducing its cross-sectional area. Therefore, the reason that so many river channels remain in their channelized form is that channelization becomes more than the simple 'pulse' disturbance caused by the reconfiguration of the channel. Instead, channelization schemes often integrally incorporate bank protection works where erosion is expected. In addition, the erosion within and upstream of the channelized reach will cause eroded sediment to be deposited within or downstream of the scheme, possibly exacerbating the prospect of flooding adjacent to these areas. Therefore, most river management agencies will subsequently undertake periodic maintenance works to dredge the channel of deposited sediment and/or use *ad hoc* remedial works to add bank protection where signs of erosion are observed. These are costly activities: it is estimated in the UK that more than £10 million are spent every year just in removing deposited sediment in channelized rivers (Sear *et al.*, 1995).

Bridge crossings

Other impacts on the river channel, albeit on a smaller spatial scale than major dams and channelization, arise from the location of mills, channel margin installations and works, and the interposition of bridge crossings. Careful site selection is necessary for crossings, with hydrologic evaluation of the chosen sites in relation to likely flood frequencies to be conveyed through the bridge structure, and consideration of hydraulic effects in relation to potential scour around the supports of the structure proposed, but in addition there can be effects immediately downstream of bridge structures (Gregory and Brookes, 1983). Although bridge crossings are designed to convey adequate flood flows without scour around bridge piers, less attention has traditionally been given to their downstream impact on river channels or to the impacts of on-going channel changes on the long-term safety of the bridge. In the New Forest, UK (Gregory and Brookes, 1983) changes occur in channel width downstream up to two and occasionally four times their original values, channel capacity can be at least twice the capacity above the bridge, and the downstream effects could persist for 20 times the channel width. As there are probably more than 50 000 bridges in England and Wales, and in Hampshire there is an average of one bridge for every 4.5 km of river channel, the effects on channels immediately downstream of bridge crossings should not be underestimated. Likewise, an extensive survey of hydraulic conditions at bridge crossings in the USA (Brice, 1981) prompted proposals for mitigation of the erosional effects. Culverts under road crossings can affect the channel above and below the crossing, and many small creeks in Australia frequently show deposition above the culvert and scour downstream. This is probably the most widespread morphological impact of barriers to flow, although it may not be clear whether erosion below the culvert is caused by clear-water scour, or whether an erosion head has worked upstream and is held by the culvert (Rutherfurd, 1999).

Desnagging

At the reach scale, removal of coarse woody debris (CWD) from river channels, a process often referred to as desnagging, is a particular form of channel modification. This form of clearance, undertaken to increase the velocity of water and sediment through desnagged channels, has effects similar to dredging and plant cutting, for example, by annual maintenance operations in Britain and North America (Brookes, 1985). The higher velocities, greater stream power values and attendant changes of channel morphology and ecology produced have been demonstrated from a range of areas (Gregory and Davis, 1992) involving bank erosion, bed scour and the introduction of new forms of channel bank collapse (Gregory and Davis, 1993). It was once imagined that the removal of CWD in rivers was of benefit to fish passage, but the overall impact on fish habitat of removing CWD is usually overwhelmingly negative when impacts on the channel morphology, the pool–riffle sequence (Gregory *et al.*, 1994), cover habitat, water temperature and foodweb are taken into account. Management strategies can, according to the character of the area considered, require removal of debris, its maintenance or, in some cases, its introduction into the channel (Gregory and Davis, 1992). Clearance of riparian vegetation, especially trees, can similarly induce downstream effects, whereas invasion of exotic species, such as Tamarisk on the Colorado plateau river channels (Graf, 1988b), or reforestation of floodplains and channels can reduce downstream velocities, also bringing about morphological and ecological change.

Mining of channel or floodplain sediments

Mining of channel or floodplain sediments can also affect river channel characteristics. Channel changes sometimes occurred because waste sediments were discharged directly to the nearest stream channel for removal, whereupon the material was transported as waves or 'sediment slugs' (p. 73), causing a sequence of channel metamorphosis, channel bed and floodplain aggradation and channel incision that requires management response, in addition to problems of the downstream dispersal of toxic mining waste (Macklin and Lewin, 1997). Inchannel mining of alluvial sand and gravel for construction sediment tends to greatly increase channel change through incision-related processes, although in a few rapidly aggrading rivers, gravel extraction may reduce river channel hazards (p. 294) associated with bed-level rise (Kondolf, 1994a,b). Sand and gravel are typically mined either directly from the active channel bed or from deep pits in the channel floodplain. Mining the active channel mechanically causes bed degradation and, during subsequent storm events, a knickpoint may be generated that migrates upstream. Because channels suitable for gravel mining tend also to be laterally active, the risk associated with floodplain pit mining is that during storm events the channel will migrate into the pit, also causing the development of a large knickpoint. Subsequent to knickpoint formation, upstream bed degradation causes channel incision, possible undermining of upstream structures, changes to bed sediment composition that reduce spawning gravel habitat, lowering of regional water tables and damage to riparian vegetation (Wyzga, 1991; Kondolf, 1994a, 1997). Extractive industries have had a significant impact on the Hawkesbury-Nepean River, New South Wales, where 13 physical and 14 biological impacts (Table 4.8) could be distinguished (Erskine, 1998); it was concluded that there is a need for management authorities to be more aware of the interaction between all types of physical and biological processes in rivers, and to reframe rapidly environment protection measures when they are demonstrably inadequate (Erskine and Green, 2000). The first of two episodes of sedimentation from hydraulic gold mining in the Bear and American basins of California (James, 1999) caused aggradation in many channels, and continues to affect sediment loads and stability of sediments. Along river channels in the French Alps, management response to channel impacts has changed since the nineteenth century: it is now contended that protecting civil engineering works threatened by incision, restoring phreatic levels and preventing further

Table 4.8 Environmental impacts of extractive industries on the Hawkesbury-Nepean River

Physical impacts	Biological impacts
• Bed degradation • Increased water depths • Channel widening • Bank erosion • Channel enlargement • Channel avulsion • Turbidity plume • Bed armouring • Mud deposition on the bed • Changed hydraulics • Changed estuarine salinity • Pyrite oxidation • Changed sediment transport patterns	• Loss of macrophytes, emergents and riparian vegetation • Weed invasion • Water acidification • Fish kills • Mycotic fish diseases • Thermal, oxygen and salt stratification • Reduced plant uptake of nutrients • Reduced light penetration • Water quality barriers to fish migration • Loss and fragmentation of riparian corridor and aquatic habitat • Reduced diversity of aquatic habitat • Loss of large woody debris • Reduced abundance of native fish • Increased abundance of exotic fish

Note: Biological impacts were largely dependent upon physical impacts.
Source: after Erskine (1998); cited by Erskine and Green (2000).

mismanagement of floodplains are priorities in defining sustainable development policies (Bravard et *al.*, 1999b).

Catchment impacts including land use change

Modifications to parts of the catchment, or to the catchment as a whole, influence river channel processes and hence the river channel itself. Such effects, which have occurred for as long as human activity has altered vegetation and land use (Gregory, 1995), are not always clearly distinct from point and reach changes, and examples of both types may be present in any one area. Impacts of vegetation and land use change affect the basin in three main ways. First are culturally implemented and progressive changes of land cover including deforestation, impact of intensive grazing, ploughing of land for agriculture and eventual industrialization and urbanization. Second are processes by which these changes are accomplished, sometimes dramatic and short lived and, if involving fire, with very significant, albeit short-term, influences upon river channel morphology. Third are cases where landscape management, often including conservation measures, may have been implemented to induce changes that include river channel adjustments. Changes in channel-forming discharges, in sediment sources and in sediment transported together with vegetation changes affecting channels can all contribute directly to cause channel changes, and have effects that are superimposed upon changes of sediment yield rates through time (Schumm and Rea, 1995). Land use changes have a range of potential effects (Table 4.9) and have often occurred sequentially (Figure 4.4). with land use changes from forest, to grazing, to agriculture, to urbanization (Wolman, 1967).

The three types of impact named above are all manifested by the impacts of *urbanization*, providing some of the most obvious and dramatic influences of land use change on river channels, within or downstream from urban areas. Higher sediment yields during

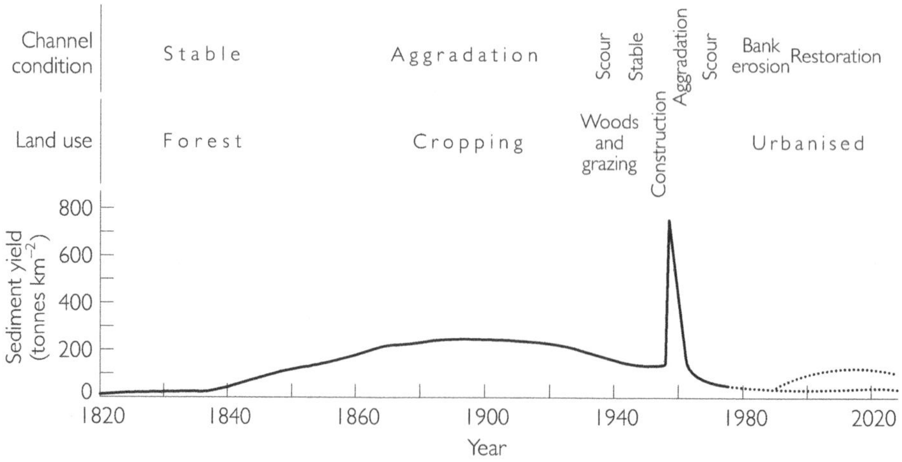

Figure 4.4 *Sequence of development of channels in urban areas. Sediment yield in urban areas may remain low or may increase under restoration management so that two lines are shown (modified from Wolman, 1967)*

building activity and the urbanization process, together with the greater runoff and peak discharges, which continue when urbanization is complete, produce substantial effects upon channels within and downstream from urbanized landscapes. Substantial runoff increases as a result of urbanization mean that channel change, often enlargement, occurs downstream of an urban area, and scour and bank erosion may be accentuated because sediment sources are no longer freely available in many urban areas. The major hydrological effects and consequent changes are summarized in Table 4.9, although there is considerable variety in urbanizing and urbanized areas. During building activity, substantially increased sediment yields accompany higher flows, leading to more extensive sedimentation on floodplains as illustrated in the Denver area of Colorado (Graf, 1975). Subsequently, when urbanization is complete, sediment transport may be at much lower levels than that prior to urbanization, although the significantly higher discharges result in enlarged channels that can be up to four or occasionally up to eight times greater than the previous channel capacities, with channel enlargement necessarily varying according to the character of the areas urbanized (Roberts, 1989; Figure 4.5B). However, the effects of urbanization do not simply result in all channels increasing their dimensions, because they may first decrease in capacity and subsequently enlarge, as shown along the Watts Branch, Maryland (Leopold, 1973). Particular areas develop distinctive patterns of erosion, as demonstrated in southeast Australia near Wollongong (Nanson and Young, 1981)

Table 4.9 Effects of basin changes on processes and channel morphology

***Why* change occurs**	***What* changes may occur**				
	Process change (Hydrology, H; sediment, S)	**Channel adjustment** (Clearly evident, *)	**Spatial effects** **Downstream extensive (#)**	**Upstream extensive (#)**	**Complete network (Σ)**
Deforestation	H+ S+	* locally, channel debris+			Gully development may occur with knickpoint recession
Grazing pressure	S+			Some gullying possible	
Crop cultivation	H+ S+	*locally	#		
Land/agricultural drainage	H+	locally			
Afforestation	S– but can increase from tracks	*	#		
Conservation methods	H– S–				
Surface mining	H+ S+	Local change			
Road construction	H+ S+				
Building activity	S+	*sediment accretion	#		Road drains complement stream network
Urbanization	H+ S–	*enlargement	#		

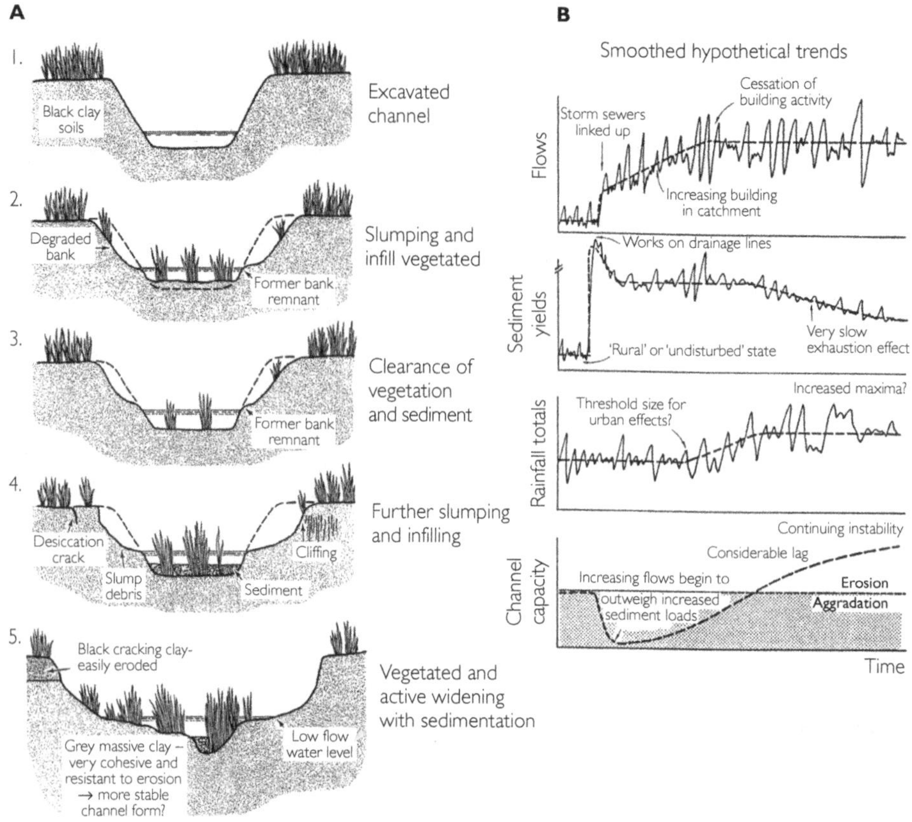

Figure 4.5 *Urbanization and channel enlargements.* ***A*** *Sequence of morphological adjustments of Avondale stream channel, downstream of Harare, Zimbabwe (after Whitlow and Gregory, 1989)* ***B*** *Hypothetical trends in flows and channel capacity after urbanization (after Roberts, 1989)*

and where dambos are converted to incised channels (Figure 4.5A) downstream of Harare (Whitlow and Gregory, 1989). Patterns of channel change in urban areas can be demonstrated by mapping indicators of channel adjustment as used downstream of urban areas in the Monks Brook basin of southern England (Gregory *et al.*, 1992), or by comparing channel dimensions above and below the points at which stormwater drains reach the drainage network, as exemplified by the pattern of urban channel erosion at Catterick Garrison, northern England (Gregory and Park, 1976). In the dryland area of Fountain Hills, east of Phoenix, Arizona, the network of ephemeral washes has effectively been fragmented into sections by stormwater inputs at road crossings (Chin and Gregory, 2001), with scour resulting immediately downstream of crossings and sedimentation often occurring immediately above. Not all urban areas are affected by channel enlargement because some urban channel networks have been modified by channelization complete with bank revetments so that they cannot adjust, although such engineering work does not always succeed, as shown in Harare (Whitlow and Gregory, 1989) and in Kansas City (Vaughn, 1990). Knowledge of channel adjustments and of temporal variations (Chapter 5, p. 136) should be kept in mind as guiding principles for planners, policy makers and citizens (Riley, 1998) when attempting a design with nature approach (Keller and Hoffman, 1977).

Over the drainage basin as a whole, each change from one vegetation cover or land use to another can potentially instigate changes in channel morphology and hence in ecology. One result of land cover change is that new sediment sources occur as soil or rock is exposed, for example through quarrying, road construction or soil erosion consequent upon vegetation and land use changes, which themselves embrace afforestation, deforestation, grazing, the incidence of fire, land use management practices, mining, and industrialization and urbanization. Any of these changes can influence the hydrology of, or the sediment production from, the basin, thus altering the channel-forming conditions and the ecological character of the river channel (Table 4.9). Deforestation, grazing, burning, agriculture or changes associated with drainage schemes (Table 4.5) can all have impacts on the channel system. In steep forested areas as in Pacific northwest USA, tree removal tends to leave slopes far more susceptible to landslides, while the installation of forest roads creates a substantial geomorphological impact (US Department of Agriculture, Forest Service (USFS), 2001; Luce and Wemple, 2001). The roads, their drainage ditches and culverts increase drainage density and act to concentrate runoff, thus increasing flood frequency. The result can be diversion of stream channels and initiation of gully erosion (Croke and Mockler, 2001). In the Klamath National Forest, Orgeon, erosion attributed to forest roads is believed to result in unit-area sedimentation rates downstream that are 58 times the background rates from landslides, and 290 times those from surface erosion (Klamath National Forest (KNS), 2002). Together, these impacts can give increased potential for channel erosion caused by runoff increases, and inchannel change by aggradation further downstream as eroded sediments are deposited.

Land use change can affect many aspects of processes that can affect channels. In the Upper Mississippi Valley, reduced infiltration caused by tillage and grazing increased peak discharges from high magnitude floods in small catchments by 200–400% above their pre-agriculture natural magnitudes (Fitzpatrick and Knox, 2000; Knox, 2001), with additional sediment production occurring in response to the increased frequency of surface runoff, exposure of bare soil and down-slope ploughing. Additional sediment can cause significant overbank deposition accompanied by a sequence of lagged responses in the river channel. In Shullsburg Branch, Wisconsin, increased overbank sedimentation increased bank heights, which led to the containment of higher flows and an acceleration of lateral bank erosion from the 1820–90 average of 0.06 m yr^{-1} to a maximum during 1925–40 of 0.70 m yr^{-1}. Bank erosion caused channel expansion that eventually reduced overbank flow frequency, especially since land conservation practices were introduced in the 1950s, but a new floodplain surface has developed. In this case, armouring of the channel bed by cobbles and boulders prevented channel incision (Knox, 2001).

If change in the magnitude and frequency of peak discharges caused by land use change is greater than critical threshold levels (see p. 125), then channel morphology may adjust incrementally or channel type may change. The complexity of the adjustments, and of the changes that can take place, must not be underestimated however, and a collection of geomorphic responses to land use change (Brierley and Stankoviansky, 2002) show how geomorphic insights into the nature and extent of change at regional, catchment, reach and hillslope scales have considerable implications for the sustainable management of landscapes. More effective integration of the differing scales of sedimentary cascades could significantly expand the range of critical tools that geomorphologists provide for future land management and planning (Brierley and Stankoviansky, 2002).

River responses to anthropogenic changes of land use over the last 15 000 years (Gregory, 1995) have included changes since the Neolithic, prior to the more dramatic changes occurring from the seventeenth century onwards. Over the last several thousand years, changes in land cover and land use instigated by human activity have led to alterations in runoff and sediment production that in turn have significantly affected river channel morphology. Thus, in many areas, the accumulation of fine floodplain sediments by alluviation was a consequence of deforestation, so that later land use changes could occasion reactions of the river channel system, as in southern Poland where extensive potato cultivation prompted the development of multithread river channels that returned to single thread once the intensity of land use decreased (Klimek and Starkel, 1974; Starkel, 1991). Size of the area modified, in relation to the total basin size, may also be significant: where agricultural land drainage affects a small area draining directly to a river with a large drainage area there will be comparatively little hydrological effect and channel adjustment; whereas if the affected area is large relative to the total basin area then the effects may be substantial.

4.3 Climate change and channel adjustment

Allogenic channel adjustment can be forced by changes of climate, although it is not easy to determine which changes of climate, on short, historic or palaeohydrologic timescales, will be significant. It is especially difficult to separate climatic from human impact effects, although a checklist can be compiled for any area (Table 4.3). The difficulty of separating climatic change from human impact effects is exemplified in Australia where a sequence of events was recognized (Nanson, 1986) and interpreted as episodes of vertical accretion and catastrophic floodplain stripping comprising a mode of disequilibrium floodplain or valley bottom development. Such episodes have been associated with flood-dominated and drought-dominated regimes arising from sequences of wet and dry years. It has been suggested (Nanson and Erskine, 1988) that some channels are in a state of 'cyclical equilibrium' whereby they widen, straighten and increase bedload during two to three decades of high flood frequency and magnitude (flood-dominated regimes, FDR); then narrow and become more sinuous in the subsequent two to three decades of lower flood frequency and magnitude (drought-dominated regimes, DDR). The alternation in the last two centuries led to the suggestion that FDRs and DDRs occur on a cyclical basis every 50 years or so (Rutherfurd, 2000). However, others have argued (e.g. Brooks and Brierley, 2000) that flood-dominated periods would not cause catastrophic channel change if it were not for the removal of catchment, especially riparian, vegetation (see also Erskine and Warner, 1988). Thus, the process of floodplain stripping, argued to represent a natural cyclical process in some floodplains (Nanson, 1986) may be encouraged by the absence of the original dense floodplain vegetation formerly occupying the floodplains (Rutherfurd, 2000).

The most dramatic climate changes were originally thought to have been associated with glaciations but recently it has been appreciated that there were other particularly significant aspects of climate change in the Holocene post-glacial period (Gregory and Benito, 2003b). Three major climatic episodes (20 000–14 000 BP, 14 000–9000 BP and 9000 years BP–present) have been shown to have major influences on fluvial system behaviour (Knox, 1995). Over short timescales, in Europe, for the period since AD 1200, enhanced periods of fluvial activity have been demonstrated between 1250 and 1550 and between 1750 and 1900 (Rumsby and Macklin, 1996) coinciding with periods of climatic transition, whereas the Medieval Warm Period AD 900–1250, the period 1550–1750 and the most severe phases of the last neoglacial (the Little Ice Age, AD 1550–1750) were associated with lower rates of fluvial activity. Floodplains are particularly affected and many in Britain have 200–300-year-old trees established when lower temperatures and larger floods occurred during the Little Ice Age. Second there have been changes in the incidence of specific events, including changes in the frequency of flooding, which can affect channel morphology by changing the frequency of channel-forming discharges. The relationship between climatic variability and hydrologic series has been explored (Hirschboeck, 1988) by *Joseph* and *Noah* effects (Mandelbrot and Wallis, 1968). Whereas *Joseph* effects are the incidence of very long periods of low river flows (or precipitation) or very long periods of high flows, *Noah* effects refer to short-term rare occurrences of extremely high flows (or precipitation), effects which can be transmitted to the river channel morphology. Flood- and drought-dominated regimes necessarily involve extreme events, requiring reconsideration of the magnitude–frequency concept (Richards, 1999).

Medium and large channel and floodplain systems in humid climates may develop a quasi-equilibrium morphology (Table 3.6), adjusted to the water and sediment delivered, so that alterations of climate inputs may result in channel and floodplain adjustments (Knox, 1984). Long historical records from the Mississippi at St Paul, Minnesota, demonstrated (Knox, 1984) nonrandom and persistent departures from average flows because, before 1895 and since 1950, meridional patterns prevailed frequently with relatively cooler and moister episodes than those between 1895 and 1950. The result was that, before 1895 and after 1950, more frequent and relatively large floods contrasted with the intervening episodes (Figure 4.6). Long-term variations in the magnitude of the 1.58-year recurrence interval 'bankfull' flood identified in western Wisconsin (Knox, 1984) also meant that, whereas many engineering structures had been designed to withstand the effects of the probable maximum flood, there could be very significant differences according to the period of record used to calculate the probable maximum discharge. Four distinct major environmental episodes characterize the magnitudes and frequencies of floods in the upper Mississippi River valley since about 25 000 years BP, each episode associated with changes over the basin, reflected in the channel morphology, with the fourth episode initiated by the introduction of Euro-American agriculture in the early decades of the nineteenth century (Knox, 1999).

Recognition that river channel changes can reflect change over long periods of time required research in palaeohydrology (Schumm, 1977b), defined as 'the reconstruction of the components of the hydrological cycle, of the water balance, and of sediment budgets

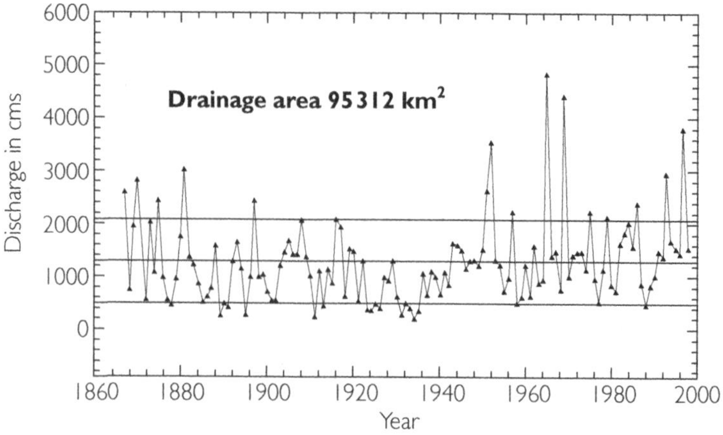

Figure 4.6 *Annual maximum flood series for upper Mississippi River at St Paul, showing large floods were relatively common during the late 1800s and since 1950 but were relatively infrequent during the first half of the twentieth century (after Knox, 2000)*

for the time before continuous hydrological records, and necessarily embracing an understanding of the way in which changes of the hydrological cycle occurred and establishing how they differed from the contemporary hydrological picture' (Gregory, 1983: 10). Palaeohydrological research (e.g. Starkel et *al.*, 1991; Gregory et *al.*, 1995; Branson and Brown, 1996; Benito et *al.*, 1998; Gregory and Benito, 2003a) provides background for river channel management in the context of global environmental change through derivation of data, as well as by analysis of mechanics of temporal change, spatial contrasts, coupling of global climate change models to hydrological models, all providing a basis for construction of new models of hydrological change (Gregory and Benito, 2003b). Reconstruction of hydrological data, especially palaeoflood events (Baker, 1994, 1996), prior to the instrumental record, can be based on slackwater deposits (SWD) enabling derivation of more effective flood frequency curves than those derived solely from the instrumental record. Such reconstructions are particularly important in relation to the estimation of design discharges for in- and near-river structures (Enzel et *al.*, 1993; Baker, 1995; House et *al.*, 2002).

Catalysed by awareness of global warming (see p. 169), attention is now focused on changes in the frequency of severe floods and droughts during the Holocene, possibly the clearest indication of climate control on river systems (Knox, 1983; Starkel, 1983; Baker, 1991; Macklin and Lewin, 1997; Knox, 1999). A consistent pattern emerging from a number of world areas is one of alternating periods of high and low frequency of the order of 10–25 years duration over the last few centuries. It now becomes possible to identify possible global patterns of clustering of flood or drought events, and progress is being made in relating these to the influence of the El Niño–Southern Oscillation (ENSO) in low latitudes and of the Intertropical Convergence Zone (ITCZ) in higher latitudes (Macklin and Lewin, 1997).

Because the past may not provide suitable analogues for present and future environmental conditions (Baker, 1996), palaeohydrological information should be employed to

Table 4.10 Examples of observations since 1984 stressing the importance of a temporal view

A Affirming the need for a temporal perspective:
- each river has a history, reflected in landscape memory
- engineers typically concerned with a particular design period, varying timescales should be considered
- engineering geomorphology – defining all strategies at the reach and watershed scale based on historical and contemporary assessment of geomorphological dynamics

B Reasons why temporal change should be considered in river management:
- variability in time as well as in space is key to the understanding of fluvial systems
- longer-term studies are required to explain river processes
- the effects of climate change still affect contemporary river systems
- late Quaternary environmental changes provide the 'initial' conditions for present-day processes, their activity rates and the resulting morphology
- landscape histories and the cumulative consequences of land use changes may assist in the prediction of geomorphological futures
- geomorphic inputs may explain the reasons for erosion and flooding problems
- long-term evolution provides understanding of the proximity of individual reaches to threshold conditions

C Specific implications:
- palaeo landforms can furnish information on levels of stability
- analysis can be aided by application of multiple stable and unstable equilibria , amending regime approaches

Source: after Gregory (2003a); see Gregory, 2003b: table 21.4.

condition future modelling, to enlighten understanding of river channel change and behaviour, and to contribute to future management scenarios (Gregory, 1998, 2000, 2003a). Research papers published since 1984 indicate three major issues that confirm the need for information from the past to inform river channel management (Gregory, 2003b): first, arguments for a temporal perspective on management (marked A in Table 4.10); second, reasons why management should be considered in the context of temporal change (B), particularly that variability, in time as well as in space, is key to the understanding of fluvial systems; complemented, third, by specific implications (C). Although investigation of the past can illuminate decision-making, interactions of climate change with thresholds and sensitivity need to be evaluated (Chapter 5, p. 130).

4.4 The world pattern of river channel changes

Changes occurring in the channel cross-section, the channel planform including the floodplain and the channel network (Gregory, 1976a, 1987a), can each be thought of in terms of changes of channel size, shape or composition (Table 4.4). Compiling data from published research for five major types of human-induced river channel changes (Gregory, 1976a, 1987a, b, 1995), indicates that substantial, but highly variable, ratio changes in channel capacity, width or depth can occur (Table 4.11). The result implies that knowledge gained about such impacts needs to be taken into account in river channel management.

The potential complexity of channel changes and the need to embrace knowledge of channel history for channel management is demonstrated by the Gila River channel in Arizona. The channel was fairly stable and narrow from 1846 to 1904, when the channel meandered through a floodplain covered with willow, cottonwood and mesquite. During this period, average channel widths were less than 46 m in 1875 and less than 91 m in 1903. However, in the period 1905–17, beginning with significant floods in 1905 that swept away all traces of the meandering channel, a braided channel formed that was more than 1 km wide in some places (Burkham, 1972). The average channel width increased to 610 m before dense phreatophyte growth and sedimentation narrowed the channel and facilitated reconstruction of the bottom land from 1918 to 1970, which saw the average channel width reduce to less than 61 m in 1964. Minor widening occurred again in 1965 and 1967 so that the average width in 1968 was 122 m (Burkham, 1972). A large flood occurred in 1983, after which further reconstruction occurred, so that by the early 1980s the river had a compound appearance apparently returning to the meander geometry of almost a century before (Graf, 1988b: 207). Overall, the morphological response of the Gila since 1920 (Hooke, 1996) has depended upon the sequence of high flow events and feedback effects on morphology, so that the morphology may not show a simple relationship to size of event or to wetness of a period. As such, care must be taken when making inferences about change from morphological evidence alone (Hooke, 1996).

In many basins and along many rivers a sequence of impacts, each with associated adjustments, has been superimposed; the separate effects of each adjustment not easily distinguished. In Australia the complex range of adjustments that can occur has been admirably demonstrated (e.g. Brizga and Finlayson, 2000), with four major types of human impact on Australian river channel morphology categorized as (Rutherfurd, 2000):

Table 4.11 Magnitude of river channel adjustments

Cause of change	Range of channel change	Number of studies used
Reservoir construction and dams	0.09–3.0; 74% of studies show reduction	57
Urban development	0.15–>10.0; 76% of studies show increase	37
Channelization and other reach changes	0. 23–>12.0; 57% of studies show increase	49
Catchment land use change	0.05–>15.0; 64% of studies show increase	53
Water transfers	0.95–>9.0; 80% of studies show increase	10

Note: Channel change is the ratio of capacity, width or depth after change to that before.
Source: developed from Brookes and Gregory (1988); Gregory (1995).

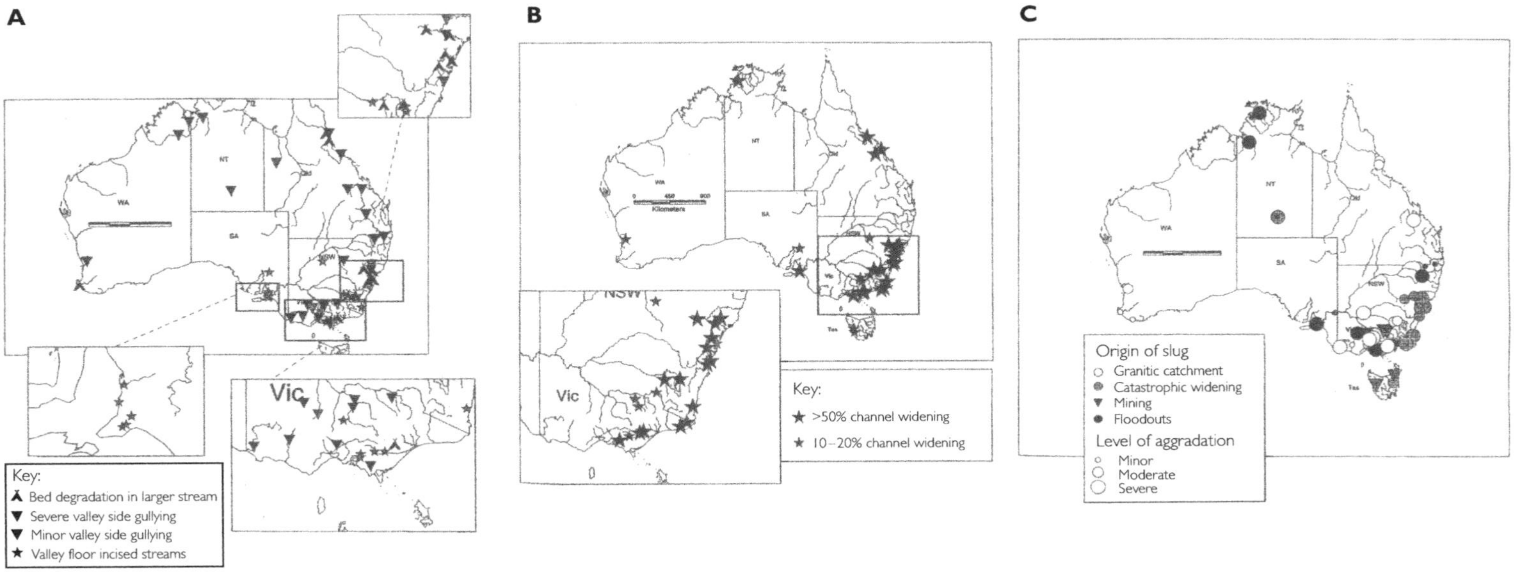

Figure 4.7 *Reported channel changes across Australia.* ***A*** *Reports of channel deepening in Australian streams;* ***B*** *reports of channel widening in Australia;* ***C*** *reported distribution of historical sand slugs in Australian streams (after Rutherfurd, 2000). Rutherfurd notes that these maps show distribution of published examples and may not necessarily reflect the true distribution of the phenomena*

1 incision of smaller streams comprising channel network extension or valley floor incision (Figure 4.7A);
2 general enlargement of larger streams, and accelerated meander migration (Figure 4.7B);
3 catastrophic channel enlargement (Figure 4.7B);
4 channel avulsions.

In reviewing the reported distribution of historical sand slugs in Australian streams (Figure 4.7(C)), Rutherfurd (2000) notes that much of the research literature has focused on the first and third of these categories. A further dependency is that many of the rivers studied are those accessible from major university and research centres, which may cause a bias analogous to that of different perceptions of river behaviour (p. 80).

Dumaresq Creek on the Tablelands of New South Wales illustrates how several adjustments have been superimposed (Gregory, 1977a). Settlers in the nineteenth century, after 1848, were responsible for clearing the eucalypt woodland so that greater runoff changed the character and pattern of river channels from headwater depressions and channels embracing chains of ponds (Eyles, 1977) to new types, with more continuous channels, and to the incidence of valley side and valley floor gullies. Completion of Dumaresq Dam in 1898 regulated flow downstream so that the channel morphology was reduced; the subsequent urban growth of Armidale ultimately affected about 20% of the basin with urban runoff. However, the impact of urbanization has not been uniform. Sediment from the urban area coupled with regulated flows led to siltation of the main channel through Armidale, the urban runoff increased once the stormwater tanks of individual properties were discontinued, and a greater urban runoff increase occurred once stormwater drains were installed and fed into the tributary channel network. With the requirement for urban stormwater management plans this is now an area for possible holistic integrated management of the basin (Gregory, 2002).

4.5 Conceptualizing river channel adjustment

The historical and catchment hydrosystem context unique to each river means that individual studies of river channel adjustments are not easily generalized into models of river channel change. Such models would ideally indicate what, why and how much change occurs, as well as answering other questions of where, how and when it will occur (Chapter 5, p. 134). This presents a challenge requiring analytical capability beyond a regime theory approach.

In addition to the conceptualization (Figure 4.4) of land use change related to channel change (Wolman, 1967), another conceptual model of river channel change over time was provided by relating gullying to alluviation, providing a link between adjustments in the headwater areas of the basin with those downstream (Strahler, 1956). Specific investigations of channel change were also employed, using space–time substitution (the ergodic hypothesis) to provide estimates of the amount of channel change that could occur over time (e.g. Gregory and Park, 1976; Gregory, 1977a). Degrees of freedom were suggested (Lane, 1955a, 1957; p. 91) to relate bed material load (Q_s), particle diameter (D), water discharge (Q_w) and stream slope (S) in the form:

$$Q_s D \approx Q_w S$$

This equality was used to deduce the implications of six types of change (Lane, 1955a). For instance, if water was diverted into the system, Q_w would increase so that the equality could be maintained only if either Q_s or D increased or if S decreased. This approach was extended by developing an approach to river metamorphosis based upon empirical relationships obtained from a number of stable alluvial channels in semiarid and humid areas (Schumm, 1969), including the approximation:

$$Q_w = \frac{w\, d\, ML}{S}$$

where w is channel width, d is channel depth, ML is meander wavelength. The approximation indicated how possible changes could occur but was limited in that there may be correlation between the component variables that are not always appropriate and comprehensive, and that several outcomes may be possible from any one situation. For example, in the case that water discharge increases as a result of catchment land use change then a positive response could occur in numerous of the parameters, including:

$$Q_w^{+} = \frac{w^{+} d^{+} ML^{+}}{S^{+}}$$

Application of this approach to specific examples led Schumm (1973) to propose complex response (p. 101) to typify how several alternative change scenarios might result from comparable environmental conditions; such complexity was analysed by experiments employing flumes and a drainage basin hardware model (Schumm *et al.*, 1987). Consideration of river metamorphosis also led to conceptualization of thresholds (p. 125) and to generalization of the potential problems (Table 4.12) faced in interpreting river channel changes (Schumm, 1991: 94). Such problems should be assessed as part of river channel management, although three of the ten, singularity, efficiency and multiplicity, may be inherently accommodated by most investigations.

River channel changes occur in response to an alteration from a notionally stable condition, originally described as an equilibrium condition (Table 3.6). An intriguing approach to the sequence of changes was provided by a rate law (Graf, 1977) in the form of a negative exponential function, similar to that used to describe the relaxation times of radioactive materials and chemical mixtures, providing a model for relaxation times in geomorphic systems (Graf, 1977, 1988a). Graphic representation of the disrupted system (Figure 4.8) includes the magnitude of response in the measured channel parameter as it progresses from steady state or equilibrium A to steady state D, through reaction time B (the time needed for the system to absorb the impact of the disruption), and relaxation time C (the time period during which the system adjusts to new conditions). The channel change depicted by this conceptual model can be autogenic, inherent in system dynamics, or allogenic, whereby it is superimposed on the system as a result of climatic forcing or human impact.

Catchment hydrosystems and most problems of river management are four-dimensional, including time as one dimension, but numerical models of surface water systems are

Table 4.12 Problems identified by Schumm (1991) that need to be considered in interpreting river channel changes

Problems of scale and place:

- *Time* – the period of time for data collection is invariably too short and we deal with physical systems that operate over varying time spans.
- *Space* – two aspects are *scale*, which involves the resolution at which the object is viewed and *size*; these are affected by the resolution possible in any investigation.
- *Location* – involves differences between places and perspectives obtained by workers according to their training.

Problems of cause and process:

- *Convergence* – may be referred to as equifinality: different processes and causes can produce similar effects.
- *Divergence* – similar causes and processes produce different effects.
- *Efficiency* – the ratio of the work done to the energy expended, referring to the impact of an event or series of events on a system.
- *Multiplicity* – multiple causes act simultaneously and in combination to produce a phenomenon, so that a multiple explanation approach should be applied to problems and multiple working hypotheses used as appropriate.

Problems of systems response:

- *Singularity* – the condition, trait or characteristic that makes one thing different from others, and is really the randomness or unexplained variation in a dataset, called indeterminacy by Leopold and Langbein (1963: 190).
- *Sensitivity* – refers to the propensity of a system to a minor external change and embraces the proximity of a system to a threshold – if it is near, and sensitive, it will respond to an external influence.
- *Complexity* – the complex response that occurs when the system is perturbed because a complex system, when interfered with or modified, is unable to adjust in a progressive and systematic fashion.

still rarely developed beyond two dimensions and limitations to the dimensionality of numerical models have recently led engineers to further de-emphasize time (James, 1999). Although it is not possible to predict channel change responses exactly because of unique combinations of environmental conditions and basin history, it was concluded (Brown and Quine, 1999) that traditional engineering approaches such as regime theory, and traditional geomorphological approaches such as hydraulic geometry, may not be suitable for the prediction of the effects of environmental change on fluvial systems. Aspects of the possible scales and magnitude of change, their spatial distribution and propagation, indicates that the response of the fluvial system is not only, or necessarily, dependent upon the extrinsic factor, but upon the configurational state of the system, in turn a product of the geomorphic history of the system, involving nonlinearity of response (Brown and Quine, 1999). The wider range of biophysical processes that are part of the configurational state

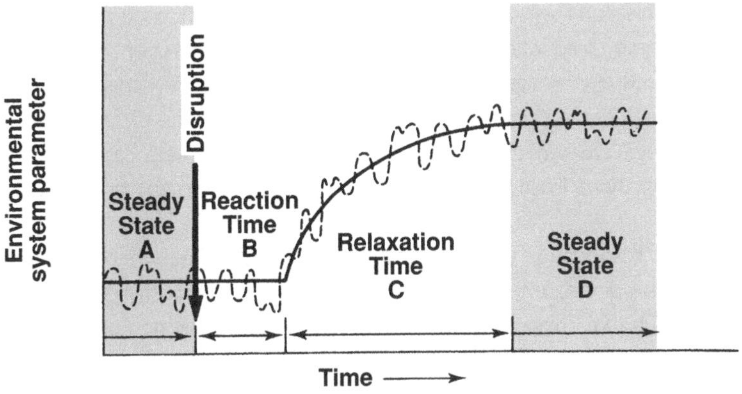

Figure 4.8 *System change between two states of equilbria (A and D). The dashed line represents observed values, the solid line is the mean condition of the system, B is reaction time and C is relaxation time (Graf, 1977, 1988a modified by Gregory, 1985)*

can affect the response of the system to environmental change in the way that the growth of aquatic weed as a result of higher stream temperatures can affect channel processes. This relates to the interest in 're-naturalizing' rivers (see Chapter 7, p. 185), coinciding with a time when environmental change is occurring at an even greater rate than ever before.

Attempts to predict the new equilibrium channel form expected after complete adjustment to a specified and finite set of environmental changes, need to allow for the way that the disturbance usually involves continual dynamic adjustment processes reflected in a series of transient states (Richards and Lane, 1997). If the transient states persist for significant periods of time, or the complex oscillatory responses to stepped or impulse changes result in overshoots beyond the likely final equilibrium state, then a methodology based on the governing process laws is required (Richards and Lane, 1997), and nonlinear dynamic modelling has been advocated generally for the investigation of Earth surface systems (Phillips, 1999).

4.6 Implications for river channel management

Impressive studies in many drainage basins have enabled demonstration of the very extensive consequences of river channel change (what) and the range of causes responsible (why), so that it has been realized that dynamic sequences of channel change should be considered in river channel management, reinforcing the increasing lobby for environmentally sound and sustainable river channel management alternatives. Awareness has been raised of the amount of change that has occurred in the past, some of which is not immediately apparent in the present system and some not easily predictable because complex response (Schumm, 1969) may give different outcomes to the same cause. Thus, circumstances that produce channel enlargement upstream and floodplain aggradation downstream in one area may produce extensive gully development in a different type of environment. In any landscape, several types of influence may be imprinted, or palimpsest, one over the other; some changes over the catchment may be manifested directly by changes in the drainage network of river channels, so that intensification of agriculture involving large-scale commercial farming is both a land use influence on channel change but equally may have direct effects upon extensions of the drainage network through the addition of agricultural drainage schemes or networks of irrigation channels. In west central Wisconsin, the incision of nearly every tributary stream in the catchment in response to land use change created enlarged channels with considerable capacities to contain floods, preventing them from inundating and depositing sediment on the valley floors (Faulkner, 1998).

Complexity of response is also exemplified by the fact that, whereas some channel adjustments can be direct, others may be indirect and less obviously recognized. The outcome of change is that river systems have effectively been segmented, for example, by dams on most rivers in the northern hemisphere with associated large-scale disruption of the physical and ecological integrity of rivers (Dynesius and Nilsson, 1994; Graf, 1999), in urban areas according to inflows of water and sediment from point sources in the urban drainage network (Gregory and Chin, 2002), and by piecemeal channelization in many lowland rivers. Despite progress by investigations of channel change a number of outstanding issues remain (e.g. Table 4.12). River channel management requires consideration of uniqueness in the river continuum and consideration of the implications of channel sensitivity and questions of where, how and when change occurs (Chapter 5), conclusions to emphasize are that:

- *any river system has several degrees of freedom*, with adjustments in a particular location (Table 4.4) depending upon local environmental characteristics and history;
- *contemporary processes are not always responsible for the present character of the channel*, the channel pattern, the channel network or the floodplain so that past processes need to be considered including not only autogenic change inherent in the system but also allogenic change that is superimposed as a consequence of climate change or of human impact;
- *channel adjustment can arise through a combination of natural forces and human activity* occurring at points or along reaches in the channel network (Table 4.3), and may affect large parts of, or the entire basin (Table 4.5). Superimposed channel change induced by climate change creates complex results that are not easily separated;
- *many river channels have been segmented* into sections each showing impacts of particular types of human activity, conflicting with the notions of continua suggested for undisturbed catchment hydrosystems (Chapter 3). Dams and reservoirs are prime examples of the cause of such segmentation;
- *understanding of changes over periods longer than 'engineering timescales' is necessary as a precursor to management.* River systems should be visualized in relation to several timescales of change; the design periods of the engineer (steady time), traditionally up to 50–100 years, should be complemented by longer time periods up to 10 000 years (Table 4.2) required to understand the evolution of the fluvial system in relation to a range of significant influences that have affected the river or the entire basin;
- *in view of the requirement for longer-term information, understanding gained from palaeohydrology potentially contributes useful information* about the history of a river and of river channel change as an input for river channel management because each river has

a history reflected in landscape memory. Late Quaternary environmental changes may provide the 'initial' conditions for many present-day processes, including the proximity of individual reaches to threshold conditions, information from palaeolandforms furnishing information on likely levels of channel stability and the potential that contemporary river systems are still responding to long-term climate change.

5
River Channel Hazards: Sensitivity to Change

Synopsis

River channel management operations frequently respond to a channel-related hazard such as bank erosion, which is often depicted as 'instability', although such instability also produces environmentally valuable river assets. Specific risks can be estimated from knowing where, when, how long-lived and how much adjustment will occur, all influencing the robustness and resilience of the habitats dependent on the channel environment (5.1). Approaches for determining the risk of hazard occurrence have the goal of understanding river channel sensitivity to change so that management can respond appropriately to the type of hazard and the risk of its occurrence (5.2). Sensitivity to change in drainage basins undisturbed by human activity is a response to drainage basin characteristics that provide both extrinsic and intrinsic controls on river channel processes that are variable in time and space (5.3). However, river channel sensitivity to change is also conditioned by the impact of human activities on runoff, sediment production and calibre, flow magnitude, duration and timing, flow hydraulics and sediment transport processes (5.4). Consequently, river channels are complex, sensitive phenomena where the risk of hazards and the risk to assets prompt six fundamental implications for river channel management (5.5).

5.1 River channel hazards, assets and risk

Expenditure on river channel management has frequently responded to a perceived channel-related hazard, that is, 'a potential threat to humans and their welfare' (Smith, 1992: 6). Such hazards are usually associated with channel adjustment, occurring either naturally or as a result of human activity, and are often portrayed as examples of channel 'instability', although they may actually reflect changes under conditions of dynamic equilibrium (Chapter 3). Hazards involve the probability of a hazardous event occurring and the loss or vulnerability associated with the hazard. The product of these two elements defines the *risk* (R) associated with the hazard although potential losses (L) (e.g. human lives) are generally weighted more strongly than their probability of occurrence (p) so that $R=pL^x$ where $x>1$ (Smith, 1992). For instance, the threat of river bank erosion to urban infrastructure is a commonly perceived river channel hazard, and action often follows to protect property deemed to be at risk whether or not the likelihood and rate of bank erosion is known. The action is generally to install bank protection measures (i.e. increasing hazard resistance), although this may cause undesirable channel responses such as increasing the erosion hazard downstream. An alternative involves 'vulnerability modification', that is, to lower the risk associated with bank erosion by moving valuable infrastructure outside of the path of the eroding bank (see p. 306).

While the hydrological hazards of flood and drought associated with flowing river water have been studied intensively, hazards associated with river channels are primarily geomorphological and have been far less clearly articulated. Geomorphological hazards are generally associated with instability (Schumm, 1988; Gares *et al.*, 1994) as the 'probability that a certain phenomenon reflecting geomorphological instability will occur in a certain territory in a given period of time' (Panizza, 1987: 225). For river channel hazards, it is fair to substitute the term 'adjustment' for instability because even change associated with natural river processes, such as the progressive migration of a meander bend, can be perceived as a hazard even though it occurs in geomorphologically stable systems. Therefore, assessment of river channel hazards requires an understanding both of natural system operation and the degree of human impact over extended time periods. Natural hazards have previously been characterized in terms of magnitude, frequency, duration, areal extent, speed of onset, spatial dispersion and temporal spacing parameters (Burton *et al.*, 1978). For river channels, hazard dimensions can be characterized in terms of *where*, *when*, *how long-lived* and *how much* adjustment occurs (Gregory, 1987a) and managers cognisant of these dimensions should understand the location, likelihood, persistence and magnitude of prospective hazards (respectively). Competing management options can then be evaluated conscious not only of *what* channel adjustment might result from different options and *why*, but also the specific dimensions of the adjustments expected to result from each option. This could provide the ideal basis for making rational management choices to minimize river channel hazards and promote river channel assets; defined as features, sites or catchments of great habitat or other environmental value, based on rarity, uniqueness, critical place or function in the ecosystem, scenic attraction or heritage value (Downs and Gregory, 1994; Cavallin *et al.*, 1994). River channel assets are usually maintained by the same geomorphic processes as river channel hazards, introducing a source both of conflict and potential opportunity in river channel management.

River channel management using hazard (or asset) assessment as a core component requires that river channels with only a small risk of adjustment are managed very differently from highly changeable channels. Therefore it is important to analyse the dimensions of channel adjustment with accuracy and precision. Various types of river channel hazard exist (Schumm, 1988), namely:

- abrupt, producing a catastrophic event, such as dramatic channel widening in response to a large storm event;
- progressive change that leads to abrupt change, for instance meander growth leading to a meander cut-off;
- progressive change that has slow but progressive results such as bank erosion.

Such hazards arise either naturally from geomorphological processes, from the inadvertent consequence of land use activity or from a response to channel management actions either in the managed reach or upstream or downstream of it (Gregory and Chin, 2002, see Table 4.5). If geomorphological assets are considered, two further causes of hazard can be incorporated (Cavallin *et al.*, 1994): direct management impact resulting in loss of asset and induced, indirect, management impact resulting in loss of surrounding assets (Figure 5.1).

Channel hazard types can be categorized (Table 5.1) according to their key variables of time, discharge, sediment load and base-level (Schumm, 1988). Major hazard categories include erosion, deposition, pattern change and metamorphosis leading to hazards such as channel incision, bank erosion, aggradation and meander migration (Figure 5.2). River channel hazards effectively involve changes in one or more of the channel's nine 'degrees of freedom' (Hey, 1982, 1997b; Chapter 4, p. 91) and could potentially be solved using the

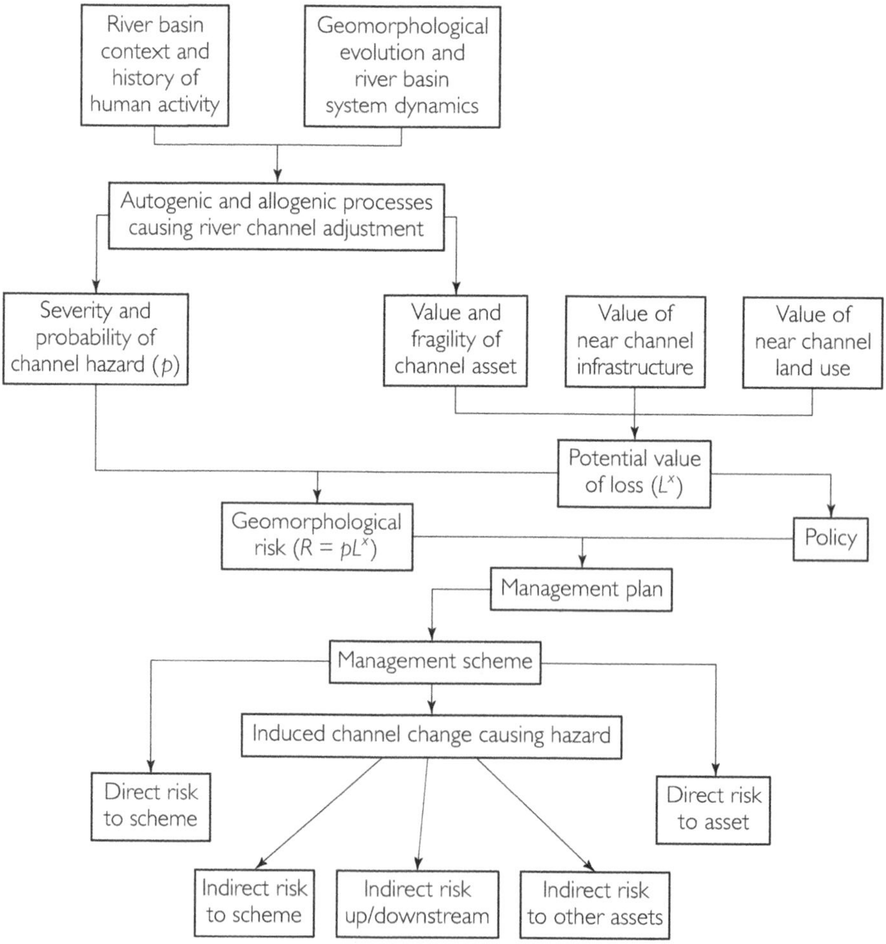

Figure 5.1 *Summary of the range and causes of river channel hazards, and their management implications (developed for river channel hazards from Panizza, 1987; Cavallin et al., 1994)*

Table 5.1 River channel hazards and their four major controlling variables

River channel hazards	Variables						
	Time	Discharge		Sediment load		Base level	
		Increase	Decrease	Increase	Decrease	Up	Down
Erosion							
Degradation (incision)		■			■		■
Knickpoint formation and migration	■	■			■		■
Bank erosion	■	■		■	■	■	■
Deposition							
Aggradation			■	■		■	
Back and downfilling			■	■		■	
Formation of berms			■	■			
Pattern change							
Meander growth and shift	■	■		■	■		■
Island and bar formation and shift	■			■		■	
Meander cut-offs	■	■		■		■	■
Avulsion	■	■		■		■	
Metamorphosis							
Straight to meandering		■		■			■
Straight to braided			■	■		■	■
Braided to meandering		■			■		■
Braided to straight			■		■		■
Meandering to straight		■		■		■	■
Meandering to braided			■	■		■	

Notes: Time is included as a variable because river channels evolve naturally as a process of time. Discharge, sediment load and base level can be influenced both by natural processes or human actions.
Source: Adapted from a table of geomorphological hazards in Schumm (1988).

governing equations associated with each parameter to indicate exactly what aspects of channel adjustment are causing the hazard, although the evolutionary complexities of geomorphic systems have thwarted attempts at a full deterministic solution to these equations to date.

In defining geomorphological risks associated with river channel management, potential losses associated with a channel hazard should be allied to the probability of hazard occurrence and its associated severity (i.e. combining the likelihood, persistence and magnitude aspects of channel adjustment). However, probabilistic approaches to river channel adjustment are somewhat rare (see Graf, 1984; Tung, 1985; Miller, 1988; Downs, 1994, 1995a; Rice, 1998) and field indicators are more commonly used to evaluate the extent to which a channel may be unstable. Such indicators have been used to evaluate relative stability in relation to highways and bridges (Brice, 1981; Shen *et al.*, 1981; Simon *et al.*, 1989), in straightened and incised channels (Schumm *et al.*, 1984; Brookes, 1987b; Simon and Downs, 1995) and to indicate river bank stability (Thorne, 1993). More comprehensively, hazard evaluation in channels should include assessment of where, when, how long-lived and how much adjustment will occur, all elements of the sensitivity of the river channel to change as developed in the remainder of this chapter.

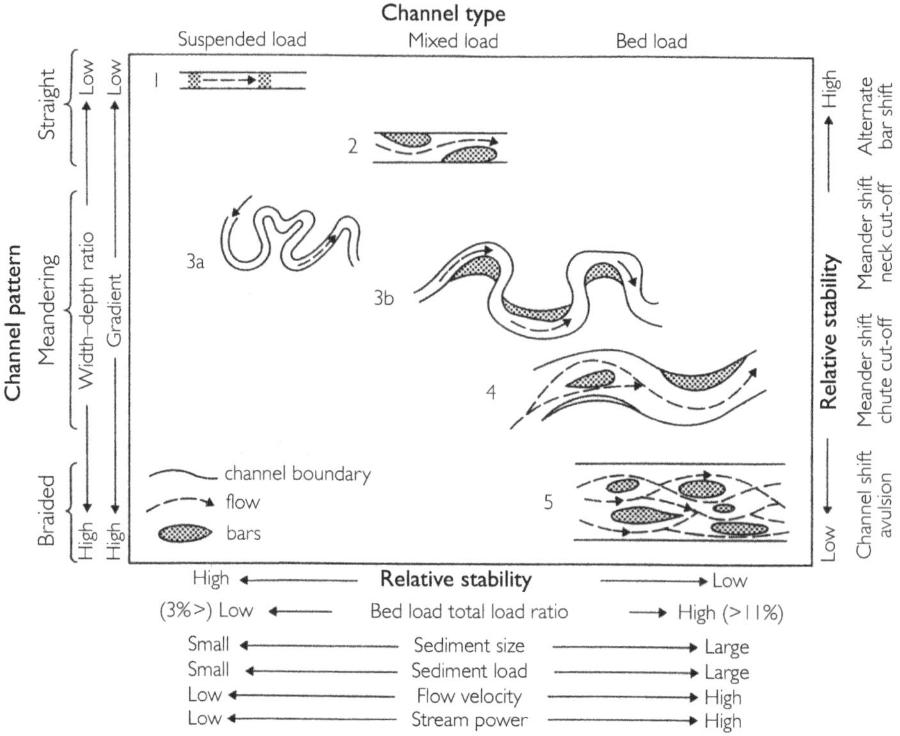

Figure 5.2 *A classification of river channel hazards (right axis) according to relative channel stability (Shen et al., 1981). Hazards indicated are a sub-set of those in Table 5.1*

5.2 River channel sensitivity to change

In associating river channel hazards (and assets) with adjustment, risk becomes an expression of the sensitivity of the channel to change. Sensitivity is a system response characteristic (Schumm, 1991) that describes a ratio between the magnitude of channel adjustment and the magnitude of change in the stimulus causing the adjustment (Downs and Gregory, 1995). Sensitivity is a nested phenomenon because the sensitive component can be changed by a stimulus that is itself sensitive to other stimuli. For instance, channel erosion downstream of urban development may be caused by increased shear stresses acting on the channel bank during storm flows, but these increases may be proportional to the increase in peak runoff, which is a function of the increase in extent of impervious area.

The concept of geomorphological sensitivity is defined as '... the likelihood that a given change in the controls of a system will produce a sensible, recognizable and persistent response. The issue involves two aspects: the propensity for change and the capacity of the system to absorb change' (Brunsden and Thornes, 1979: 476). Implicit in this definition is the notion that the risk of response (i.e. adjustment) is variable both in space and time. Spatial dimensions involve the capacity of the system to absorb change through its strength, morphology, structure, filtering capacity and existing system state (Brunsden,

1993) and is conditioned by thresholds in geomorphic response that define the location, type and magnitude of a hazard (i.e. *where* and *how much*). Identified thresholds (Schumm, 1979) include:

- *extrinsic*, describing the threshold condition at which the landform responds to an external influence, for example, the way in which a river channel cross-section may change abruptly from a single thread to a braided channel following a large storm event;
- *intrinsic*, where geomorphic response is caused by exceeding a threshold in an internal variable without the need for an external stimulus, such as where long-term weathering reduces the strength of hillslope materials until slope failure occurs, potentially providing a significant change in sediment production to an upland river channel;
- *geomorphic*, in which abrupt landform change occurs as a consequence of progressive intrinsic or extrinsic adjustments, for example, where progressive river bank erosion eventually causes a meander cut-off which prompts further changes related to bed-level adjustments in the channel.

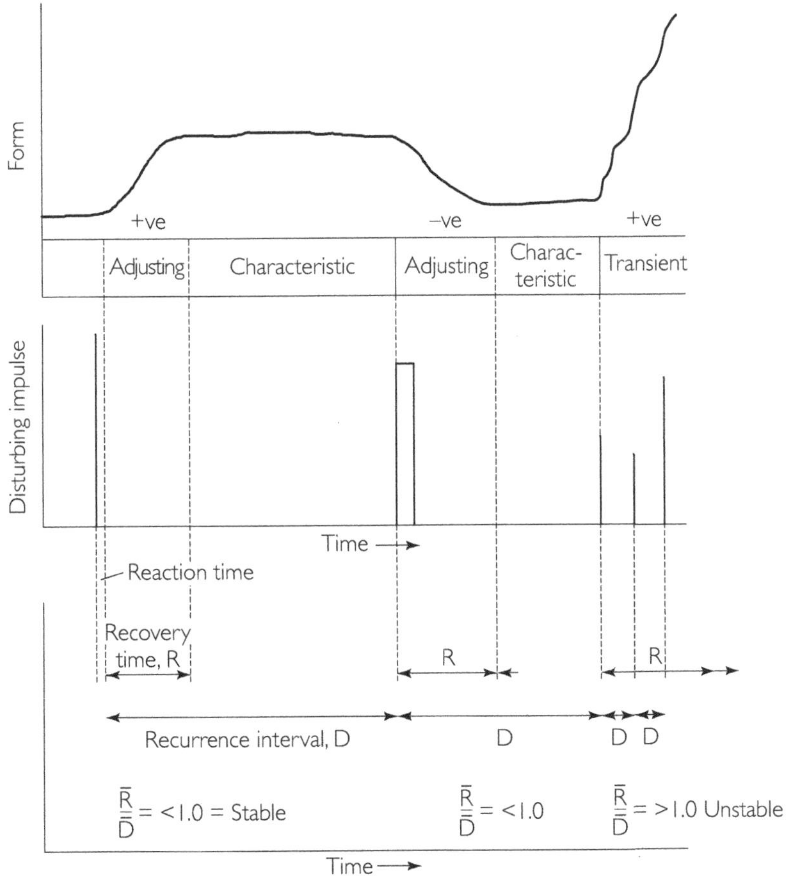

Figure 5.3 *Temporal dimensions of river channel sensitivity indicated by channel stability measured as the ratio of the mean relaxation to recurrence times (Brunsden and Thornes, 1979)*

Table 5.2 Four dimensions of river channel sensitivity and their related channel changes, related to river channel hazards and assets

Dimension of sensitivity	Interpretation of channel change	Hazard example	Impact on asset
Location	The ratio of disturbing to resisting forces – change occurs *where* the disturbance exceeds the barrier to change, that is, the ability of the channel to resist change.	Periodic migration of channel meander following large flood events causes homes near bend apex to be threatened.	Periodic migration of channel meander following large flood event rejuvenates ecosystem, creating new surfaces for colonization and input of coarse woody debris for wooded riparian areas.
Likelihood and type	The relationship of disturbing forces to a specified threshold condition in channel behaviour – the proximity of the channel to the threshold condition determines *when* (relatively) and *what style* of adjustment will occur (other things being equal).	In channelization, the balance of changes in discharge, slope and depth determine the type of change. Rivers raised above the stream power threshold of >35 W m^{-2} may threaten incision-related hazards, those below 15 W m^{-2} may suffer deposition-related hazards. The stream power prior to channelization will determine the extent of change required to cross the threshold.	Steady flows downstream of dams cause channels to metamorphose in form and frequently to stabilize (as a consequence of low stream power and greatly reduced sediment supply), being detrimental to habitats that require periodic disturbance.
Persistence	The longevity of morphological change according to the channel's time-dependent ability to recover from the disturbance – encompassing the recurrence interval of events causing change and the relaxation period to a new or former equilibrium condition (*how long-lived?*).	Large flood events following ground disturbance in an upland catchment release large volumes of sediment causing receiving single-thread channel to become braided and consequently much wider, threatening floodplain land uses. The interval between large floods determines the hazard longevity.	Periodic large flood events causes riparian ecosystem to be re-set through extensive channel change, floodplain sediment deposition and recruitment of riparian trees for inchannel habitat downstream. Provides space for early colonization and prevents senescence of ecosystem. Interval will determine balance of species.

Magnitude	The quantity of change in the channel per unit change in one or more of the controlling variables – ideally, *how much* is defined relative to the initial input and output factor magnitudes to facilitate comparisons between the effect of different stressors.	Urban development causes quantifiable increase in impervious area that increases runoff and can be related to the enlargement ratio of the downstream channel which may threaten riparian land uses.	Prospect of anticipating the impact of human activity on downstream assets, for example, urban runoff can increase the near-bed flow velocities and the potential for bed armouring, thence to likely changes in macroinvertebrate composition and to the resultant suitability for higher-order species (e.g. fish).

To describe the hazard likelihood and its persistence (i.e. *when* and *how long-lived*), temporal dimensions of sensitivity are based on the propensity to change and are defined by the ratio of the event's recurrence interval to its relaxation time (Brunsden and Thornes, 1979; Chorley *et al*., 1984; Figure 5.3). Recurrence interval is the average length of time separating events of a particular magnitude sufficient to cause change, and relaxation time is the elapsed time between the start and completion of the landform response (Graf, 1977). The temporal dimension therefore reflects the frequency of individual events nested within patterns of longer-term environmental changes; channels can consequently be described either as *sensitive* in the face of external change or *robust* (or resilient, Gordon *et al*., 2001). A highly robust channel is likely to present only a low risk of being hazardous in its current state. The spatial dimension explains why apparently similar landforms may exhibit different rates of change (Thomas, 2001) and why apparently similar river channels may show different responses to the same event or change in driving processes (Schumm, 1985a; Schumm, 1991). In this sense, *hypersensitivity* denotes an apparently disproportionately large channel adjustment resulting from a small hydrological change, whereas *undersensitivity* denotes a disproportionately small response (Brown and Quine, 1999). Together, the spatial and temporal dimensions of sensitivity account for the 'complex response' of geomorphological systems (Schumm, 1979).

Understanding the sensitivity of river channels to change is an integral component in successfully evaluating the risk of geomorphological hazards in river channel management (Downs and Gregory, 1995; Gilvear, 1999). In a river basin untouched by human activity, the sensitivity of the channel to change is dictated by properties of the drainage basin and aspects of climate, whereas in catchments already affected by human activity, river managers need to accommodate additional hazards caused by past activities, indirect hazards caused by surrounding contemporary activities and the direct and indirect hazards potentially resulting from the proposed management actions (see Figure 5.1). Specific investigations are limited both by the data available for analysis and interpretation (Allison and Thomas, 1993) and by the evolutionary nature of river channel systems that negates the prospect of time-invariant geomorphological sensitivity analysis in the manner of, for instance, testing mechanical instruments (the original scientific basis of 'sensitivity'). Therefore, multiple definitions of sensitivity will always be required to describe the location, likelihood, persistence and magnitude of river channel hazard and so to describe the *where*, *when*, *how long-lived* and *how much* dimensions of risk of hazard occurrence (Downs and Gregory, 1995; Table 5.2). These are as explored in Sections 5.3 and 5.4.

5.3 Sensitivity of undisturbed systems

The sensitivity of a channel hydrosystem undisturbed by human activity results from competition between storm events that act to send 'waves of aggression' through the system (i.e. pulses of kinetic energy; Brunsden, 2001), and the ability of the system to resist change. System resistance occurs in the form of 'barriers to change' that impart spatially and temporally variable thresholds in the response of the channel to the storm events (Brunsden, 1993). The barriers determine the amount of alteration in individual environmental components necessary to change processes that alter channel morphology, including aspects such as where the river system will be most and least affected and the length of time required for the system to recover to a quasi-equilibrium state (Knox, 2001). Barriers to change also impart time-dependent aspects of natural sensitivity, for instance, the extent of weathering of materials or the amount of channel bed armouring can determine the impact of individual storm events.

The effectiveness of individual storm events and the magnitude of barriers to change are regulated by drainage basin characteristics that provide extrinsic and intrinsic control (Brown, 1995) on the processes of river channel adjustment. Such characteristics encompass the rock types of the drainage basin, their disposition and arrangement, and features related to climate history, contemporary climate and the legacy of palaeohydrological events within the catchment (see p. 108). They provide extrinsic controls on the potential energy available for erosion, the resistance of channel materials to erosion, the quantity and dynamics of sediment supply, transport and deposition, and intrinsic control on channel processes according to elements of the pre-existing channel and floodplain morphology (Table 5.3). At any point in time, drainage basin characteristics combine to establish sensitive locations within the river system and the likelihood of change of a given type, and can dictate the persistence and magnitude of changes so determining hazard severity or asset fragility and the ability of the system to recover to its former condition following a disturbance or establish a new condition.

Because a combination of characteristics act together to influence channel sensitivity to change, and because of the likely superimposition of human impacts, explicit examination of how drainage basin characteristics affect channel sensitivity is rare although certain generalizations can be made (Table 5.3). For instance, alluvial channels are usually more sensitive to change in high relief-ratio drainage basins, especially where continued uplift of the headwaters is occurring, and the speed of onset of change may be more rapid where storm flows are routed downstream efficiently by dendritic channel networks. Such rapid response may increase the risk of channel hazards but also means that the abundant sediment supply and effective transmission of storm events can allow rapid channel recovery. Conversely, low relief catchments without active tectonics are more likely to be 'over-relaxed' (Brunsden and Thornes, 1979), that is, largely incapable of morphological adjustment under current climate conditions. Bedrock channels generally provide significant resistance, resulting in insensitive channels, especially in headwaters where stream power is low (Merritts and Vincent, 1989) as a function of discharge, and therefore present only minimal channel hazard. River basins composed entirely of fine sedimentary rocks or clays may yield only fine sediment throughout the channel network, resulting in narrow, cohesive channels with limited lateral mobility and overbank flood sediment as the main channel hazard. Conversely, gravel and sand-bedded channels may be highly mobile, laterally, presenting a hazard where human habitation exists near eroding meander bends.

The importance of relative position in the catchment is represented by the tripartite division of drainage basins into zones of sediment production, transport and deposition (Schumm, 1977b) although real river systems may exhibit multiple occurrences of the three zones (Newson, 1997). Never the less, the way in which certain processes are intrinsically affected according to their position is significant for planning management operations. For instance, in mid-catchment positions of temperate drainage basins, high stream power caused by the peaking of the slope:discharge product is the basis for preferential sediment transport and lateral migration processes, with higher rates of adjustment likely in banks with low cohesion (Lawler, 1992). In Japan, river channel hazards can be differentiated according to channel gradient. In large rivers the threshold between high turbidity torrent flows and overbank flooding of fine sediments occurs at channel gradients of approximately 0.0010 (Ohmori and Shimazu, 1994), whereas further upstream, the divide between torrent flows and high density boulder-strewn debris flows occurs at approximately 0.0800. Debris flows are generally carried through the entire tributary where the downstream channel profile is linear or fits a power function curve, whereas boulder deposition occurs in the middle reaches of channels with exponential downstream profiles. Locally, the sensitivity of channel avulsion on alluvial fans in southern Arizona, USA, is favoured primarily where sediment deposition during preceding small floods caused lowering of the relative bank height (Field, 2001): the avulsion invariably reoccupies and enlarges existing small channels on the fan surface, demonstrating the importance of floodplain architecture in determining hazard-prone locations (Figure 5.4).

Drainage basin topography in combination with antecedent weather events is another important determinant of natural channel sensitivity. In the southern Sierra Nevada, California, recent debris flows have been triggered either by intense rainfall from large, long-lasting frontal rain events that require little antecedent wetness, by rain-on-snow events that act to melt significant volumes of snow, or by snow melt (DeGraff, 1994). Their hazard potential is small at present but may become more significant as population increases. Thirty years of monitoring and periodic surveys in the Howgill Fells in north west England indicated that the channel receives sediment primarily from footslope gullies, causing the channel to be wide and braided immediately downstream of the gullies and narrow and single-thread elsewhere (Harvey, 2001). However, during extreme events, such as the June 1982 convectional storm (greater than 100–year recurrence interval) improved channel–hillslope coupling resulted in more widespread braiding; subsequently the channel progressively recovered, demonstrating the system to be sensitive yet robust (as the recovery time is less than the recurrence interval of major events). In October

Table 5.3 Some generalizations of the sensitivity of undisturbed river channel hydrosystems, according to properties imparted by drainage basin characteristics and climate-related factors

Feature	Implication for sensitivity of adjustment processes
Drainage basin characteristics	
Bedrock structure	May guide the formation and evolution of drainage patterns, and local channel morphology.
Bedrock lithology	Determines grain size and hardness of channel sediments.
Soil properties (and weathered regolith)	Determines amount and calibre of sediment available to the channel.
Relief	Determines energy potentially available for erosion and sediment transport and affects morphological resistance. Weathering and denudation of the land surfaces will typically reduce relief over time in the absence of base level changes.
Tectonic activity	Changes relief. Uplift in headwaters increases available energy, uplift in lower basin decreases. Lateral shearing activity may expose sediment sources.
Topography	Influences sediment delivery to the channel through hillslope processes, extent of direct hillslope–channel coupling and likelihood of landsliding.
Position	Influences sediment sources and calibre. Headwaters provide sediment sources from hillslopes and temporary sediment storage, intermediate zones deliver sediment overbank to floodplain storage and to point bars but also derive floodplain sediment from bank erosion, the lower catchment delivers fine sediment to floodplain storage.
Drainage patterns and basin shape	Determines the efficiency of runoff routing: dendritic drainage patterns for instance, tend to route runoff more effectively than parallel patterns.
Downstream floodplain fining	Influences channel morphology, flow and sediment production processes.
Climate-related	
Eustatic sea level change	Influences catchment relief and the energy available for sediment erosion and transport. Sea level rise decreases available energy and *vice versa*.
Post-glacial isostasy and subsidence	Uplift following the glacier recession from the land surface: increases relief and the chances of channel incision. Subsidence may increase sediment deposition.

Feature	Implication for sensitivity of adjustment processes
Long-term sediment source and supply	Formerly glaciated catchments may have large volumes of eroded sediment which has been re-worked; but now have lower supply rates. In unglaciated mid- and low-latitude catchments, greater aridity led to formation of sand dunes and loess which can now provide the largest sediment source.
Sequence of flood events and sediment sources	Floodplain metamorphosis occurs in response to changes in sediment sources and in the sequence of flood events and is reflected by the composition and type of floodplain.
Contemporary precipitation and seasonality	Influence geomorphic process rates. Rates are often higher in semi-arid catchments and those with a distinct dry season as the sparser vegetation offers lesser resistance to erosion when precipitation occurs.
Catchment vegetation	Change through time will influence the volume of sediment derived from hillslopes as well as runoff rates.
Floodplain vegetation	Can influence channel processes. Formerly forested lowland catchments were conducive to an anastomosing channel pattern, generating overbank processes and rapid channel changes.
Antecedent weather events	Prior to a flood event can determine the effectiveness of the flood event.
Long-term flood regime changes	Reflected in the contemporary valley and floodplain morphology that provides background for contemporary channel processes. Evidence of former palaeofloods may remain. In formerly glaciated catchments, upstream floodplains were extensively terraced whereas downstream floodplains continue to aggrade with progressively finer sediment, changing channel morphology.
Flood regime variability	May be responsible for creating a majority of channel forms, especially in semi-arid environments where more extreme variability occurs. Variability may be quasi-cyclic such as in the influence of ENSO or quite subtle where changes in air mass dominance occur.
Floodplain architecture	Sediment calibre and depositional patterns recorded in the floodplain stratigraphy from past channel adjustments provide topographic influence and influence local sediment supply affecting current adjustments and possibly armouring of channel bed.

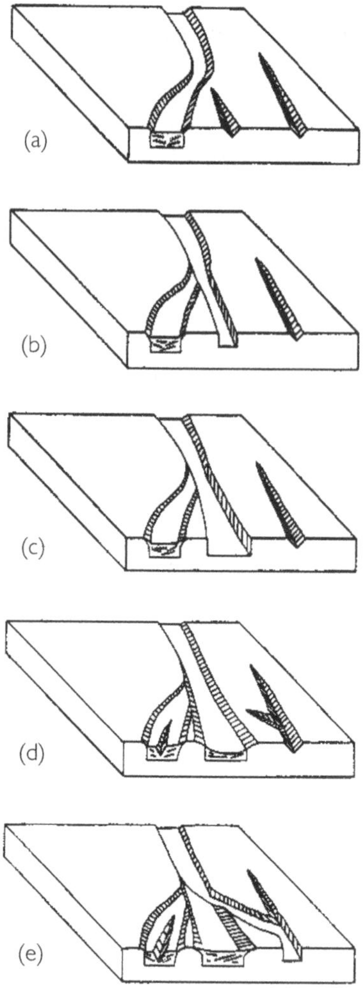

Figure 5.4 *Conceptual model of channel avulsions on alluvial fans in southern Arizona. View is looking upstream. Deposition within the main channel encourages overbank flows that preferentially re-occupies relic channels as avulsion occurs (Field, 2001)*

1998, the wettest week for 30 years saw debris flows almost reach the channel and would have caused system transformation if coupling had occurred. Such transformation is more likely under climate change projections and would cause rates of valley floor aggradation not seen for a millennium (Harvey, 2001).

Climate variations can be important controls affecting valley alluviation and the natural alternation between cycles of deposition and incision, especially in semi-arid environments where vegetation type often controls sediment supply from hillslopes and is highly sensitive to precipitation amount and seasonality. Examination of floods and droughts over the last few hundred years has indicated a consistent pattern of alternating periods of high and low flood frequency with a periodicity of 10–25 years (Macklin and Lewin, 1997):

a pattern potentially of great importance to river managers as rates of channel activity are intimately linked with these periods (Rumsby and Macklin, 1994). In environments where channels are incising into their floodplains they may expose sediments deposited under different climatic (and vegetation) regimes, presenting a channel hazard different from those near to the floodplain surface.

The legacy of the river basin's sequence of palaeohydrological events as preserved in the architecture of the contemporary floodplain can condition contemporary hazards. For instance, in drainage basins previously subject to large glacial meltwater floods with low sediment loads, channel gradient will have been reduced by upstream erosion and downstream floodplain aggradation (Brown, 1995). As contemporary process rates are much lower than in glacial periods, the majority of sediment transport now involves reworking of material rather than primary production, so that channel sediments get progressively finer in time. For instance, it is estimated that channel perimeters in the British Midlands in the early Holocene were composed of approximately 60% sand and gravel, whereas now the average is closer to 30% and will be expected to get finer over time (Brown and Keough, 1992), affecting channel sensitivity to change and the type of hazards requiring management. Understanding process rates under both contemporary and past climates is important both for setting management priorities and determining appropriate management responses. To be sustainable, management operations should assess future channel sensitivity to change in relation to projections of climate change. High energy environments may be more sensitive to small changes in climate than low energy environments, and river channels affected by climates with a highly pronounced variability between seasons or years may be geomorphologically more responsive to subtle climate variations than in humid areas (see p. 108).

5.4 Sensitivity and human activity

In populated catchments, the effect of human activities adds additional complexity to the sensitivity to change of the channel hydrosystem. Important activities include both direct influences on sensitivity occurring at points and reaches within the hydrosystem related to existing water resources management and channel engineering and maintenance operations, and indirect influences stemming from land use and land use changes distributed across the drainage basin (p. 92). Channel adjustments (and, therefore, hazards) caused by water resources management usually relate to flow regulation caused by large dams or water abstraction for irrigation or urban water supplies. The intensity of the impact is dictated by factors including the magnitude of flow regulation, the resistance of boundary materials, the quantity and calibre of sediment delivered (Petts, 1979), the effect on individual flood events (Petts, 1982) and reservoir operating rules. Changes to channel sensitivity caused by channel engineering and maintenance arise as a result of local alterations to flow velocities, discharge capacity and sediment transport processes. The indirect impacts on sensitivity of land use and land use changes are associated with the consequence of accelerated runoff and greater runoff volume and changes in the rate and calibre of sediment delivery from hillslopes. Human activities potentially obscure the channel's natural sensitivity to change and can be fundamental in determining the location, likelihood and type, persistence and magnitude of river channel hazards.

Location of channel sensitivity changes

The precise ***location*** of channel sensitivity changes relate to the type of impact. Associated with water resources management, sensitivity is generally affected downstream of the dam or point of water abstraction (or return), although ponded water upstream of a reservoir may cause deposition-related hazards. Downstream of the dam, flow regulation can subdue the sensitivity of the channel to natural flow variability but the reduction in sediment supply caused by trapping of sediment behind the dam will act to instigate a progressive change in the channel morphology through incision of the channel bed and deposition of coarse materials locally at the confluence with unregulated tributaries (p. 92). The downstream impact continues until the influence of regulated mainstem channel flows are mitigated by the incoming tributaries (i.e. until the proportion of the catchment regulated by the reservoir is small relative to the unregulated component). Impacts on Lower Mangrove Creek in New South Wales are dictated by the fact that instantaneous peak discharges have been reduced by 94% and sediment trapping efficiency is near to 100%, causing channel changes for at least 16 km downstream (Sherrard and Erskine, 1991). Downstream channel hazards associated with regulated flows may include accelerated bank erosion and the undermining of bridges and other near-river infrastructure related to channel incision, especially if the impact of regulation is reinforced by instream gravel mining (Kondolf and Swanson, 1993; Kondolf, 1997), and a reduction in the extent and variety of channel habitats subsequent to hydrological changes and metamorphosis of the channel bed and floodplain. As the channel morphology reaches equilibrium with the imposed changes in flow and sediment transport over time, the channel becomes markedly less sensitive and is essentially paralysed. This can result in serious degradation of channel habitat assets that require periodic disturbance to maintain species diversity (Kondolf, 2000).

Channel engineering has invariably involved straightening the channel, which locally increases the channel gradient potentially causing the formation of an active knickpoint or knickzone within the channel reach. Channel hazards related to knickpoint erosion can then be propagated through the reach and upstream of it (Parker and Andres, 1976) while, immediately downstream of the engineered reach, channel erosion and enlargement is followed further downstream by deposition of the eroded material (Brookes, 1987b; Simon, 1989; Figure 5.5). The location of the hazard will vary through time as a function of the channel's recovery potential. The sensitivity of the channel in terms of its propensity to change varies according to relative change in its stream power, related to the product of the discharge per unit width and the channel slope (Bagnold, 1966; Rhoads, 1987). Projects in which the channel is straightened without increasing the channel width, such as for agricultural drainage in moderate to steep gradient channels, result in greater potential energy and the prospect of erosion-based hazards related to deepening and increased gradient of the channel, such as in the Bluff line streams of Mississippi (Patrick *et al.*, 1982; Schumm *et al.*, 1984). In lowland areas such as southern England and Denmark, where channelization is often accompanied by increases in channel width, potential energy may be decreased overall and deposition of material transported into the channelized reach can progressively increase flood hazards and may require operational maintenance (e.g. Brookes, 1987a).

Hazards arising from land use changes associated with agriculture, commercial forestry and urban development generally occur downstream of the area of ground disturbance

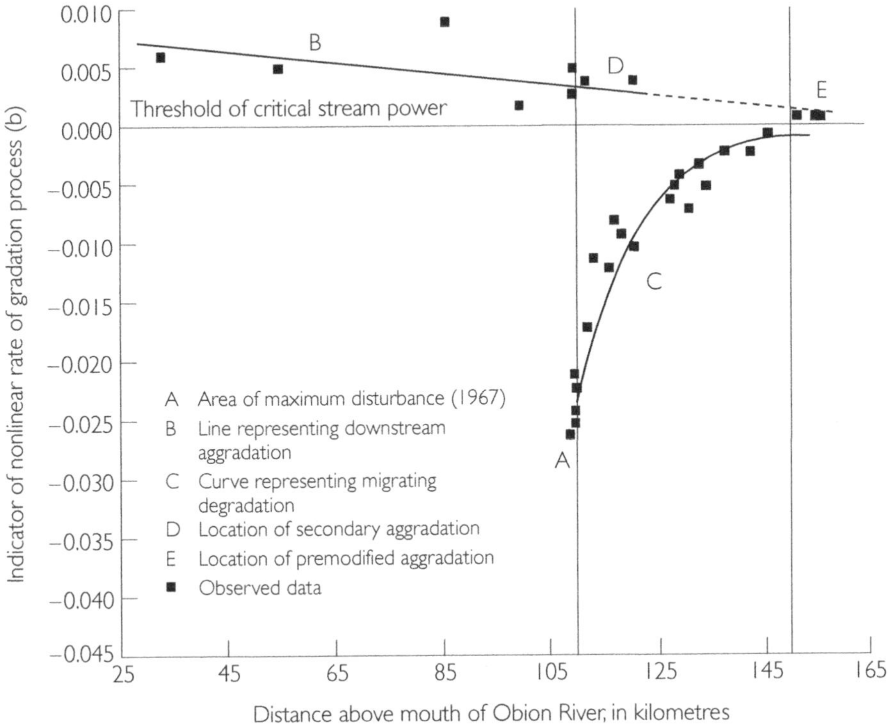

Figure 5.5 *Illustration of bed-level response to channel straightening and re-sectioning on the Obion River, West Tennessee (Simon, 1989) Responses are a guide to the location of potential hazards*

and clearing. Ground disturbance through forest clearance or cropping directly loosens soil, making it easier to erode, and clearance accelerates flows to the channel, increasing runoff volume and peakedness and increasing runoff and erosion potential in downstream channels via sheetwash and the development of rills and gullies where flow concentrates. Ground clearance also reduces the volume of plant infiltration and evapotranspiration promoting near-surface runoff and allowing the ground to saturate more readily, increasing the prospect of sediment yield to nearby channels via mass movements (e.g. landslides or the possibility of gully development). In urbanizing areas, sediment yields often increase dramatically during the construction phases and then decrease to below background levels as the urban infrastructure caps the ground surface. Conversely, under cropped agriculture or forestry, periodic ground disturbances continue to elevate sediment yields through a variety of processes. In cropped areas subject to agricultural intensification since the mid-twentieth century, elevated levels of fine sediment delivery to channels has also occurred because reduced crop rotation practices since the Second World War have reduced the organic residues that promote soil cohesion, leading to increased rates of soil erosion, leaching and gullying from fields (Macklin and Lewin, 1997). In pasture, intensive trampling by cattle in wet fields can cause the breakdown of clay soils and also the creation of impermeable surfaces that promote additional runoff and the prospect of enhanced erosion where flow concentrates. The removal of native vegetation also tends to result in additional runoff in small and medium size storm events causing channel erosion or, in steeper terrain, debris flows and torrents. Such additional runoff is often aided by the construction of artificial drainage networks such as agricultural fields and subsurface drains and urban storm drains that increase drainage density, reduce infiltration and consequent baseflows, and greatly accelerate storm flows causing permanent regional changes in flood magnitude and timing. As such, channel hazards related to land uses tend to increase the channel propensity to change by increasing the intensity of 'waves of aggression' (Brunsden, 2001) without adjusting the 'barriers to change' (Brunsden, 1993). Erosional channel adjustments occur close downstream of the land use and aggradation-related hazards such as flooding occur further downstream as the eroded sediments are re-deposited. However, during ground disturbance by tilling (agriculture), felling (commercial forestry) and construction (urban development), the initial response may be deposition of a large volume of sediment close by, that eventually travels downstream, possibly as a sediment wave or 'slug' of sediment (p. 73; Chapter 3). In steeper terrain, channel erosion may form knickpoints that then begin to erode upstream, so that land use changes may cause upstream hazards over time. A further hazard is that sediments derived from soil on hillslopes are invariably finer than those in the channel, resulting in a decrease in the average grain size of channel sediments. Downstream deposition of the finer material may cause a physical water quality hazard, and a chemical hazard if residual herbicides and pesticides are bonded with clay particles. Deposition may also result in changes in the species composition of in-channel and riparian flora and be hazardous to invertebrates and fish such as anadromous salmon that rely on interstitial water movement through the gravel bed for spawning success (Kondolf, 2000).

Likelihood and type of change

The likelihood and type of changes depend upon the ways in which human activity can change available energy in the channel relative to the proximity of the channel to thresholds in channel behaviour. In regulated rivers, the abrupt and permanent reduction in flows and change in sediment supply characteristics downstream of the dam make channel change extremely likely (except in bedrock channels with high resistance) unless flood flows and sediment are regularly released from the reservoir to mimic natural flood flows. If not, the impact on the channel is similar to a rapid and significant climate change, to which the channel will respond by attempting to create a new quasi-equilibrium condition with the prevailing flow and sediment regimes. Thus channel metamorphosis is likely (Schumm, 1969; Petts, 1979) whereby formerly braided channels may become single thread as a consequence of reduced coarse sediment supply (e.g. Kondolf, 1997), anastomosing channels may become single thread as a result of reduced flows and flow variability, and single-thread channels may narrow as reaction to a supply of finer sediment. In the Lower Mangrove Creek, New South Wales mid-channel bars and channel benches dominated by sand developed and became vegetated, creating a narrower, deeper channel and new floodplain surface for at least 16 km downstream of the dam (Sherrard and Erskine, 1991). The balance between scour and aggradation in regulated rivers is largely a function of the dam's relative capacity to trap sediment and impound water and thus depends upon reservoir size and climate characteristics (Petts, 1982; Figure 5.6A).

A large literature on channel adjustments following channelization (see Brookes, 1988)

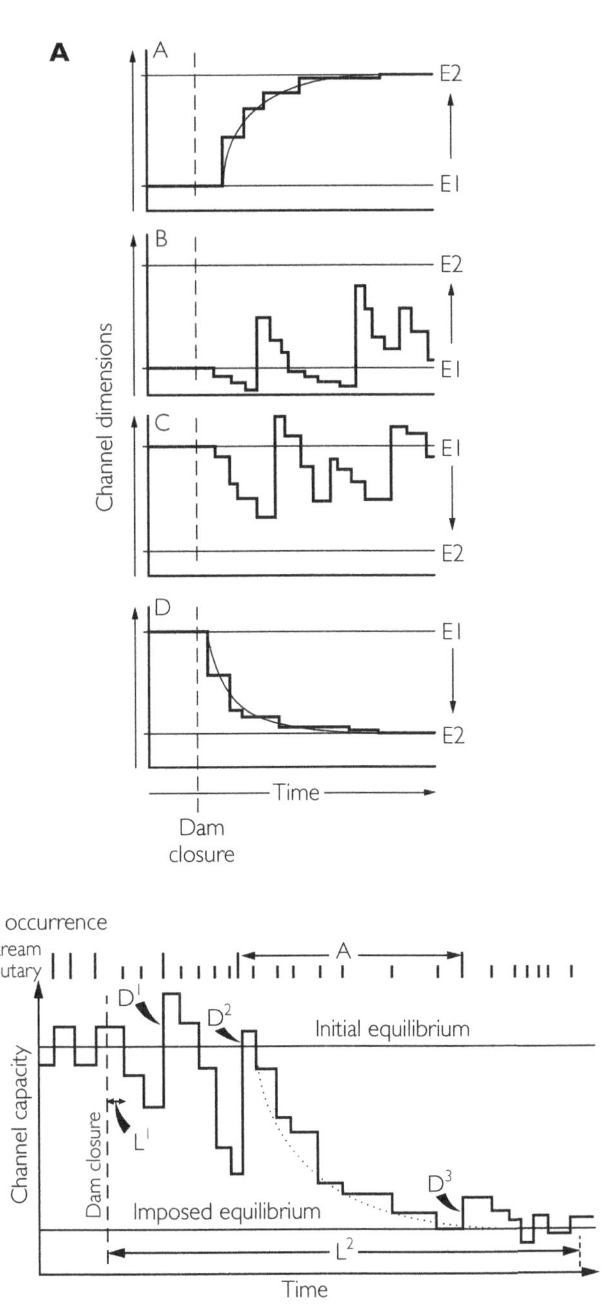

Figure 5.6 ***A*** *Hypothetical response of channel morphology to flow impoundment behind a dam. Potential responses involve a change from an initial equilibrium state (E1) to an equilibrium state (E2). Simple scenarios involving only degradation and scour (a) and aggradation (d) are the extremes in a continuum of adjustment prospects, two of which are indicated in b and c.* ***B*** *Hypothetical response in time of a regulated river below a major tributary sediment source. The relaxation path reflects the frequency of major reservoir releases (D1, 2, 3) modified by tributary sediment injection and vegetation establishment during* ***A*** *L^1 is reaction time and L^2 is relaxation time (Petts, 1982; see definitions p. 117)*

suggests that the likelihood of adjustment is high, with the type of change depending on the change in the energy available to the channel as stream power (Bagnold, 1966; Brookes, 1987b; Rhoads, 1987) relative to resistance of the channel perimeter. For channelized rivers in lowland England and Denmark (Brookes, 1987a, b) indicated that a specific stream power of 35 W m^{-2} was a lower threshold above which erosive adjustments occurred, whereas below 15 W m^{-2} adjustment by deposition was probable (Figure 5.7). Clearly, in the latter case, a significant upstream supply of sediment is required to allow deposition: Thorne (1993) used the term 'moribund' to describe lowland channels in England that have insufficient energy or sediment supply to recover from their channelized state. Recovery following channelization may also involve channel pattern changes, particularly the recovery of a sinuous course (e.g. Lewin, 1976; Brookes, 1987a; Thorne, 1999). For instance, following straightening for land drainage in 1932, Big Pine Creek, Indiana, began to re-establish its sinuosity to reduce the channel gradient, allowing the channel's pool–riffle sequence to reform, followed by bank erosion and reductions in channel capacity (Barnard and Melhorn, 1982; Figure 5.8A).

The widespread evidence for fundamental changes in river systems following the introduction of intensive agriculture in areas subject to rapid European settlement such as in the USA (Fitzpatrick and Knox, 2000; Knox, 2001), and Australia (Fryirs and Brierley, 1998, 1999), suggest that the likelihood that agriculture has altered the sensitivity of river channels worldwide is great. In areas of severe disturbance such as in Mississippi state, soil erosion led to severe valley aggradation and resulted in the formation of a distinct sedimentary layer up to 5-m thick in the upper floodplain, termed 'post-settlement alluvium' (Happ *et al.*, 1940). The likelihood of a hazard may also be conditional on the extent of land use change. In the humid USA, channel changes caused by urbanization have been estimated using a combination of various hydrological models and a risk analysis of channel change based on stream power and bed material size (Bledsoe and Watson, 2001). Levels of imperviousness of 10–20% are sufficient in many cases to at least double pre-urban estimates of the two-year discharge and to increase the frequency of moderate flood flows. Together, these changes caused a marked increase in the time-integrated sediment transport capacity (especially where the majority of sediment load is suspended sediment) and a much increased risk of channel change by incision or braiding. Forecasting the precise type of channel change is difficult because the response is conditioned by local factors such as disturbance during construction (Nanson and Young, 1981; Neller, 1989), position in the drainage network (Neller, 1988; Ebisemiju, 1989), local lithology and channel slope (Neller, 1988; Booth, 1990; Bledsoe and Watson, 2001) configuration of the foul sewerage network (Roberts, 1989; Bledsoe and Watson, 2001) and road drainage network (Chin and Gregory, 2001; Gregory and Chin, 2002), local mitigation measures such as channelization (Neller, 1988; Gregory *et al.*, 1992), and flow storage devices such as detention basins (Bledsoe and Watson, 2001).

Persistence of change

The persistence of changes depends fundamentally on whether the hydrological and sediment changes are essentially permanent (i.e. 'press' disturbances; Underwood, 1994), such as with river regulation, or whether, such as with channelization, changes are one-off 'pulse'

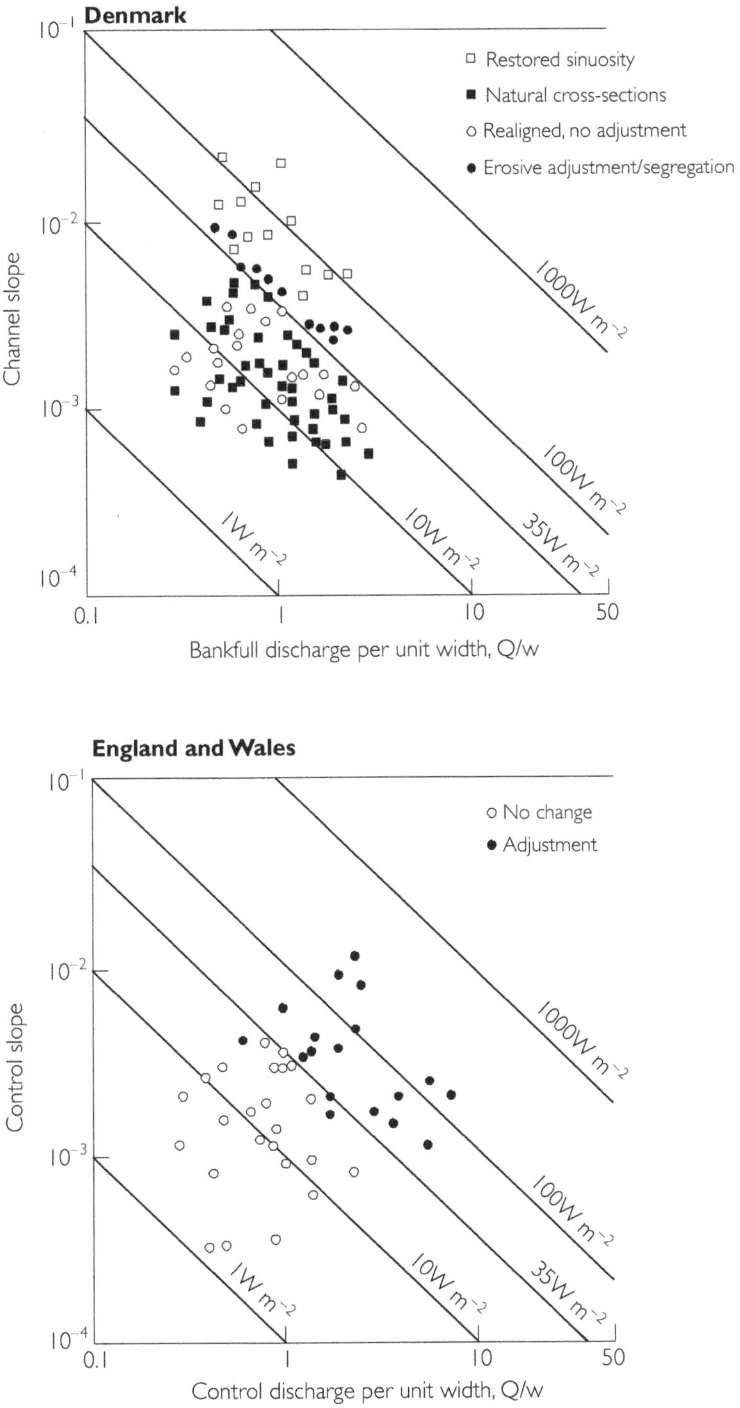

Figure 5.7 *The likelihood and type of channel change caused by channelization in England and Wales and Denmark (Brookes, 1987a, c), according to the resultant stream power of the reach*

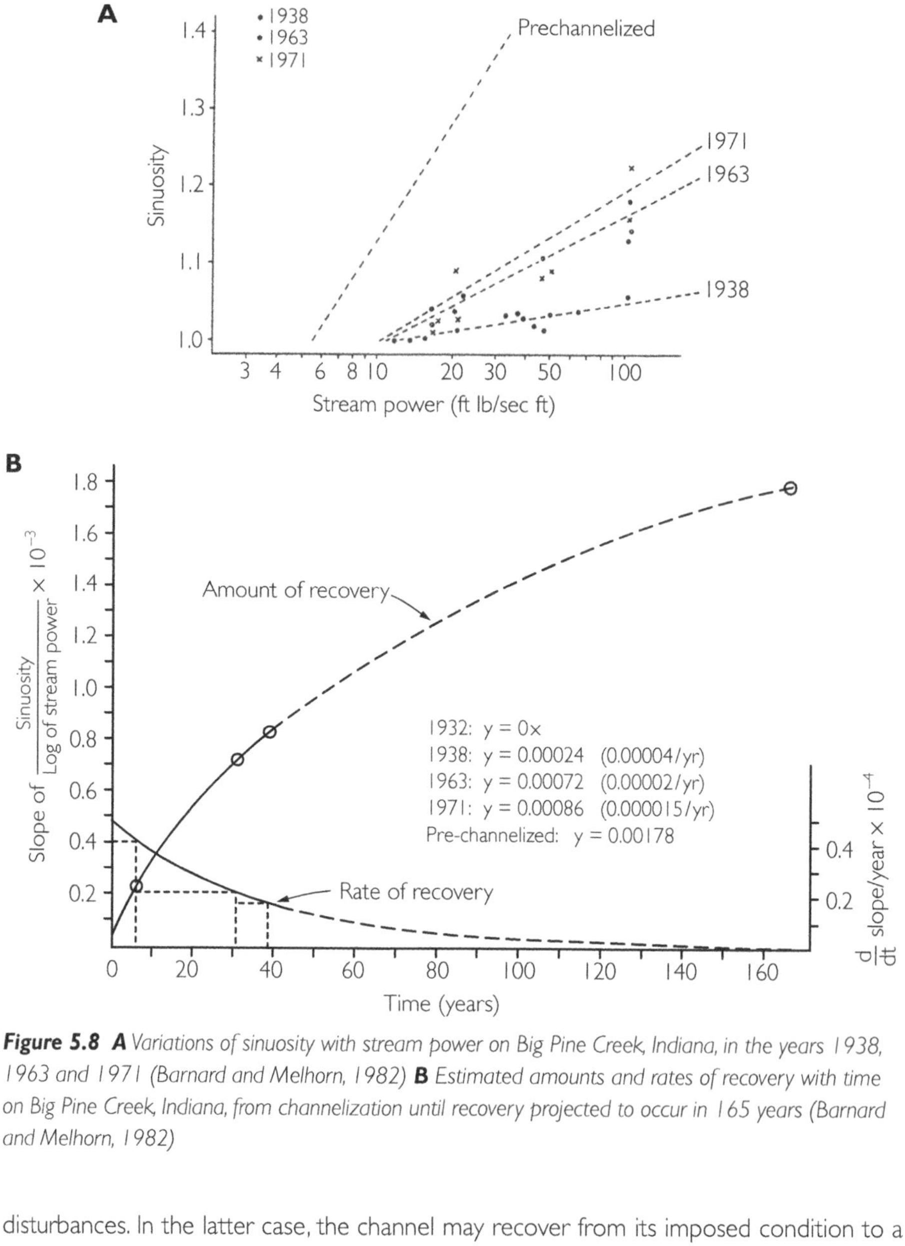

Figure 5.8 ***A*** *Variations of sinuosity with stream power on Big Pine Creek, Indiana, in the years 1938, 1963 and 1971 (Barnard and Melhorn, 1982)* ***B*** *Estimated amounts and rates of recovery with time on Big Pine Creek, Indiana, from channelization until recovery projected to occur in 165 years (Barnard and Melhorn, 1982)*

disturbances. In the latter case, the channel may recover from its imposed condition to a morphology that is closer to equilibrium with the prevailing flow and sediment transport characteristics (in the absence of hard revetment of the channel banks and/or bed). The adjustment period (Figure 5.3B) consists of a reaction time until changes begin and a relaxation time for the adjustment to be completed in the absence of other channel management or land use changes. The relaxation period is often characterized as the channel's recovery potential (Fryirs and Brierley, 2000). The reaction time for changes caused both by flow regulation and channelization can be quite short because the hydrological and sediment changes occur in-channel, whereas many channel hazards prompted by land use changes are manifested only after significant storm events when effects on the land surface begin to influence the channel, so there may be significant time lag between the land use change and hazard occurrence. In the Bollin-Dean basin, UK, changes in runoff arising from increased agriculture and land drainage in the early twentieth century and an increased urban population from the 1920s did not instigate river channel adjustments until triggered by a sequence of floods in the 1930s (Mosley, 1975).

The relaxation time required for regulated rivers to equilibrate with their post-regulation flows and sediment transport varies according to the post-regulation magnitude of destructional mainstream flows, constructional tributary inflows, the regional climate and vegetation (Petts, 1982; Figure 5.6B). The permanent hydrological change means that the channel morphology is likely to reach a new equilibrium condition. Conversely, channelized rivers without revetments and with sufficient energy can recover to a state that is similar to, but not the same as, the former condition. The conceptual summary of the longitudinal parts of adjustment processes is known as the 'Channel Evolution Model' (CEM; Schumm *et al.*, 1984; Simon, 1989; Figure 5.9) and its stages describe the complex evolutionary response of an oversteepened channel (i.e. steep relative to the equilibrium long profile). Typically, the high shear stress generated on the bed of the straightened reach allows the

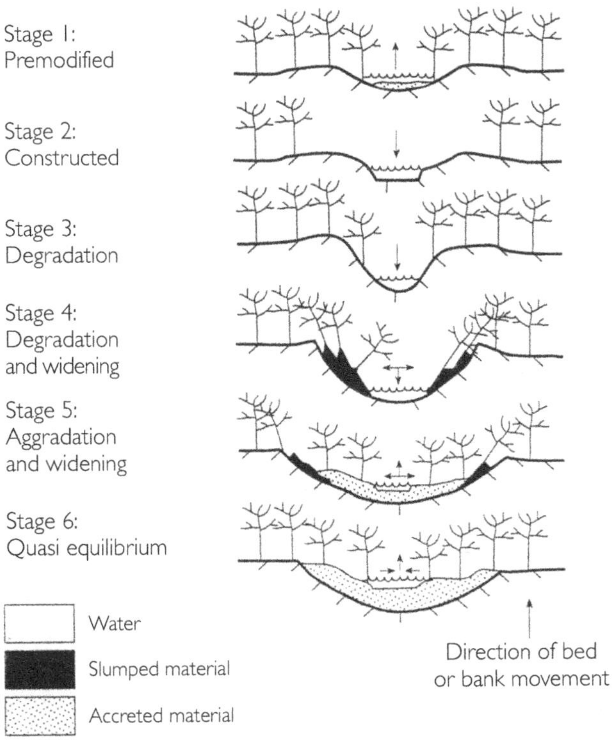

Figure 5.9 *Six-stage model of channel evolution following channelization. All stages are depicted for sites at or just above the limit of channel work. Arrows above the channel indicate the direction of degradation or aggradation (Hupp and Simon, 1991)*

knickpoint to migrate upstream by eroding the channel bed. The reach incises to the point that its banks are unable to withstand their own weight when saturated and mass failures occur following flood events causing the channel top width to increase markedly. Sediment begins to accumulate in the still widening channel. Eventually, large volumes of sediment are received in the formerly eroding reach from erosion occurring upstream and aggradation occurs within the incised channel until a channel of a cross-sectional area similar to the pre-disturbance channel will re-form, set within a compound section and in which the former floodplain is now a terrace. The relaxation period of channel gradient adjustment may also include recovery of a sinuous planform in the late stages of adjustment (Thorne, 1999) creating a change in potential channel hazard to one associated with lateral migration subsequent to channel widening and bed incision. In Big Pine Creek, Indiana, the relationship between channel slope and sinuosity in the pre-channelized river and at various dates following channelization as channel recovery proceeded, suggested that the channel could fully recover its original form in 165 years (Barnard and Melhorn, 1982; Figure 5.8B): the drainage basin specific conditions associated with each channelization scheme prevents a generalization of recovery time.

The relaxation time downstream of land use changes is a function of the relative change in hydrology and sediment transport characteristics and is not well generalized. One exception is in mid-Wales where extensive studies of forestry impacts enabled development of a generalized conceptual model of suspended load and bedload sensitivity to the commercial forestry cycle of afforestation, mature trees and felling (Leeks, 1992). The relaxation period for most downstream channel changes resulting from land use change will involve progressive channel enlargement by bank erosion and bed incision driven by storm events, and may be relative to the effectiveness of the installed network of drainage ditches and storm drains in urban areas. The recovery process may be more complex in headwater and low-order channels. In these cases, increased runoff resulting from land use changes may cause zero-order and swamp-type headwaters to develop distinct channels that will accentuate hydrological changes downstream causing additional erosion. Where flow concentrates at a distinct channel head, this can form a knickpoint that will retreat upstream in time (Kirkby, 1980; Dietrich and Dunne, 1993). The resulting sequence of changes becomes a complex response to knickpoint incision similar to that for channelization (Schumm *et al.*, 1984). Because of the potential for land use changes to fundamentally alter the channel network and drainage density, reverting to pre-disturbance land uses (e.g. allowing agricultural land to remain fallow until native vegetation re-establishes) may not result in complete channel recovery, or may require hundreds if not thousands of years for recovery to occur (Brierley and Murn, 1997). Therefore, the persistence of channel sensitivity changes may be long-lived and distinct in type from those associated with flow regulation (direct 'press' disturbance) or channelization ('pulse' disturbance).

Magnitude of change

The time-independent relative magnitude of adjustment is a further prospective measure of channel sensitivity to change consequent upon human activity. Such a sensitivity analysis is approached using a measure of the proportional adjustment of specified channel

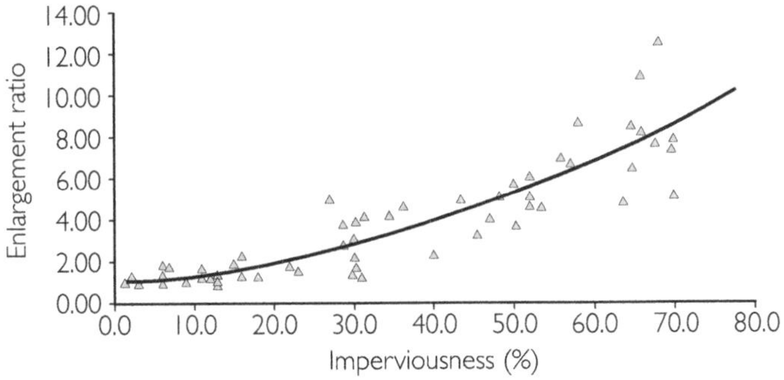

Figure 5.10 *Projection of 'ultimate' increase in channel enlargement ratio as a function of impervious cover for 60 reaches of erodible alluvial streams in Maryland, Vermont and Texas. Method uses contemporary and historic channel surveys and aerial photography and hydraulic models. Time to ultimate enlargement is estimated at 50–75 years. From Caraco (2000) with permission of the Center for Watershed Protection*

parameters relative to the proportional change in some parameter of the human activity causing the channel change (Table 5.2). In this way, the severity of the potential channel hazard could be forecast as part of assessing the environmental impact of the activity. For instance, adjustments resulting from flow regulation might be understood relative to peak flow reductions, channelization by the relative change in stream power in the reach and land use impacts by the areal extent of the use change. This level of analysis is largely untested to date (p. 226), partly owing to the temporal and spatial dependency implicit in channel sensitivity to change (p. 124). However, several studies have attempted to determine the hazard associated with channel enlargement following urbanization. For instance, in urbanized alluvial streams in Maryland, Vermont and Texas, a relationship has been developed between the projected enlargement ratio in channel capacity and the percentage of impervious urban cover in the catchment (Bledsoe and Watson, 2000; Figure 5.10). The challenge in providing a process-based rather than associative sensitivity analysis lies in ascribing explicit mechanical linkages first between channel morphology adjustment and related changes in hydraulics and sediment transport, and second, between the hydraulic and sediment transport changes and some causative parameter in human activity. A similar study linking the change in specific stream power caused by increasing imperviousness to the risk of channel instability concluded that the risk was a function of proximity to a geomorphic threshold, the degree of channel entrenchment, the bank and floodplain conditions, properties of time-integrated sediment transport and the resilience of the stream type (Bledsoe and Watson, 2000, 2001), underlining the non-trivial nature of this task.

5.5 Implications for river channel management

Options for river channel management during the twentieth century often tended to assume that channel morphology would be insensitive to climatic variations during the lifetime of the management solution (e.g. 50–100 years) so that the prospect of channel adjustment could be ignored. However, results from palaeohydrological research linking long-term climatic shifts to evolutionary channel adjustments (Table 4.10), and evidence that even quite subtle variations in climate associated with decadal-scale changes in air mass dominance may influence channel activity (Rumsby and Macklin, 1994), have shown channels to be potentially sensitive to small variations in climate even within the lifetime of a management project. Further, some channel adjustments previously attributed to human activity are being re-evaluated. For instance, the marked increase in arroyo formation (i.e. trenched channels) in the semi-arid southwestern USA was usually ascribed to the impacts of overgrazing, but studies have discovered evidence for the existence of palaeoarroyos from periods in the Holocene that cannot be linked to agriculture (e.g. Balling and Wells, 1990; Webb *et al.*, 1991). Instead, some arroyo formation may be linked to periods of increased precipitation intensity such as in southern Utah where flood magnitudes in the period 1866–1936 are unmatched over 500 years (Webb *et al.*, 1991). Clearly, the potential impact on the channel of climate variations or fluctuations through the life of the management project should be considered in assessing the most suitable option.

Of course, the past, present and potential future human impacts arising from land use change, flow regulation and previous channel management operations may also have significantly influenced river channel sensitivity and should be assessed and accommodated by the preferred management solution. Human influence may be of long-standing, such as on the River Tanat in mid-Wales where land use differences along with site differences have made the Tanat far more sensitive to periods of valley aggradation and degradation than a neighbouring section of the River Vrynwy for over 4000 years (Taylor and Lewin, 1997). Not surprisingly, channels disturbed more recently show an even clearer link between channel activity and human disturbance. In North Fish Creek, Wisconsin, discharge peaks and sediment loads during large flood events increased markedly as the catchment was deforested and developed for agriculture in the late nineteenth century (Fitzpatrick and Knox, 2000), and the legacy of incised channels in the upper catchment continue to promote downstream flooding and sedimentation despite several decades of agricultural decline. Likewise, a 'dramatic' acceleration in rates of channel incision and bank erosion occurred in just a few decades following European settlement of the Cobargo catchment, New South Wales, and the upper catchment may take thousands of years to recover (Brierley and Murn, 1997). Channel erosion responses during the 1993 flood-of-record in the Raccoon River basin, Iowa, are demonstrated to have been conditioned by land use patterns (Prestegaard *et al.*, 1994) in combination with processes determined by catchment location, so that intensive modern agricultural practices can make fluvial systems far more sensitive to change than in the historical past (Knox, 2001). Moreover, the potential for river channel sensitivity to have been significantly altered by river engineering and flow regulation is easily demonstrated and implicit in research studies over several decades (e.g. Gregory, 1977b; Petts, 1984; Brookes, 1988). Overall, while long-term records of river channel change may be 'climatically driven but culturally blurred' as contended for Britain (Macklin and Lewin, 1993), it appears that the last several hundred years of channel change may be interpreted as 'climatically enacted but culturally prepared' by increasing populations and intensive floodplain land uses in many river basins, especially those subject to rapid European colonization during the eighteenth and nineteenth centuries. The prospect is that many river channels are now far more sensitive to change in morphology than they were prior to human disturbance (Knox, 2001).

It is clear that contemporary river channel sensitivity to change is influenced not by one single factor, but by a combination of overlapping factors providing numerous 'stressors' distributed across the catchment. These stressors include natural background characteristics, the influence and legacy of environmental changes (progressive, fluctuating and cyclic), superimposed impacts stemming from contemporary human activity and legacy impacts stemming from prior human activity. Understanding river channels as sensitive phenomena demands an understanding of system response based on the influence of natural and human influences, in both small and large magnitude events (Allison and Thomas, 1993), and provides the logical basis for determining the appropriate location, timing and type of management to subdue river channel hazards and promote channel assets. As many river channels will not be static over the timeframe of management projects, and channels are more sensitive to small changes in sediment supply and runoff than was previously imagined, an appreciation of river processes over longer timeframes is essential to guard against unanticipated changes following project implementation. Further, knowledge derived in one river environment cannot necessarily be transferred to another (Macklin and Lewin, 1997). Fitzpatrick and Knox (2000: 105) conclude that:

> *... it is not possible to predict a geomorphic response simply by assessing changes in runoff and sediment loads. To understand the response and recovery of North Fish Creek to historical land use change, other factors, including the historical sequence of flood events and their relative magnitudes, the downstream spatial variability in sensitivity of channel morphology to erosion and sedimentation, and physiographic diversity in the watershed, need to be assessed.*

It follows that management should be based on local environmental assessment set in the catchment context, rather than on the application of an overarching 'one size fits all' engineering solution as was the practice during much of the twentieth century (see Chapter 2). Specific implications for river channel management of river channel sensitivity can be summarized as:

Managing the present

- *Complexity of contemporary system activity*: the risk of contemporary river channel hazards (*where*, *when*, *how long-lived* and *how much*) should be understood as a complex response to numerous interacting climatic and human influences, spatially distributed across the catchment. Threshold-driven channel responses and the possibility of significant time lag (i.e. reaction time) between the influence occurring and the river channel responding may make it difficult to pinpoint the exact cause of channel change. Scientific environmental assessment that seeks to incorporate the dynamics of channel adjustment is required (Chapter 8).
- *System activity as asset*: drainage basin characteristics and human influences that determine river channel sensitivity to change can condition not only river channel hazards but also river channel assets, that is, components of environmentally valued river landscapes. Consequently, best management practice requires an inventory of the channel morphology and dynamics (Chapter 8) as the basis for identifying and understanding

the template of present-day habitat values, zones of high conservation value, the robustness and resilience of the habitat and special features of the system such as the role that periodic disturbances play in maintaining habitat value (Chapter 3). Adopting this understanding is the basis for 'design with nature' management solutions (Chapter 7).

Managing in the context of the past

- *Long-term inheritance – 'system memory'*: river channel sediment and morphology hazards and assets may be related, in part, to the inheritance of features of the river's floodplain derived over the Late Quaternary period (125 000 years) and especially the last 10 000 years (Macklin and Lewin, 1997). Contemporary processes may be reworking sediment sources derived from former (possibly more extreme) periods of physical weathering, water runoff and/or sediment transport. Therefore, channel management should seek to understand past mechanisms, patterns, magnitudes and rates of river instability in the context of prior climates, as the basis for a successful solution (Chapter 8). This understanding may act as the basis for forecasting future channel conditions but allowance should be made for the prospect that human activity has conditioned today's alluvial channels to be more sensitive to change than at any other historic period.
- *Influence of antecedent events*: evidence that river channels are sensitive to smaller climatic fluctuations than previously imagined requires managers to assess the regularity, cyclicity or episodicity of recent storm events, and to ascertain their effectiveness in mobilizing and transporting sediment that affects channel erosion and deposition. These events may have created short-lived phases of channel activity that do not require management, or may result in impacts (potentially time-lagged) that are truly indicative of channel trends. Long-term records of hydrology and sediment transport records are vital (Chapter 8).

Managing for future conditions

- *Incorporating future actions*: channel management solutions should be planned cognisant of likely future channel hazards brought about not only by existing influences in the catchment, but also by planned future human actions that may affect channel sensitivity to change, including the impact of the management solution itself. This is central in attempting to achieve sustainable management solutions and requires catchment-based planning (Chapter 6) linked to regular monitoring of the implemented solution as the basis for later actions.
- *Incorporating climate change and variability*: reliable predictions of future climate changes are required as the basis for forecasting future river channel changes (Chapter 4). This task is extremely challenging because geomorphological change is a tertiary impact that requires downscaling of regional estimates of precipitation and temperature changes from which estimates of evapotranspiration and groundwater conditions provide the basis for predicting future vegetation assemblages. From this information, runoff and sediment yield may be estimated from which to summarize sediment transport and channel geomorphological response. Further, climate change will locally alter patterns of water demand and flood risk that may influence policy priorities for water resources management with additional implications for river channel management.

In summary, river channel sensitivity to change is complex, based on stressors related to drainage basin characteristics, evolutionary environmental changes and present and past human activities. These stressors dictate changes in flow magnitude, duration, timing, variability and hydraulics, and sediment production, calibre and transport rates that combine to cause channel adjustments that give rise to channel-related hazards and assets. Although river channel management actions often occur in response to perceived channel-related hazards and assets, human perception of the channel hazards is limited in at least three ways: in always assuming stability, leading to the conclusion that any change is not natural; in always assuming instability, implying that change will never cease; and in always assuming that changes will be major (Schumm, 1994). Methods for rational management of channel hazards and assets ideally require an environmental assessment of river channel sensitivity to change to define the hazard dimensions of where, when, how long-lived and how much channel adjustment occurs (Table 5.2, and see Chapter 8). In this way, different hazards/assets could be accommodated differently according to the hazard dimensions and the human value placed on potential losses.

6
The River Basin Planning Context

Synopsis

The drainage basin was used as a management unit initially to facilitate single purpose water resources projects, but evolved as the basis for multipurpose (6.1) and then integrated schemes, resulting in a diversity of terms (6.2) to describe integrated river basin management (IRBM). Recent trends influencing IRBM and its implementation are legal, organizational and technological (6.3). River channel management in IRBM has historically been underplayed but is gaining in importance amid growing concern to integrate land and water management objectives and to preserve and restore environmental values (6.4). Future developments in IRBM will provide new challenges for river channel management especially with regard to future resource demands, global change and sustainability (6.5).

Successive phases of river use (Table 1.2) and associated engineering methods, including significant modifications of river channels (Table 2.1), have been related both to subduing river channel hazards (Chapter 5) and to initiatives in water resources management. As scientific advances prompted gradual acceptance that water resources management should be undertaken at the drainage basin scale, management and planning required what is now thought of as integrated river basin management (IRBM). In order to provide a context, this chapter shows how basin-scale management and planning developed in a complex way, how the river channel is being increasingly recognized as a fundamental component in achieving IRBM and the implications this has for river channel management.

6.1 Development of river basin management

River channels were regarded as rather incidental to water resources management projects until the end of the nineteenth century. As the range of river uses increased (Tables 1.2 and 2.3), management methods were developed in response to particular problems, usually related to water resources, in specific locations. Management of perennial water sources for local agriculture and domestic water supplies, and the opportunistic use of seasonal floods and rains for agriculture, were succeeded by rivers used for navigation and power together with informal regulation of seasonal floods for irrigation agriculture and drainage of wetlands (Petts, 1990). Prior to the eighteenth century, river regulation was more of an art than a science (Petts, 1999) but was achieved in the nineteenth century by river training, followed in the twentieth century by river discharge regulation achieved by large dams, by a chain of major dams or by run-of-river impoundments created by navigation weirs and locks, all intended to maintain water levels, or by channelization schemes (Petts, 1999). River channels were therefore managed as necessary in relation to water supply, flood control, water quality control and water as a power source (McDonald and Kay, 1988b) with an emphasis upon the solution of a single problem, undertaken with little consideration of the effects that the 'solution' would have upstream or downstream of the project and elsewhere in the basin.

Several reasons therefore collectively encouraged the search for an alternative approach. First, the consequence of dealing with individual problems, often specific to a particular stretch of river, was to create impacts elsewhere, such as pollution or downstream flooding. Second, legislation had been established in relation to separate issues such as water rights, riparian rights, water resources, river channel movement, pollution, fisheries or navigation (e.g. Hodges, 1976) with insufficient coordination. Third, inefficiency arose from dealing separately with water resource provision, flood control, navigation and erosion control, all leading to the need for a more integrated approach. In addition, the extent of human impact, in regulating the flows of rivers, in creating new inundated areas, as well as in increasing demand for water supply and irrigation, and in protection from floods and droughts all created new pressures that required a more integrated approach.

River basins are ideal planning units for an integrated approach (Boon, 1995) because they are clearly bounded, physically functional, hierarchical in scale and culturally meaningful (Newson *et al.*, 2000). Initially utilized from the end of the nineteenth century for single-purpose projects such as the Suez Canal (1869–) and the Tisza regulation, subsequent use of the basin framework (McDonald and Kay, 1988) included other single-purpose schemes such as the Indus barrages (early twentieth century), the Aswan Dam (1902–), which produced multipurpose benefits; and followed in the mid 1930s by multipurpose projects such as the Tennessee Valley Authority (TVA). Growth of urban areas increasingly necessitated basin-based water supply and flood control; in Australia early basin-based efforts were stimulated by the growing cities of Sydney and Melbourne, whereas in the 1880s the US government sponsored regional planning basin efforts on the Mississippi system to address flood control and navigation (Hooper and Margerum, 2000). Use of the basin framework was reinforced in the early 1900s in both the USA and Australia when severe droughts with impacts on farm productivity increased awareness of soil erosion and promoted the adoption of regional-scale, basin-based conservation and land management practices.

An excellent illustration of the goals of early multipurpose approaches to water resources management is provided by the Tennessee Valley Authority (TVA), established as an independent federal agency in 1933 to,

> *improve the navigability and provide for the flood control of the Tennessee River; to provide for the reforestation and the proper use of marginal lands in the Tennessee Valley; to provide for the agricultural and industrial development of said valley; to provide for the national defense by the creation of a corporation for the operation of Government properties at and near Muscle Shoals in the State of Alabama (Viessman and Welty, 1985; cited in National Research Council, 1999).*

The Tennessee River and its tributaries (comprising 1% of the US land area) were developed by the TVA into one of the most controlled river systems in the world. Although the initial brief did not explicitly include provision of electrical power (Street, 1981), hydroelectric power (HEP) generation soon became an integral element in the agricultural and industrial development strategy, with the series of 60 major dams and reservoirs bringing changes in water quality, aquatic habitats, fisheries, hydrology and water uses; and having consequences for land reclamation from gullying, for soil conservation, for use of fertilizers, for demonstration farms with crop diversification and for cooperatives for marketing. This approach to basin management, and the type of organization, was not replicated elsewhere in the USA (Muckleston, 1990) and has been evaluated not just as a power system but as one devoted to total regional development (Owen, 1973) and as a massive electricity generating authority which, in 1980, was a far cry from what was originally envisaged in 1933 (Saha and Barrow, 1981). The TVA has evolved partly in response to national trends by working to understand the impact of changing hydrological conditions and seeking to facilitate multipurpose operation of the system of dams and reservoirs (National Research Council, 1999). Recently, emphasis has been placed on water quality, and on integrating local residents, businesses and government agencies into watershed protection efforts (National Research Council, 1999). Created to ameliorate environmental problems, the TVA subsequently influenced the approach of other organizations throughout the world by attracting supporters who saw it as the model for integrated basin management and critics who regarded the heavy dependence upon dam building as a weakness (Newson, 1992).

From a very different background, but also illustrating a multipurpose approach, is the Huang He (Yellow River) basin of China (average estimate 77 × 10^6 km^2; Renssen and Knoop, 2000) with its inherent problems indicated in traditional sayings about the river's water quality ('he who falls into the Huang He will never be the same again'), flooding (the Huang He is 'China's sorrow') and soil erosion ('once the skin is gone where can the hair grow?'). The Huang He has the largest sediment load of any of the world's rivers, carrying nearly 10% of all the sediment transported to the oceans from the surface of the globe (Walling, 1981), averaging 1.69 billion tonnes annual sediment production 1919–96 (Changming, 2000). As a result of the high sediment load the lower reaches of the river channel have silted, many sections of the river bed have risen within their constraining embankments (Figure 6.1) and the 'suspended river' means that the river bed in the lower reaches is generally 3–5 m, and sometimes 10 m, higher than the floodplain behind the levees (Changming, 2000). The Huang He's history of disastrous floods, including that of 1881 and of nine major course changes, including the deliberate break of the dikes on the southern bank in 1938 which flooded 54 000 km^2 of land resulting in the death of 900 000 people, prompted a river-based management strategy (Figure 6.1). Project-driven ('dig the beds deep, keep the dikes low') since the seventh century BC, the river had a series of dams constructed in the twentieth century with very significant downstream effects on the river channels (e.g. Chien, 1985). A programme for the complex utilization of the river received formal approval on 30 July 1955 and embraced 46 major dams on the main river; many small structures on the tributaries, particularly for electricity generation; the expansion of irrigation and the extension of the navigable sections from 160 to 3610 km (Smil, 1979); and 85 medium-scale reservoirs reported to have been constructed on the main tributaries in the middle basin with the primary objective of silt retention (Greer, 1979). The multipurpose plan for permanently controlling the Huang He and exploiting its water resources had objectives of flow control to reduce flooding, development of hydroelectric power for irrigation and industrial purposes and the extension of navigable reaches of the river. Increasing irrigation in the upper and middle courses has led to drying of the lower reaches for up to 226 days per year since 1990, giving problems of managing water allocation throughout the basin. Other basin problems include degradation of ecosystems as a result of deforestation, grassland degradation, desertification, salinization and alkalization, and water pollution. There has also been severe soil erosion in the middle reaches and a flood hazard created in the lower reaches: the lower Yellow River ceased flowing in its lower 704-km reaches on 21 occasions from 1972 to 1998, and in the 1990s the lower reaches remained dry for increasingly longer periods every year. It therefore was realized that planning should be undertaken for the Yellow River basin as a whole, that the Yellow River Commission (YRC) should have complete control, requiring clarification of Yellow River Law and of the relative responsibilities of the YRC and basin water management agencies (Changming, 1989, 2000). Other rivers in China such as the Yangtze (Chen *et al.*, 2001) continue to require a multipurpose approach, particularly with the advent of a plan to divert 10% of the Yangtze discharge to northern China.

6.2 Progress towards integrated approaches

Although the TVA and the Huang He contrast markedly in the problems addressed, they both illustrate the advantages of progressing from single purpose to multipurpose schemes of basin management, but also the potential pitfalls of an approach centred on achieving one primary technological function; a truly integrated scheme should allow for interaction between multiple scheme objectives. Moving in this direction, four essential stages were recommended by the United Nations for integrated river basin planning (United Nations, 1970):

- Stage 1: a preliminary stage, usually in response to an immediate problem such as erosion or flooding. Appreciation of the broader implications of the problem is the first step in defining the bounds of the basin system;
- Stage 2: the existing physical, socio-economic and administrative system is studied in detail and a development strategy formulated;
- Stage 3: small-scale pilot projects are implemented to investigate the likely success of the large strategies proposed in Stage 2;
- Stage 4: completion of the project by physical structures and management of the overall scheme.

Although these stages were intended to be the basis for an integrated approach they were criticized because they did not include any post audit provision, and were largely technocentric and centred on provision of major engineering structures (Saha, 1981). Conversely, river basin management was already being viewed as a vehicle for comprehensive regional development involving the management of land and water together under a unified administration (White, 1957).

A

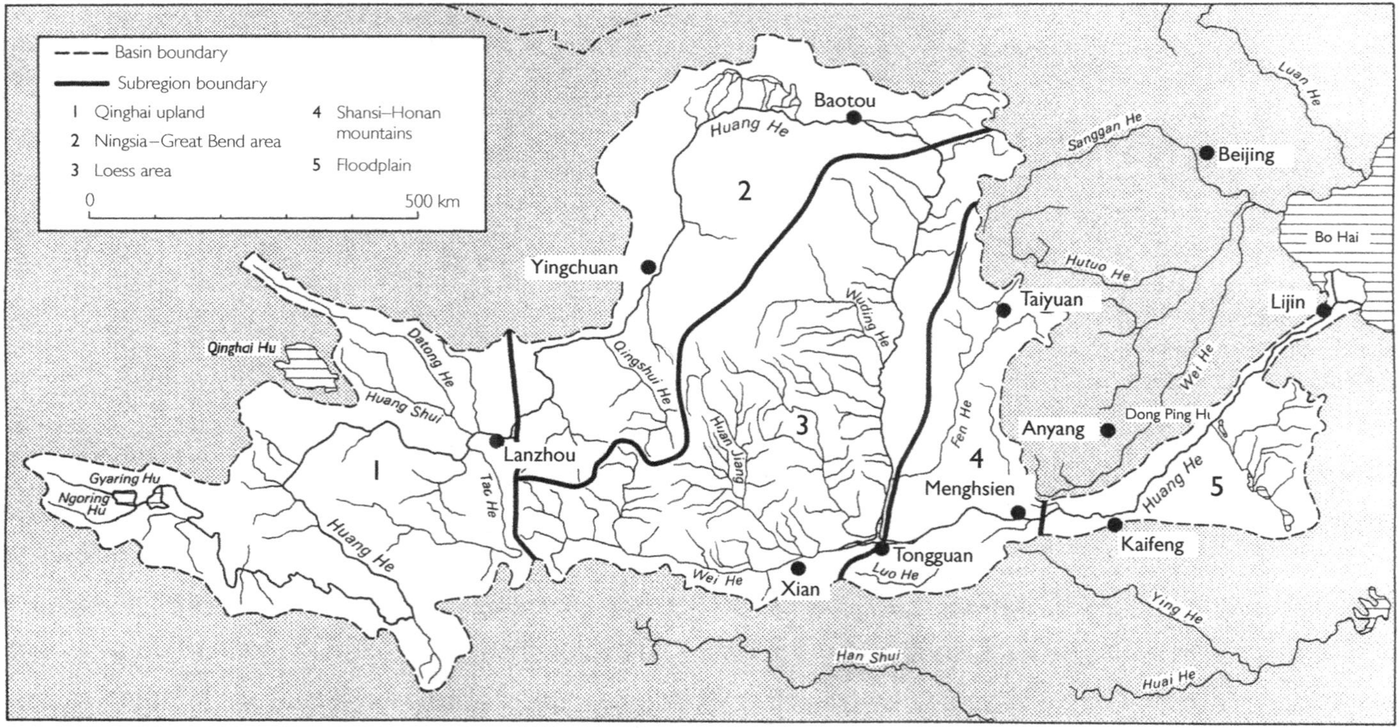

Basin boundary
Subregion boundary
1 Qinghai upland
2 Ningsia–Great Bend area
3 Loess area
4 Shansi–Honan mountains
5 Floodplain
0
500 km
Baotou
Huang He
Beijing
Luan He
Sanggan He
Bo Hai
Yingchuan
Hutuo He
Taiyuan
Lijin
Wuding He
Qinghai Hu
Datong He
Qingshui He
Wei He
Huang Shui
Huan Jiang
Fen He
Dong Ping Hu
Anyang
Lanzhou
Gyaring Hu
Ngoring Hu
Tao He
Menghsien
Kaifeng
Tongguan
Luo He
Xian
Ying He
Han Shui
Huai He
1
2
3
4
5

B

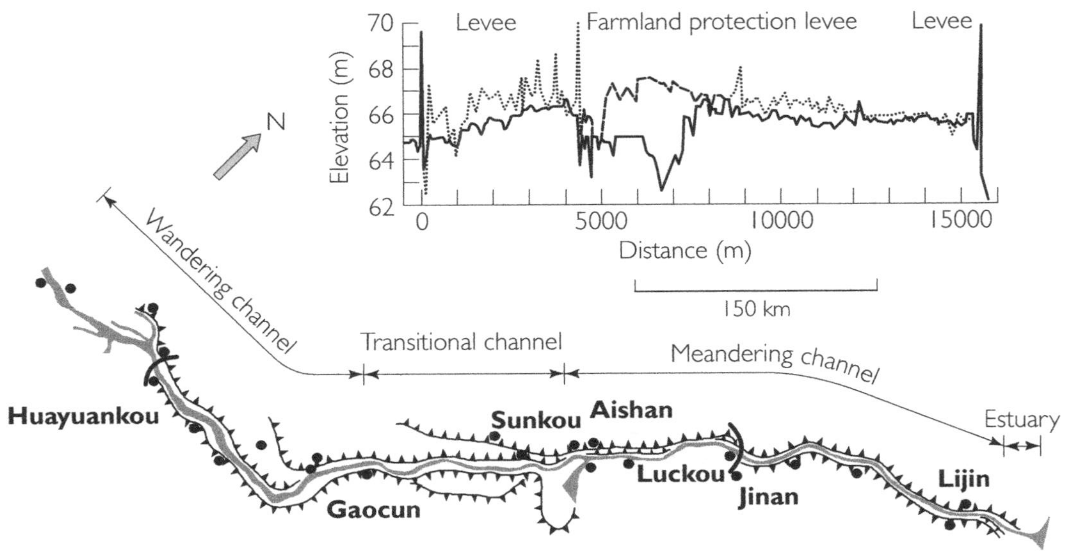

Figure 6.1 ***A*** *Regions of the Huang He Basin (after Greer, 1979).* ***B*** *Channel types of the floodplain region (after Ning Chien, 1985). Inset shows section across the channel (after Mai et al., 1980)*

Defining integration

A number of different terms have been used in order to connote the idea of an integrated approach to river basin management, reflecting the different disciplinary training of practitioners involved, the types of environment to which the approach is applied and the organizational framework used. Diversity is inevitable because of the different priority requirements for management that exist between river basins. Terms have evolved as the idea of unified management has been progressively enriched (Lord, 1982). By analysing accounts of 36 management schemes from the 1970s and 1980s, the development period of a new phase of integrated approaches, it was shown (Downs *et al.*, 1991) that the terms used to describe the schemes encompassed the multi-objective nature of the project, the unit area forming the focus of the scheme and the project aims (Table 6.1). The multi-objective nature of the scheme usually reflected recognition of feedback processes arising from manipulating the catchment hydrosystem and the 'inevitability of expanding responsibilities' in management (Coy, 1981): the term 'integrated' was most frequently employed in the sense of 'complete by addition of parts'. The management area was most often described as the 'river basin' and aims were generally described as 'management', which connotes the process of implementing and undertaking the management plan, although other descriptors included 'planning', relating to the controlled design and development of the basin and 'development', usually concerned particularly with future allocation of water resources.

As the move towards integration in water resources management continued, differences persisted between schemes and the level of true integration was called into question (Downs *et al.*, 1991), either because not all relevant components of the hydrosystem were included or because interactions between components were not fully recognized. The quest for real integration was articulated in various ways: as a systems approach by the World Bank, or as an ecosystem approach (Marchand and Toornstra, 1986) adopted for river basin management (Newson, 1997). Treating the entire river basin as an ecosystem (Marchand and Toornstra, 1986) necessitated a series of guidelines involving:

1 preservation or improvement of the spontaneous functions fulfilled by the river;
2 conservation of the natural values of the river basin;
3 conservation of the river basin's extensive exploitation functions;
4 development of sustainable intensive exploitation functions;
5 improvement of the overall health situation in the river basin;
6 guiding principles for regional planning.

Each guideline was associated with a set of specific management recommendations often with implications for river channel management, for example, the first requires:

(a) restoring erosion/sedimentation processes, through countering increased silt loads caused by upstream erosion (improvement of watershed management);
(b) preserving genetic diversity, through conserving natural areas and threatened species;
(c) preserving the self-purifying capacity of the river through combating pollution (water treatment plants, at-source anti-pollution measures).

Table 6.1 Examples of terms used for specific countries or areas (developed from Downs, Gregory and Brookes, 1991)

Date	Term employed	Area or country to which the term was related
1978	River basin management	FR Germany
		England and Wales
1981	River basin planning	Ogun/Oshum, Nigeria
	Basin management	Acelhuate, El Salvador
	Floodplain management	Hunter valley, NSW, Australia
	Integrated river basin management	Atchafalaya, USA
1983	River basin management	Thames Ontario, Canada
	Watershed management	Central Ontario, Canada
	Comprehensive basin planning studies	Various, Canada
	River basin management	Alberta, Canada
	Comprehensive water quality management	Stratford/Avon, Ontario, Canada
	River basin planning	Newfoundland, Canada
	River basin management	Thames, England
1985	River basin planning	Gongola/Sokoto, Nigeria
	Integrated river basin development	Nile, especially Sudan
	Basin-wide planning	Huang, China
	Ecosystem modelling	Nam Pong, Thailand
	River basin development	Han, China
	Ecosystem approach	Great Lakes, St Lawrence, Canada
	River basin management strategy	Tisza, Hungary
	Basin management	Colorado, USA
1986	Total catchment management	NSW, Australia
1988	Integrated river management	Zambesi, Zambia and others
	Basin management	Murray Darling, Australia
	Catchment management	New Zealand

Source: Developed from Downs *et al.* (1991).

The catchment ecosystem approach (Figure 6.2) focuses not only on interaction between the components but could also embrace any possible damage and provide a basis for future management by including regulation and economic instruments to form a template for water management in the future on an international scale (Newson *et al.*, 2000).

International interest in IRBM was demonstrated, not only by the systems approach advocated by the World Bank that sponsored work on international aspects of river basin development such as the RAMSAR convention (1971) for the intergovernmental 'management' of wetlands that produced recommendations about integrated basin management, but also in the guidelines contained in key paragraphs of Agenda 21 inter-

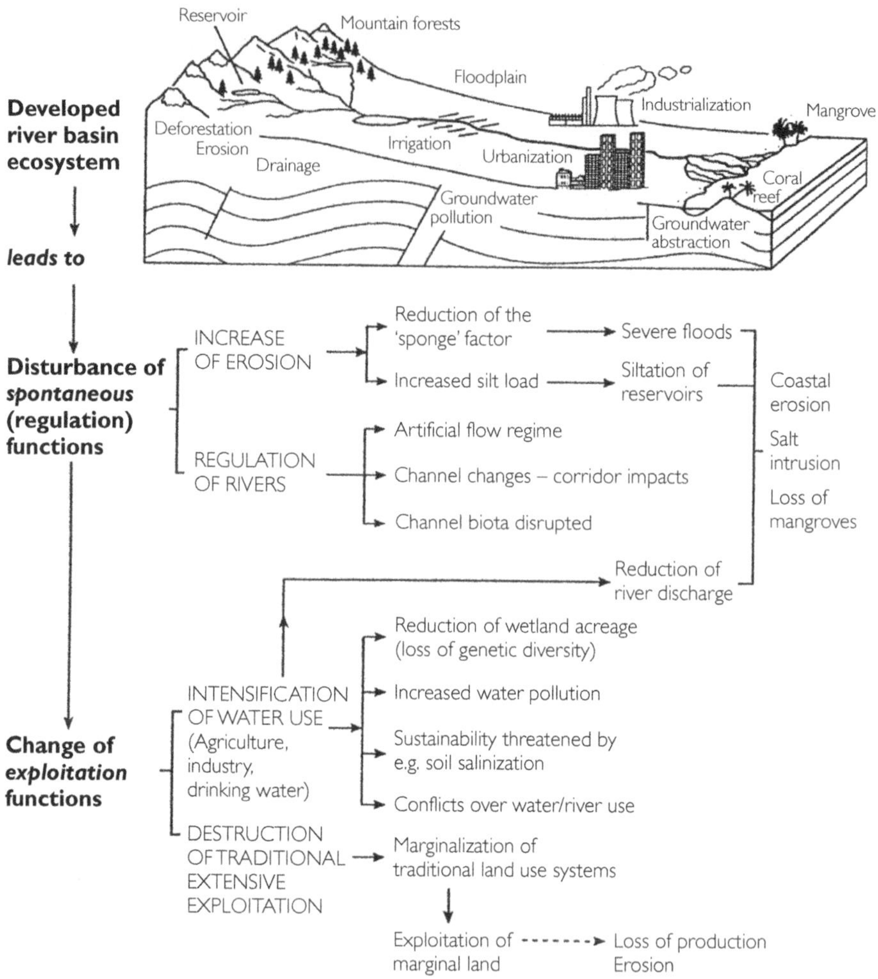

Figure 6.2 *The catchment ecosystem (from Newson, 1997, developed from Marchand and Toornstra, 1986; in Newson* et al., *2000: 249)*

nationally established in 1992 after the Rio Earth Summit (Table 6.2). Agenda 21 refers to an integrated and holistic approach to the management of water resources, placing a requirement on individual countries to establish national action programmes for water management, based on catchment basins or sub-basins, including integration of water resource planning with other resource management programmes.

The notion of 'holistic' river basin management, signifying management where 'the whole is more than the sum of the parts', was also recommended after reviewing management schemes up to 1990 (Downs *et al.*, 1991). It was suggested to encompass a fully comprehensive approach incorporating implications of river channel change, energetics of the fluvial system, embracing not only aspects of connectivity between water resources but

Table 6.2 Extracts from Agenda 21

Key paragraphs
• Water resources must be planned and managed in an integrated and holistic way to prevent shortage of water, or pollution of water sources, from impeding development. Satisfaction of basic human needs and preservation of ecosystems must be the priorities; after these, water users should be charged appropriately.
• By the year 2000 all states should have national action programmes for water management, based on catchment basins or sub-basins, and efficient water use programmes. These could include integration of water resource planning with land use planning and other development and conservation activities, demand management through pricing or regulation, conservation, reuse and recycling of water.

also the consideration of potential impacts throughout the drainage basin, and a dynamic approach to management and planning. Holistic basin management is also argued to be necessary because problems, their symptoms, causes and engineering solutions (structural and nonstructural) should be viewed in the context of the whole catchment (Gardiner, 1988a, 1990, 1991a) (Figure 6.3A) and because options for a solution should be examined in relation to the dynamics of all important influences including environmental, economic, institutional, political and social factors, with a view to facilitating adaptive management in response to future change (Gardiner, 1991b). The holistic view of river basins might also reform river basin management in the less economically developed world (Newson, 1988: 69) although the scope of such a holistic approach needs to be thought through (Mitchell, 1990). It is likely to require two stages whereby, at a strategic level, a comprehensive viewpoint implies scanning the widest possible range of issues and variables; and at an operational level an integrated approach is more focused by concentrating upon those issues and variables judged to be the most significant.

Progress in achieving integrated management at the national level is illustrated by the watershed (i.e. drainage basin) thinking proposed as a new strategy for the USA watersheds (National Research Council, 1999). Developed from the watershed protection approach of the US Environmental Protection Agency (USEPA, 1993) watershed management is presented as:

> *... an integrated, holistic problem-solving strategy used to restore and maintain the physical, chemical, and biological integrity of aquatic ecosystems, protect human health, and provide sustainable economic growth.*

In focusing on drainage basins or watersheds, the distinguishing characteristics of such thinking (National Research Council, 1999) are founded upon problem specification, identifying the primary threats to human and ecosystem health; stakeholder involvement, embracing those most concerned or most able to take action; and integration of actions, once the solutions are determined. This focus upon problems, stakeholder views (see Chapter 7) and integrated solutions is viewed as a way of achieving lasting agreements necessary to restore or to prevent further degradation of the watershed.

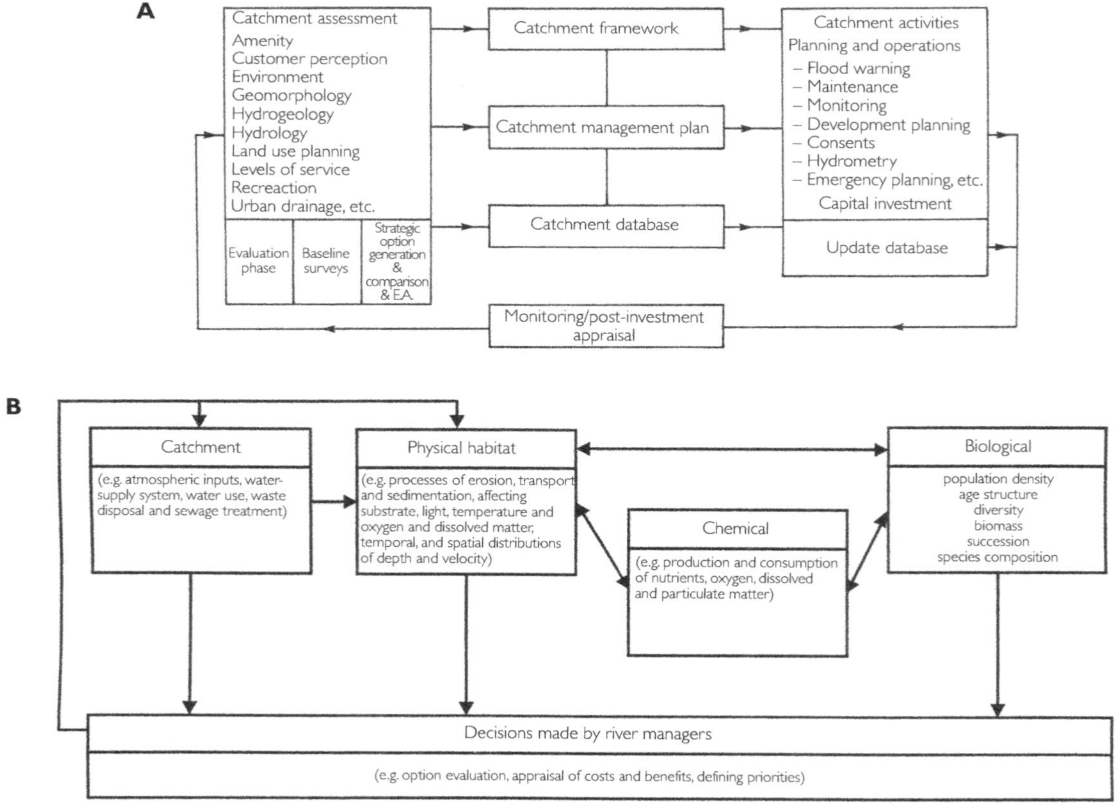

Figure 6.3 ***A*** *Holistic approach to river management as suggested by Gardiner (1991a) (in Figure 2.1);* ***B*** *Subsystems which should be considered in relation to river restoration (from Brookes and Shields, 1996c, figure 14.1, adapted from Statzner and Sperling, 1993)*

It is now usually agreed that the drainage basin or watershed is the ideal functional unit for water resources management, although integrated basin management and planning will continue to vary from one area to another according to features of the physical environment, including climate, and the approach to management, be it focused in hydraulics and concerned primarily with the distribution of water resources, or with hydrology, concerned primarily with water collection (Newson, 1997). The terms 'integration' and 'watershed approaches' are still often used in a variety of ways (Hooper and Margerum, 2000) by different people (Table 6.3). However, an integrated approach is not a quick fix but rather a long-term process (Margerum, 1996) because it takes time for stakeholders, organizations and individuals to come together, to reach consensus and to begin implementing a strategy. In addition it requires planned change so that government organizations must change policies and priorities and allocate new resources, and individuals must change their practices and actions (Hooper and Margerum, 2000: 512) to support the content requirement of integrated schemes.

Table 6.3 Aspects of watershed typology usage

Aspect	Activities and approaches included
Education and information	School water watch programmes Watershed information systems Information materials and brochures
Monitoring and planning	Ambient monitoring Watershed studies and assessments Local and regional management planning
Permitting	Discharge permits by watershed Watershed modelling Compliance monitoring
Soil conservation	Farmer information and education Farm planning Restoration and rehabilitation Incentives and demonstration projects
Stormwater control	Land use planning Stormwater management Plan assessment and review Site design
Integrated watershed management	Goal-based approach to management Strategic regional-scale river basin management (cross-jurisdictions) Combination of several or all of above approaches

Source: Based on Hooper and Margerum (2000).

Content

The main components of integrated river basin management can be visualized in several ways. Seven main functions were suggested (Newson *et al.*, 2000) namely water resources provision, flood defence, pollution, fisheries, conservation, recreation and navigation. In the USA (National Research Council, 1999), commonly included functions are water supplies, water quality, flood control, sediment control, navigation, economic development with HEP, biodiversity, fish and other aquatic biota, habitat preservation and recreation. Therefore, components of schemes of drainage basin management are primarily related to water resources, river channels, land resources and leisure aspects (Downs *et al.*, 1991), as supported by analysis of 21 examples of integrated basin management (Table 6.4). The components involved most frequently still relate to provision of water and its management for industrial, agricultural and domestic purposes, and to flood control. In less economically developed countries, emphasis is often upon support for development, particularly for energy and agriculture, so that HEP generation and soil conservation are frequently

Table 6.4 Major components of schemes of integrated river basin management

Provision objectives	**Components included**
Water supply – quantity and quality	Water supplies • industrial demand • domestic demand Water quality • pollution • sediment control Hydroelectric power and economic development Fisheries Flood control/flood defence
River channel	Navigation Channel management • channel stability • biodiversity • aquatic preservation
Land	Land resource management • agriculture • forestry • soil conservation • irrigation • urbanization Preservation and conservation of landscape
Leisure	Recreation Angling

included, whereas in more economically developed nations objectives related to conservation and recreational provision assume a greater importance. As fully integrated basin management should be concerned with both water and land issues (Haas, 1981; Saha and Barrow, 1981; Pantulu, 1983; Cunningham, 1986), truly integrated basin management can be visualized as comprising the following generic functions, although not all will be equally prominent in individual cases:

- hydrologic regulation for domestic and industrial water supplies including navigation;
- water quality control of physical and chemical 'pollutants';
- river channel management to moderate channel hazards and promote channel assets (see Chapter 5);
- land degradation control for agricultural and conservation benefit;
- land use regulation for public safety, reduction of public burden and conservation benefit;
- ecosystem preservation and diversification for conservation and recreational benefit including angling;
- economic development including measures for social justice.

6.3 Implementing integrated river basin management (IRBM)

IRBM has become a national and international reality, with support from international organizations and requiring several elements (Heathcote, 1998: 403):

1. adequate expertise for multi-objective planning and evaluation procedures, especially in economic, social and environmental areas;
2. adequate resources of time and money for implementation;
3. consideration of a wide range of alternatives to solve observed problems;
4. a flexible, adaptable plan, reviewed and amended at regular intervals;
5. representation of all parties affected by the plan and its implementation;
6. sufficient authority to enforce conformity of execution with construction and operating plans.

Although implementation strategies must vary between river basins, certain difficulties are inherent to many schemes (Heaney, 1993), including:

- a perception that IRBM is a static process so that adoption of a restrictive master plan seldom finds acceptance by all to conform to a single course of action;
- problems where the basin watershed does not coincide with political boundaries;
- weaknesses in the database of many planning models so that results are not realistic and have little credibility;
- the great complexity of schemes especially if environmental impacts are included;
- the slow progress of the planning processes so that people become impatient while waiting for action.

In addition, the responsibility for effecting management can rest at different national, regional or local levels. Whatever approach to IRBM is adopted and implemented in a particular basin, and focusing particularly on the way in which river channel management is incorporated (see Section 6.4), certain essential requirements must be met despite the inherent difficulties that exist. In this regard, recent trends have been influenced by legal and legislative positions, as well as by stakeholder views, technological developments and the broader intellectual environment context.

As the river basin came to be used as a planning and management unit in the 1990s to achieve the integration of patterns and processes of both natural and social systems, integrated water resource management, management of natural resources and regional planning (Newson *et al.*, 2000), it was proposed that the whole catchment area 'should be considered as the natural unit for integrated ecosystems-based water management' (UN Economic Commission for Europe, 1993). Any approach must be embodied in, and implemented within, existing *legal and organizational frameworks* that encompass many institutional arrangements including (Mitchell, 1990) legislation and regulation, policies and guidelines, administrative structures, economic and financial arrangements, political structures and processes, historical traditional customs and values, and key participants and actors. Legal frameworks have existed for many years originally deriving from behavioural codes based on compensation or punishment approaches, and from written law codes dating back to the twenty-second century BC (Heathcote, 1998). Water legislation usually developed historically and cumulatively, concerned especially with a dominant aspect such as water resources, with variations from one country to another according to the level at which legislation was implemented, how much was included and how environment was incorporated.

Legislation has evolved incrementally with few opportunities to revise completely the legal basis for the administration of water resources. An exception occurs in the comprehensive revision of water law since 1994 in South Africa, whereas most countries have a diversity of legislation as illustrated by examples contrasting Australia, Finland, the USA and South Africa (Palmer *et al.*, 2000). Law has to be applied to water quantity and water supply, water quality, conservation legislation and related provisions. The three main stages in legislation are policy development, legal drafting and implementation (Palmer *et al.*, 2000), all requiring a good ecological understanding of river functioning, together with adequate monitoring, referencing and classification of riverine ecosystems, as well as an adequate account of terrestrial–aquatic links and accepting the catchment as the unit of legislation and management, to facilitate truly integrated river basin management. In decision-making it is also vital to understand the social and economic values that interact with environmental values because legislation can succeed only if its requirements are practicable and reflect values accepted, or at least not actively opposed, by society (Palmer *et al.*, 2000).

Organizational structures, as well as legislation, vary substantially from one country to another, being a barrier to, or an avenue for, success (National Research Council, 1999) and often evolving on a project-specific basis according to the challenges in individual basins. Project-driven structures, combined with the fragmentation of responsibility between federal, regional and local levels, have given rise to numerous problems and difficulties. For instance, in the USA, 19 federal agencies have responsibility over 15 areas of water-related management (National Research Council, 1999; Table 6.5). Once responsibility is vested in a particular organization, or resides at a particular level whether federal, regional or local, it is not easy to effect change and, if reorganization is undertaken to try to achieve greater integration, it does not eliminate boundary problems but merely redefines where the boundaries will be (Mitchell, 1990). In Europe the statutory framework consists of inter-

Table 6.5 Water-related responsibilities of federal agencies in the USA

Agencies	Water related responsibilities														
	A	**B**	**C**	**D**	**E**	**F**	**G**	**H**	**I**	**J**	**K**	**L**	**M**	**N**	**O**
Department of Agriculture															
• Farm Services Agency		○									○	●			
• Forest Sevice		○	○			●	●	●	●		●	○	○	○	○
• Natural Resources Conservation Service		●								○	○	●	○	○	
• Agricultural Service												○	○		○
Department of Commerce															
• National Marine Fisheries Service	●	○			○			○	●				○		
• National Oceanic & Atmospheric Administration	●								●				●	●	
Department of Defense															
• US Army Corps of Engineers		●	○	●	●					●	○	○	○	●	○
Department of Energy															
• Federal Energy Regulatory Commission				●											
Department of the Interior															
• Bureau of Land Management		●	○			○		●		●	●	○	○		
• Bureau of Reclamation			○	●		○		○	○		○	○	○	●	●
• Fish and Wildlife Service		●	○			○	●	●	●		●		○		
• Geologic Survey	○	●	●					○	○	○	●	○	○	○	●
• National Park Service	○	○				●	●	●	●		●	○	○		
• Bureau of Indian Affairs		○						○	○		○	○	●	○	
Department of State															
• International Boundary Commission		○											●	●	●
Other federal units															
• Environmental Protection Agency	○	●	●			○	○	○	○	○	●	○	●		○
• Tennessee Valley Authority		○	○	●	●	○		○	○	●	○	●	●	●	●
• Bonneville Power Administration		○		●	●			○	●	○	●	○	●	●	●
• Federal Emergency Management Agency														●	

Notes: Solid circle represents significant responsibilities; open circle represents some related responsibilities.

Key to water-related responsibilities: A, oceans and estuaries; B, wetlands; C, research and dissertation data; D, hydropower; E, navigation; F, recreation; G, preservation; H, wildlife; I, fisheries; J, flow regimes; K, ecological diversity/recreation; L, erosion/sediment control; M, water quality; N, flood risk management; O, water supply.

Source: After National Research Council (1999: 168), originally developed from O'Connor (1995).

national conventions and EC legislation, mainly as directives. Whereas the former are voluntary, directives are binding and must be implemented in national legislation by member states (Iversen *et al.*, 2000). The potential complexity of legislation is illustrated by Scotland, which has a web of international, national, regional and district agencies involved with legislation related to pollution control, abstraction and compensation flows, fisheries and nature conservation (Brown and Howell, 1992). Therefore, within a single nation, IRBM has to reconcile the existing organizational and legal frameworks and also to overcome ways in which particular responsibilities are traditionally allocated to national, regional or local levels, and in which particular discipline groups have dominated the development of organizational structures such as the field of sanitary engineering and water quality issues.

As rivers cross, or form, political borders, international endeavours to protect ecosystems and species have faced the same challenges as approaches to watershed management at national levels (Korhonen, 1996). International collaboration is required when large basins transcend political frontiers. Although the physiographic drainage basin unit is usually the ideal management unit, interbasin transfers to resolve water resource problems make the basin focus less definitive. Canada possesses 9% of world runoff to serve a population of only 25 million people, but the water is in the wrong places, with seasonal deficiencies in the urban areas of the east, the prairies and the west (Newson, 1997). Whereas the management approach of the 1970s was characterized as reactive – dealing with problems as they arose – the water policy released by the Federal Government after 1988 was more anticipatory and preventative in addressing the management goals of improving water quality and quantity. Sometimes, for economic reasons, certain planning decisions have to be taken at national level so that not everyone agrees that the planning and management of land and water resources is most appropriately achieved within a basin framework (e.g. Winpenny, 1994).

Legal and organization issues of particular significance to river channels involve what aspects of river channel need legislation and how existing legislation is directed, with the location of boundaries and ownership being especially problematic. Legislation took a new turn following greater environmental awareness in the 1960s, after which most countries instigated new environmental legislation exemplified by the National Environmental Policy Act (1969) in the USA (NEPA) and the Canadian (Federal) Fisheries Act (1970). Older agencies with narrow mandates were evolved into Environmental Protection Agencies. In the USA, 14 Acts comprise the water-related legislation with national (federal), state or local agencies having jurisdiction over particular aspects of water and water-related activities. Increased concern for the environment also marked a new era in policies to reduce water pollution in the USA and Australia, with local government and citizens in Australia and the USA refocusing their efforts on the watershed and the need to integrate not only point and nonpoint source pollution but also the range of land- and water-activities that occur within a watershed. Arising from environmental concerns was the development of environmental impact assessments (see Chapters 7 and 8), a field considered to be one of the more successful policy innovations of the twentieth century (Heathcote, 1998), now utilized in more than 100 countries.

IRBM also involves aspects of land management and planning and, in Australia, the concept of integrated (or total) catchment management has been well accepted with a large number of catchment groups becoming established, and catchment planning being advanced in some areas. New South Wales, for instance, first implemented its Total Catchment Management Policy (TCM) in 1994 with the aim of sustainable use and management of the state's natural resources, particularly soil, water and vegetation (Schofield *et al.*, 2000). In the context of total catchment management, defined as 'the coordinated and sustainable use and management of land and water vegetation and other natural resources on a water catchment basis so as to balance utilisation and conservation' (Sharpin *et al.*, 1999), the New South Wales government instigated a stormwater management planning programme to mitigate the environmental impacts of stormwater runoff from urban areas throughout the state. Thus urban areas, and stormwater management in particular, have developed integrally with basin management policies.

There is no one single 'correct' outcome of the IRBM process and it is now generally appreciated that a dialogue is required to involve *stakeholder views*. Whereas the original paradigm in natural resources management was sustained yield, the new paradigm is based on two principles – ecosystem management and collaborative decision making (Cortner and Moote, 1994). This paradigm shift involves a move away from decision-making solely by technical experts towards procedures that promote consensus building among stakeholders, with technical expertise and analysis of alternatives provided by agencies. Members of the public have participated in river conservation since the nineteenth century but, since the Rio Earth Summit and Agenda 21 (Table 6.2) in 1992, some nations and international agencies have increased their efforts to identify and work with citizen groups concerned with river conservation (Showers, 2000), so that the success of individual projects now depends partly upon public involvement and participation (Kalbermatten and Gunnerson, 1978; Newson, 1997). Active public involvement is perhaps one of the greatest challenges facing river workers at the beginning of the twenty-first century (Wade *et al.*, 1998).

IRBM has benefited also from *technological developments* relating to spatial data analysis (computing, remote sensing, GIS) and to real-time decision-making (global positioning systems, real-time forecasting/analysis, the information highway). Succeeding the great expansion in data available on water and environmental conditions, developments in information technology have aided the digital processing of remotely sensed data and the integration of remotely sensed information with other data sources to provide geographic information systems (GIS). The time elapsing between collection and collation of data and integrated basin approaches has been reduced by use of global positioning systems (GPS) and handheld computers that can enable immediate analysis by reference to existing data bases and GIS.

New information technologies and the use of the Internet to gather and share data and to educate could facilitate development of a new paradigm to utilize and protect water and other natural resources by implementing sustainable land use practices appropriate for aquatic systems, thus overcoming some difficulties presented by the array of disjointed policies that exists at federal and state levels, with local efforts for integrated planning remaining relatively ineffective in most US states (Brown *et al.*, 2002).

The *intellectual environment* in the early twenty-first century is revolutionary in the importance accorded to water and to its management. Increasing water-supply demands are very uneven throughout the world, where one in four people have no access to safe water (Kirby and White, 1994). The UN has identified water as 'the most critical environmental issue of the twenty-first century' (Bonnell and Askew, 2000) suggesting that the time has come for hydrologists with physical and biological training to reconnect with colleagues who are policy makers and lawyers, environmental managers, members of citizen groups and watershed residents (Endreny, 2001). Depending upon the environment of the river basin, priorities in basin management therefore include increasing demands for water both to enhance existing supplies and to institute regular and guaranteed supplies in developing areas as a matter of social justice. As such, recognition of the finite nature of water resources and the need for conservation of water supplies is of increasing concern.

6.4 The role of river channel management

River channel management must obviously be an integral part of IRBM but perhaps has not been as central as one might expect, because the pertinence of river channel change to other aspects of the ecosystem was not well perceived throughout much of the twentieth century. For instance, content analysis of 21 IRBM schemes up to 1990 (Downs *et al.*, 1991), showed how issues related to river channel stability and management were among the least frequently discussed components, with acknowledgement of river channel change and the integrity of the fluvial system being seldom apparent, despite the clear *implication* that river channel management would be required to meet many of the other objectives (see Table 6.4).

More recently, environmental awareness and river conservation concerns have focused greater attention on the role that the river channel, its stability and potential for change, plays within IRBM, partly influenced by the success of river restoration approaches (Newson, 1997; Chapter 7). Aspects of river channel management have been given more attention: for example, the first of five principles of repeated patterns of stream management (Petersen *et al.*, 1987) was that 'watershed management is the goal, riparian control the starting point', but changing people's thinking is becoming at least as important as guiding new scientific knowledge (Douglas, 2000). The holistic approach to river management (Gardiner, 1991b; Figure 6.3A), river restoration approaches placed within the catchment context (Figure 6.3B) and integration of geomorphological factors with river ecology (see Table 3.7, p. 63) all need to be achieved with reference to different timescales of management (Brookes, 1994; Brookes and Shields, 1996c). Therefore, it is not simply a matter of including river channel management, and particularly restoration, as an extra item within the integrated basin management agenda, because restoration strategies are themselves a catchment-wide agenda (Sear, 1994), so that the river channel is truly an integral part of IRBM. As a consequence, the paradigm of 'working with the river not against it' (Winkley, 1972; see Chapter 7) can be adapted to become 'working with the river in the integrated holistic river basin context' (Downs *et al.*, 1991).

Three requirements that could further progress river channel management in IRBM (Newson *et al.*, 2000) are: first, to prevent damage to intact river ecosystems by having an active conservation movement backed by legislation and finance; second, to have means of prioritizing actions so that the most costly remedies are reserved for the cases where they are warranted; and third, setting the objectives of 'restoration' more clearly (see Chapter

7, pp. 184–90). In US watersheds that have been degraded by human activities but have potential to recover the characteristics and features that make them functionally similar to a pristine system, watershed restoration has been suggested to represent the preferred management strategy of IRBM (National Research Council, 1999), using either natural or passive restoration where the watershed is allowed to recover naturally, or by active restoration involving human intervention to accelerate the recovery process (Chapter 9).

River channel management at the watershed scale still remains a major challenge (Naiman *et al.*, 1998) because it requires integration of scientific knowledge of physical and ecological relationships within a complex framework of cultural values and tradition in order to achieve socio-environmental integrity (Chapter 11). One means of progress could be to strengthen the coupling between land and water in the minds of all politicians, planners, engineers and participants who live within the basin to achieve a system-wide, holistic approach to river basin development strategies (Newson, 1994). In such an approach, the basin sediment system is the focal point of the land–water coupling, leading to alternative strategies being entertained: a selection of which are given in Table 6.6.

Table 6.6 Impact matrix of basin development taking a longer term geomorphological viewpoint

Development problem	Geomorphological status of basin	Outcome/risk
Dam construction (sedimentation)	a. soil erosion active	Sedimentation
	b. erosion Holocene or historical	Sedimentation from valley-floor or colluvial stores
	c. soil erosion controlled	(see below)
River regulation (downstream channel stability/habitat)	a. dam above active alluvial zone or silt flushing	Channel change by direct impact or by tributary inputs
	b. dam with flood control/ HEP	Habitat effects via bed structure and flooding regime
	c. dam near coast	Coastal erosion
Land use plans • erosion control • forestry • irrigation • urbanization (sustainability)	a. land on slopes with active tectonics	Failure unless small-scale
	b. land on active or semi-active alluvium/colluvium	Changed hydrology releases sediment from storage
	c. floodplains	Longer-term function conserved, or loss of habitat

Source: Developed from Newson (1994).

6.5 Future approaches: implications for river channel management

The complexity of integrated river basin management means that a reactive approach has been usual with emphasis placed on treating the symptoms instead of finding the source of the problem and curing the disease (Diplas, 2002). Instead, a much deeper understanding of natural phenomena and processes is needed, together with a large-scale, more interdisciplinary, integrated and holistic methodology, requiring a paradigm shift in tackling land use change and watershed issues (Diplas, 2002). The complete list of elements to be considered may not be fully known although many are suggested in Table 6.7.

The context for both basin and river channel management is now more globally conceived, with increasing awareness of the implications of global change. Whereas the catchment ecosystem may be depicted in terms of developed, disturbed and changed states (Table 6.7) there is now the further need to explore how a completely holistic approach can be sustainable in the future and can accommodate the implications of global

Table 6.7 Reasons for differences in watershed management summarizing the aspects and categories of factors involved that account for differences that can occur from one area to another (see also Diplas, 2002)

Aspect	Categories	Reasons for differences from one area to another
Subjects	Water, land	One particular subject may be emphasised, e.g. water resources, land conservation (see Table 6.4)
Approach	Discipline, techniques	Particular academic training may provide a dominant influence (e.g. engineering, economics). Techniques adopted (e.g. GIS, GPS, DEM) may be influential (e.g. Table 6.6)
Involvement	Stakeholders	Range of stakeholders consulted (e.g. public participation and involvement)
Strategy	Single or multipurpose, integrated, holistic, sustainable	Variations (see Table 6.1) according to character of area and history of watershed management (p. 157)
Framework	Legal, structural, political	Administrative structure may affect implementation (e.g. Table 6.5)
Decision-making process	Integrated decision-making	Extent to which decision-making is truly integrated (p. 162) and considers all aspects
Subsequent action	Adaptive management	Degree to which adaptive management implemented and results are heeded (see p. 193 and p. 273)

change. This prompts consideration of the finite nature of resources and of greater emphasis upon water resources as a 'blue revolution' (Calder, 1999) and, with the adoption of Agenda 21 (Table 6.2) in many parts of the world, of sustainability. Sustainability is accepted as a desirable, if not essential, requirement in environmental management but cannot easily be implemented in those areas where provision of water is still a fundamental requirement, or in locations where the politicians and decision makers do not heed the messages being relayed (see discussion in Newson, 1997; Chapter 11; Newson *et al.*, 2000). Agenda 21 policies emerging after the 1992 UN Conference on Environment and Development, provided a boost for the river basin concept (Newson, 1997) by bringing together the biophysical and social aspects of river basin development. More recently, the European Union Water Framework Directive (European Community, 2000) that obliges European countries to establish integrated river basin management, was transposed into national law by the end of 2003 and requires river basin management plans.

New methods offered to achieve sustainable development by progressive integration of ecological processes operating at micro, meso and macro scales, mean that river managers are continually adjusting their actions in response to monitoring data that alert them to changing environmental and economic conditions and social preferences (Leuven *et al.*, 2000). The recent switch from technological to ecological river management (Nienhuis *et al.*, 1998) has progressed in terms of sustainable development, which means that resources should be used without degrading their quality or reducing their quantity (Smits *et al.*, 2000). Sustainable management requires a carefully considered strategy for the entire watershed including (Leuven *et al.*, 2000):

- conservation or restoration of the natural flow regime and the hydromorphological dynamics of rivers;
- allowing space for rivers;
- adapting river functions to natural river dynamics.

In IRBM it is now desirable to balance sustainability with environmental change (Naiman, 1992); a landscape ecology approach (ten authors in Naiman, 1992) could enable understanding of complex interactions by including models that simulate the effects of changes altering landscape patterns and effects of landscape pattern, with implications for species persistence, invasion of exotics and resource supplies. The conflict between development and ecological sustainability has been summarized for Europe and rapidly developing countries (Meybeck and Helmer, 1989; Petts, 1994), with river degradation resolved into a developing phase followed by a restoration phase (Figure 6.4). However, further work is necessary to deal with potential impacts on river channels and with new dimensions of risk. The river basin is also essential as a framework that helps in cases of water resource conflict resolution (Uitto and Wolf, 2002).

For the management of hydrologic resources in the USA it is suggested that a nested hierarchy of hydrologic management organizations (Table 6.5) is preferable, with responsibilities for each organization dependent on the watershed scale of its responsibility, exemplified with respect to three spatial scales (National Research Council, 1999). If such lessons, principles and recommendations are distilled with particular reference to river channel management then it is tentatively concluded that:

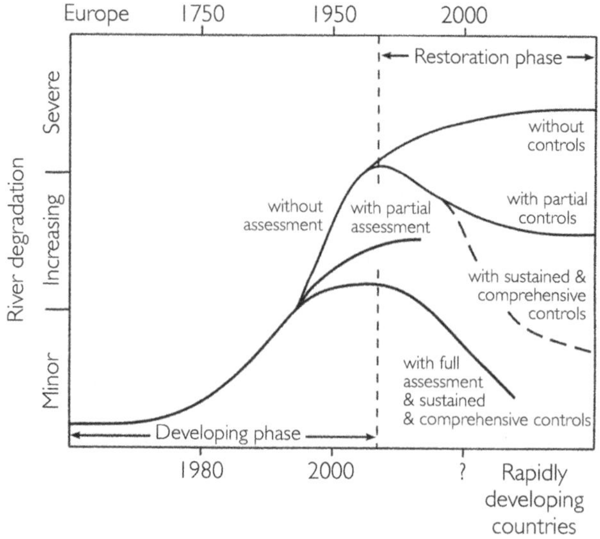

Figure 6.4 *Model for river degradation and restoration in Europe and in rapidly developing countries in relation to level of controls and environmental assessment (after Petts, 1994, developed from Meybeck and Helmer, 1989)*

- *the river basin, drainage basin or watershed is the most effective management unit,* having already been the basis for single purpose, multipurpose, integrated, holistic and sustainable approaches – all increasingly demonstrating the need to recognize the interactions between components and involving a sustainable approach. Schemes of IRBM initially emphasized water resources but, as they have become more holistic, river channel management is being incorporated;
- *river channel management is starting to be incorporated with IRBM.* Although the management of particular concerns such as water resources, pollution, soil erosion was separately organized at first, more integrated approaches became necessary because of the interrelationships between the separate aspects and the inefficiency of separate management systems. Management of channel reaches should be undertaken against the background of the management priorities and characteristics of the basin and network as a whole;
- *implementation strategies for IRBM* must be adapted according to the environmental problems and impacts in specific locations and according to whether control is national, regional or local. They include:
  - a nested hierarchy of clear responsibilities assigned to particular administrative levels;
  - dynamic ecosystem management, to minimize deleterious impacts and foster a sustainable approach;
  - channel and watershed restoration;
  - impacts on channels and watersheds as an integral part of a risk strategy;
  - as a result of technological developments and inclusion of stakeholder views, barriers to holistic management approaches are being lowered.

A re-ordering of priorities is beginning to take place as the role of the river and the river channel becomes more central to holistic basin management, and should be increasingly widespread as the twenty-first century progresses. A nested hierarchy of management organizations will continue to be required, each having a clearly specified role informed by the range of options/possible scenarios. This needs to be reconciled with a broader, more detailed context for river channel management as explained in Chapter 7.

7
A Foundation for Twenty-first Century River Channel Management

Synopsis

Several linked themes converged by the end of the twentieth century to re-define the practice of river channel management (7.1). Sciences were prompted to address the challenges faced in dealing with the catchment hydrosystem (7.2), which occurred in parallel with a worldwide increase in environmental awareness and politicization of the 'state of the environment', including river environments, leading to the need for environmental assessment (7.3). It became apparent that river channel management was as much about undoing the legacy of previous approaches as it was about responding to new management objectives (7.4) and that new methods were required to achieve the revised objectives (7.5). This chapter, therefore, reviews the shift from the 'old' to 'new' approaches to river channel management.

7.1 Retrospect and realization

Chapters 3–6 have examined several important aspects of relevance to channel management. Conclusions from these preceding chapters are translated into four requirements that have been fundamental to the redefinition of channel management practices and procedures during the latter part of the twentieth century and are indicated in Table 7.1. In outline, these requirements include:

1 new scientific approaches to the study and understanding of river channels, especially the increased use of cross-disciplinary studies to respond to the management challenges posed by the complexity of catchment hydrosystems, the deleterious effects of earlier management methods, the complexity of river channel adjustments and the need for a more integrated approach to management (Section 7.2);

2 new perceptions of the relations between rivers and people as quality-of-life concerns extended to environmental awareness and 'sustainability'. These arise from assessment of the impact of past human activities on river channels used as a guide to the likely consequences of future human activities, from greater public acknowledgement of the river basin as the logical boundary for management, and management recognition of the important role of stakeholders in decision-making (Section 7.3);

Table 7.1 Implications of conclusions from retrospect and realization (Chapters 3–6)

Retrospect/realization	Requirement			
	New scientific viewpoints	Revised perception of environment	Revised approaches including re-managing rivers	New methods to 'design-with-nature'
Chapter 3				
Catchment hydrosystems are made up of distinct components that form a continuum	•			
Physical and ecological fluvial system processes are a necessary basis for understanding river channel systems	•	•		
Understanding fluvial systems requires contributions from a number of disciplines	•		•	
Channel capacity, shape and types of channel patterns at a location in a catchment reflect the combined effect of discharge and sediment delivered from upstream, of the local characteristics of sediment, slope and vegetation, and more recent flow history			•	
A range of floodplain features relate to their processes of formation, including the river corridor and its associated ecological patterns	•		•	
Drainage networks are dynamic			•	
Connectivity of fluvial systems involves longitudinal, lateral and basin connections	•	•	•	
Stream power and equilibrium are especially helpful themes in analysis of connectivity dynamics within the overall catchment context	•			
Chapter 4				
Any river system has several degrees of freedom, adjustments in a particular location depend upon local environmental characteristics and history		•		•
Contemporary processes are not always responsible for the present character of the channel, the channel pattern, the channel network and the floodplain system, so that past processes need to be considered		•	•	
Channel adjustment can arise through a combination of natural forces and human activity	•	•		

Table 7.1 *continued*

Retrospect/realization	**Requirement**			
	New scientific viewpoints	**Revised perception of environment**	**Revised approaches including re-managing rivers**	**New methods to 'design-with-nature'**
River systems should be visualized in relation to several timescales of change	•	•	•	•
Many river channels have been segmented into sections each showing impacts of particular types of human activity			•	•
Understanding of changes over periods longer than 'engineering timescales' is necessary as a precursor to management.			•	•
In view of the requirement for longer-term information, understanding gained from palaeohydrology potentially contributes useful information			•	
Chapter 5				
River channel hazards occur as a complex response to spatially distributed natural and human factors, and require assessment as the basis for determining river channel management options	•		•	
River system activity also creates channel 'assets' in addition to hazards: identifying the character of assets is the basis for designing sustainable management options		•	•	•
Hazards and assets will be related in part to long-term inheritance of channel features – 'system memory' should be understood		•	•	•
Antecedent events may obscure longer-term trends: ideally the two should be separated to ensure effective management solutions	•			•
Management options should be designed cognisant of channel sensitivity to future human actions, as well as past and present actions		•	•	•
Management options should be designed to pre-empt channel sensitivity to future climate changes		•	•	•
Chapter 6				
The river basin, drainage basin or watershed is the most effective management unit for water resources planning		•	•	
River channel management has belatedly been incorporated into river basin management		•	•	•

Retrospect/realization	Requirement			
	New scientific viewpoints	Revised perception of environment	Revised approaches including re-managing rivers	New methods to 'design-with-nature'
Implementation strategies vary according to the environmental problems and impacts in specific areas and according to whether control is national, regional or local. They include hierarchical responsibilities, dynamic ecosystem management, channel and watershed restoration, appreciation of risk, technological development and stakeholder views	•	•	•	•

3 new approaches to management following an appreciation of river channel changes in terms of hazards and assets including their sensitivity over extended timescales, the need to 're-manage' rivers previously impacted by engineering approaches and now obsolete or counterproductive to the overall management challenge, and the desire to achieve a catchment-scale of system intervention (7.4);
4 new methods of system manipulation were required to implement the new approaches, including the need to manage each river system according to its own environmental characteristics, utilizing nonstructural methods and accommodating short-term, long-term and complex channel adjustments (7.5).

7.2 New scientific perspectives: responding to the catchment hydrosystem challenge

As knowledge of river channel processes and environments increased in the late twentieth century, the basis from which researchers sought to understand river environments also changed. Beginning in the 1960s, the 'quantitative revolution' in the environmental sciences provided the means for researchers to examine in detail the environmental impacts of river channel management methods (Petts *et al.*, 2000). The result of these endeavours was to improve the understanding of the mechanics of river processes and to demonstrate that nearly all rivers had been significantly impacted by human activity, especially in regions of intensive or long-standing settlement. However, by the 1980s, researchers with a concern for the value of river environments began to find that reductionist approaches, tied closely to one discipline, were limiting their ability to undertake comprehensive studies of river channels: biologists, ecologists, geomorphologists, hydrologists and hydraulic specialists found increasing need to use information from neighbouring disciplines to make sense of the their results. Several consequences were apparent.

First was the development of multidisciplinary research investigations, often involving teams of single-discipline-trained scientists and including projects at the river basin scale. This trend was encouraged in part by themed research topics that necessitated cooperative interaction between scientists, ranging from the International Union for Quaternary Research's Global Continental Palaeohydrology Commission (GLOCOPH; Benito *et al.*, 1998; Gregory and Benito, 2003a) to more inclusive approaches to river management: for instance the California Bay-Delta Authority (CBDA) is pursuing a science-led approach to resolving the tension between urban, agricultural and environmental water needs in California (e.g. http://www.science.calwater.ca.gov/vision.shtml (last accessed 19 October 2003)). Another significant impetus was provided by investigation of global environmental change utilizing Global Climate Models (GCMs) and the desire to understand the implications of results from these models for river channel management. Continuing challenges relate to downscaling GCMs to provide reliable local hydrological information (Schulze, 1997; Wilby and Wigley, 1997), while further issues arise because river flows are a second-order environmental change following from predicted changes in precipitation, temperature and vegetation growth (Arnell, 1996; Arnell and Reybard, 1996). River channel *changes* can be viewed as a third-order concern requiring understanding of fluvial erosion and land–water sediment transport interactions and deposition (Newson and Lewin, 1991) so that, in predicting future scenarios for channel change as the basis for understanding channel hazards, comprehension of climate change scenarios is critical to understanding. Also, as indicated in Chapter 6, technological advances in developing spatial databases using remote sensing data and Geographical Information Systems have provided consistent, detailed data over large spatial extents to permit integrated, basin-wide research studies to be practicable for the first time, leading for instance, to the prospect of catchment-scale models of sediment transport behaviour (review in Downs and Priestnall, 2003). These various studies are but part of an evolving scientific perspective towards multidisciplinary studies of river channel environments that is entirely appropriate to understand the complexity of the catchment hydrosystem (Chapter 3).

The second consequence has been the creation of hybrid sub-disciplines at the boundaries between traditional disciplines, to respond to the challenge of studying the river environment in a more holistic way. For instance, since the 1980s there has been increasing interest in the interaction between ecology and hydrology. Therefore, 'ecohydrology' stemmed in part from an engineering concern for the flow resistance caused by aquatic plants on river channel flow and reflected in the 'roughness' factor in the Manning's equation, but was for a while primarily concerned with the relationship between mire vegetation and hydrological change (Baird, 1999). Ecohydrology has been defined as '… the study of the functional interrelations between hydrology and biota at the catchment scale' (Zalewski, 2000) and '… the science which seeks to describe the hydrologic mechanisms that underlie ecologic patterns and processes' (Rodriguez-Iturbe, 2000: 3). Interest in the relationships between plants and flowing water developed to include an expanded concern for the influence of plants on flowing water, both in river channels and on floodplains, and the influence of flowing water on vegetation growth (Large and Prach, 1999). The latter is of particular interest in rivers where flow volumes have been regulated to cause an interruption to the normal process of riparian tree recruitment (e.g. Mahoney and Rood, 1993, 1998, Figure 7.1). Ecohydrological knowledge is therefore potentially critical in planning channel management activities, both to ensure that management does not

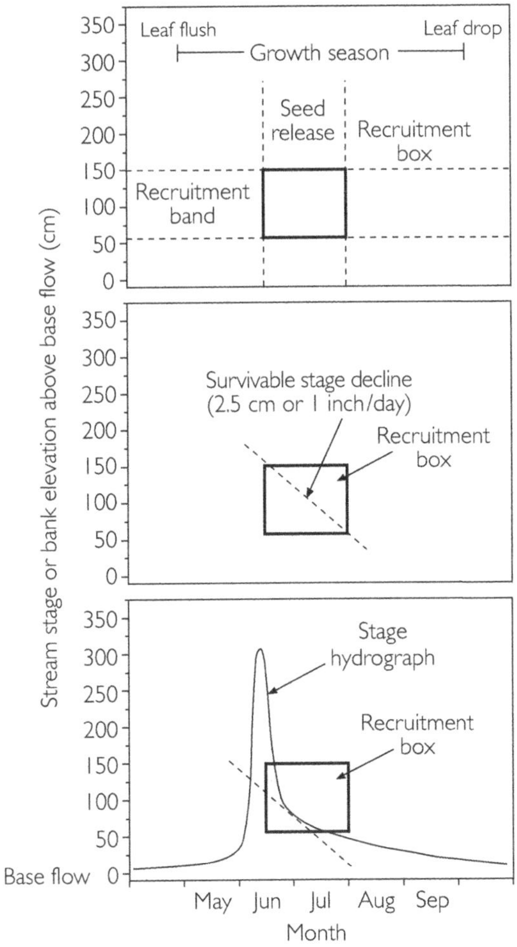

Figure 7.1 *Ecohydrology: the survival box concept (Mahoney and Rood, 1998). The ability of riparian cottonwood seedlings to survive in the rapidly changing river levels that often characterize dammed rivers of western North America is reliant on river stage declining by a maximum of 2.5 cm day*$^{-1}$ *during the critical period of 1 June to 7 July, to enable the growing roots to maintain functional contact with the receding water table*

adversely affect hydrology–ecology interactions such as tree recruitment, and also that management actions are not compromised by factors such as aquatic weed growth in lowland flood channels. A related hybrid sub-discipline has come to be known as 'ecohydraulics' (or 'habitat hydraulics') based on the principle that the energy budget of an organism is determined by the relative difference in speed between the organism and the medium in which it lives (Statzner *et al.*, 1988). At a physical level, ecohydraulic research has been concerned with the relationship between flow patterns, channel morphology and habitat availability (Figure 7.2). The Instream Flow Incremental Methodology developed by the US Fish and Wildlife Service (Bovee, 1982) to predict changes in fish and invertebrate habitat based on the interrelation of physical parameters exemplifies this

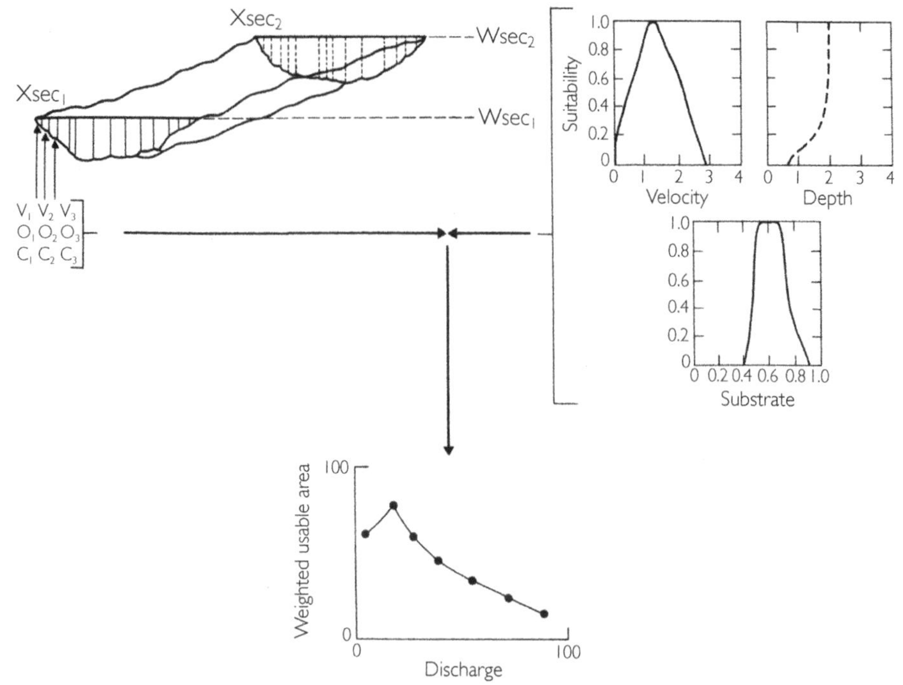

Figure 7.2 *Basic notions of ecohydraulics illustrated by the PHABSIM procedure (Bovee, 1982). Using biological research to define the habitat suitability for individual biota as a function of velocity, depth and substrate, hydraulic calculations in a river are made as the basis for predicting the 'weighted usable area' for the biota as a function of varying discharge (Petts and Maddock, 1994)*

approach, although it is often criticized for, among other things, the wide range of output scenarios possible from the same dataset according to options chosen within the program (e.g. Gan and McMahon, 1990; Kondolf *et al.*, 2000). Biologically, ecohydraulics is concerned with assessment of habitat through understanding the behavioural biology of fish, for example, based on foraging and bioenergetic models to predict position choice and territory size for drift feeding fish, noting that the answers will vary with fish life stage (Hughes, 1999). Overall, ecohydraulics is intimately linked with management efforts to improve fish species diversity through providing improved habitat diversity, and so is becoming of central concern in river channel designs.

Third, was the way that scientific researchers interested in environmental management, including river channels, increasingly found their management concerns to be linked with those of specialists from the social and economic sciences. 'Cultural' strands include a series of inherited ideas, beliefs, values and knowledge that together constitute the shared basis of human action (Gregory, 2000: 255) and are required because values 'circulate and mutate, are foregrounded or backgrounded, adopted or excluded on the basis of a very wide range of social, cultural, economic and political priorities and commitments' (O'Brien and Guerrier, 1995, in Guerrier, 1995: xiii). While the incorporation of cultural values into scientific investigation was traditionally seen as at variance with the notion and philosophy of science, because applied investigations aim to solve river channel management issues, they implicitly include value judgements so that contributing scientists should be cognizant of the cultural milieu in which they operate. Scientists making contributions directed towards a sustainable future should be aware that some commentators are sceptical of current values, that 'We cannot expect much success at sustainable development until we acknowledge that both science and politics are based on values which are unsustainable. The challenge is to provide a confident vision of a society far removed from the ones we know; we require a vision that is nothing short of an alternative future' (Redclift, 1995: 16). Such concerns have lead to a broadening of the scope of economic analysis to include the various nonmarket goods and services provided by river ecosystems and, in so doing, incorporating the complex issues surrounding the *non-use* valuation of ecological resources (Huppert and Kantor, 1998; Figure 7.3). Non-use value can include *existence* (or *passive use*) values placed on preserving natural environments for their own sake, the bequest value of holding environmental assets for future generations and the *preservation* or *intrinsic* value on obligations to ecological resource itself (Turner, 1993b). These specific, but challenging, interpretations of values are themselves but part of the philosophical challenge faced in adapting economic approaches to environmental valuation. Future challenges for environmental economics include issues of sustainability in terms of intergenerational equity, reconciling economic and ecological approaches to sustainability, an ethical shift away from individualistic moral reasoning and the conservation of cultural capital (Turner, 1993a).

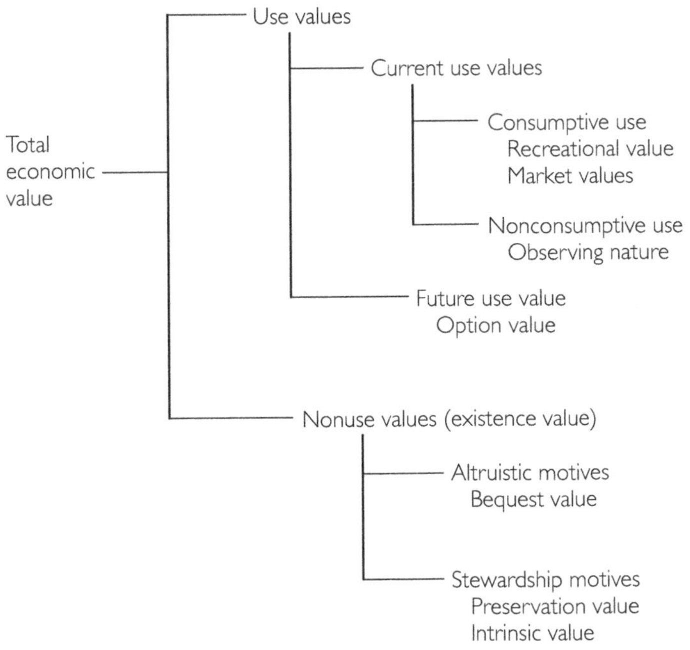

Figure 7.3 *Behavioural and motivational sources of economic value suitable as the framework for river channel management. Total economic value is the sum of use values plus non-use values (Huppert and Kantor, 1998)*

Such parallel and interwoven scientific developments are providing new perspectives for river channel managers to envision how river basin planning initiatives link with the operation of the past, present and estimated future changes in the river channel system.

7.3 New perceptions: environmental awareness and the concern for sustainable rivers

Changing scientific attitudes towards river channel management must be appreciated in the context of a growing public awareness that environmental issues are a key component for human quality-of-life. The eventual result of this awareness was a new perception of the relationship between people and rivers whereby earlier views of rivers as hazards and as drains became complemented by the perception of rivers as environmental assets worthy of protection. This change is introduced in terms of growing environmental awareness, national and international legislation.

Growing environmental awareness

Rachel Carson's (1962) book *Silent Spring* is often heralded as initiating *environmental awareness* with its repercussions for politics, especially the creation of Green Parties in many countries, education and lifestyles. Public attitudes to nature were examined, for instance, by Nash's (1967) work *Wilderness and the American Mind* and, in time, the same author was part of an evolving concern for the fundamental relationship of nature to human occupation of the Earth (Nash, 1989). Environmental ethics is 'concerned with the moral relations that hold between humans and the natural world' (Taylor, 1986), and was viewed initially as a rearguard action against those who see the predominant values in society weighted towards narrow self-interest (Cahn, 1978). In questioning the relationship of humans and nature, new concepts arose such as 'deep ecology', which contends that all species have an intrinsic right to exist in the natural environment (Naess, 1973), in contrast with 'shallow ecology' where nature is valued only from an anthropogenic viewpoint (Lemons, 1987, 1999). The management implications of deep ecology include viewing environments such as rivers from a biocentric rather than an anthropocentric perspective with a focus on preservation of unspoilt wilderness and the restoration of degraded areas (Guha and Martinez-Alier, 1997). This is in contrast with the historical antipathy of western philosophy to conservation, environmentalism and their values (Attfield and Belsey, 1994), which led Europeans to encourage human alienation from the natural environment and to an exploitative practical relationship with it (Calicott, 1995). By the 1980s, the 'green movement' had caused 'a shift in the political landscape, as environmental issues, once dismissed as a fringe movement of tree huggers, became the concern of millions of Americans' (Frodeman, 1995: 121), and of millions of others besides.

Coinciding with the first pictures of the Earth from space in 1968, Hardin's article 'The tragedy of the commons' drew attention to areas of land and water that could be regarded as commons but that had been brought to the brink of environmental catastrophe by individuals acting, quite appropriately, but alone (Hardin, 1968). This also raised awareness about sustainability issues and limits of the world's resources, a theme prominent in the *The Limits to Growth* (Meadows *et al.*, 1972). Alongside these texts were produced a series of influential popular works highlighting the environmental plight of rivers, including *The River Killers* (Heuvelmans, 1974), *Mountain in the Clouds* (Brown, 1982), *Cadillac Desert* (Reisner, 1986), *Taming the Flood* (Purseglove, 1988) and *Silenced Rivers* (McCully, 1996). Such books questioned the methods and priorities of the lead agencies responsible for river management in their respective countries and sought to publicize knowledge of the environmental state of the rivers affected by these actions. Themed academic texts also strove to examine scientifically the impacts of management methods on rivers, focusing on the effect of dams (Petts, 1984) and on capital works and maintenance of river channels themselves (Brookes, 1988).

In some countries, such as the USA and Australia, where river channel management is not governed by a single water management authority but by a plethora of regional or 'single purpose' management agencies (p. 162), the result of popular interest in the state of rivers has been the creation of 'Friends of the River' groups in the USA and agency-backed community-based planning in the 'Rivercare' programme of Australia (e.g. Boyd *et al.*, 1999). These groups run by local, concerned citizens, may carry out river maintenance activities through volunteer actions and often form the basis of riparian stakeholder representation with regard to actions proposed by a managing agency. In New South Wales, Australia, Rivercare planning on the Camden Haven River brought about considerable community involvement in workshops and 'riverwalks', designed to inform and seek community input on planning for river channel management, and resulted in a Rivercare plan with priorities that are the basis for prioritizing funding applications (Boyd *et al.*, 1999). Community involvement is commonly of three forms (Showers, 2000): (i) participation by consultation, (ii) self-mobilizing and interactive participation and (iii) functional participation in the absence of legislation. Community-based actions have spawned a further series of popular texts in the USA concerned with the practicalities of citizen action groups for river management such as *How to Save a River* (Bolling, 1994) and *Entering the Watershed* (Doppelt *et al.*, 1993), with summaries of scientific information suitable for a broad-based readership interested in 'watershed' (i.e. catchment) management (Schueler and Holland, 2000) and in the removal of small dams (The Aspen Institute, 2002; The Heinz Center, 2002). Thus, in parallel with the environmental 'green' movement has been the creation of a riverine 'blue' movement, defined by Calder (1999: xv) in *The Blue Revolution*, as 'although supported by technological advances, is more a philosophical revolution in the way we respect the world's environment and one of its most precious assets, water'.

National legislation

General interest in environmental issues led to a *national* response in terms of legislation directed towards conservation management of landscapes. For instance, in the USA during the depression era, Civilian Conservation Corps Projects and consequent, related, legislation including the Federal Aid in Wildlife Restoration Act of 1937 initiated a revised attitude to the environment. Eventually, the National Environmental Policy Act (NEPA) of 1969 encouraged '... productive and enjoyable harmony between man and his environment; to promote efforts which will prevent or eliminate damage to the environment and biosphere and stimulate the health and welfare of man; to enrich the understanding of the ecological systems and natural resources important to the Nation ...'. Notwithstanding the general impact of this legislation, it was Section 102 (c) of the Act, known as the Environmental Impact Statement (EIS) clause, which proved over time to have profound impacts for development proposals in near-river environments (Brookes, 1988). River-related environmental legislation continues to the present, for instance, in the USA in 1997, the American Heritage Rivers Initiative was passed 'to help communities alongside them revitalize their waterfronts and clean up pollution' and involved the designation of 14 rivers (National Research Council, 1999). In England and Wales, frequent re-organization of the water management agencies has provided an opportunity to trace the developing concern for environmental conservation (Table 7.2). The desirability of conservation is not articulated in legislation until the 1973 Water Act, despite centralized water resources planning focused primarily on water supply quantity and quality, flood defence and land drainage since 1930 (see Kirby, 1979; Parker and Penning Rowsell, 1980; Pitkethly, 1990; Newson, 1997). In 1973, conservation concerns were phrased passively and it was not until the adoption, in 1988, of the 1985 European Union (then European Economic Community) Directive on environmental impact assessment (85/337/EEC), that assessment was required for 'significant' river and near-river projects by developers and the water management agencies themselves. From this time, the Environment Agency

Table 7.2 The introduction and development of environmental assessment (EA) alongside river channel conservation duties in England and Wales

Year	Act	Commitment
1973	Water Act	No statutory EA but creation of ten catchment-based Water Authorities that, in addition to their traditional functional concerns, were charged '*to have regard to the desirability of preserving the natural beauty and of conserving the flora and fauna and geological and physiographical features of special interest*'.
1981	Wildlife and Conservation Act	No statutory EA but rewording of the phrase '*to have regard to the desirability of preserving* ...' to '*further and promote the conservation and enhancement of natural beauty and the conservation of* ...'.
1988	Statutory Instruments 1199 (*Town and Country Planning (assessment of environmental effects)*) and 1217 (*The Land Drainage Improvement Works (assessment of environmental effects)*)	EA is required for water resources projects where they are expected to result in 'significant' impacts on the river environment, stemming from EU directive EEC/85/337.

Year	Act	Commitment
1989	Water Act	Creation of National Rivers Authority. Definition of 'significant' environmental impacts as: '(a) … is of *more than local importance*, principally in terms of physical scale (b) … is *intended for a particularly sensitive location* … even though the project is not on a major scale (c) … is *thought likely to give rise to particularly complex or averse effects* …'.
1991	Water Resources Act	Reinforcement of '*to further the conservation and enhancement of natural beauty and the conservation of flora, fauna and geological or physiographical features of special interest*'. Section 16(1) allows conservation enhancements on their own merits for the first time: *'… it shall be the duty of the Authority to such an extent as it considers desirable generally to promote –* *(a) the conservation and enhancement of the natural beauty and amenity of inland and of coastal waters and of land associated with such waters* *(b) the conservation of flora and fauna which are dependent on the aquatic environment; and* *(c) the use of such waters and land for recreational purposes'.*
1995	Environment Act	National Rivers Authority merged with *Her Majesty's Inspectorate of Pollution* and the *Waste Regulation Authorities* to form the *Environment Agency*. The new Agency reflects the post-Rio concerns for sustainability: *'… it shall be the principal aim of the Agency … to make the contribution towards achieving sustainable development …'.*
2003	Adoption of EU Water Framework Directive 2000/60/EC	Requirement for the 'hydromorphological' quality of all inland fresh waters to reach 'good status' by 2015.

(formerly National Rivers Authority) began developing procedures for assessing river channels and habitats (e.g. Brookes and Long, 1990; National Rivers Authority, 1992) in the UK, showing the importance of *international* legislation in landscape conservation, and set to continue in 2003 with the adoption of the 2000 European Union Water Framework Directive (2000/60/EC) (see p. 206)

International legislation

International recognition of the importance of environmental matters has often focused on the global issue of sustainability. Notable in this context was the United Nations World Commission on Environment and Development report *Our Common Future* (UNWCED, 1987), which defined 'sustainable development' and advanced seven strategic imperatives and seven preconditions for sustainability to be achieved. This report stimulated environmental organizations, industry and development agencies to address sustainable issues and to reframe their strategies as necessary. Such moves were aided by *Caring for the Earth: a Strategy for Sustainable Living* (International Union for the Conservation of Nature (IUCN) *et al.*, 1991), which developed from and built upon the *World Conservation Strategy* published in 1980, by basing a strategy upon nine principles for sustainable living. The third principle, 'Conserve the Earth's vitality and diversity', like the others is accompanied by a series of actions, of which 4.5 is 'Adopt an integrated approach to land and water management, using the drainage basin as the unit of management'. The additional imperative of global change stimulated international conferences such as the United Nations Conference on Environment and Development (UNCED) in Rio de Janeiro in June 1992. The fallout from the Rio Summit was a series of internationally agreed action items such as Agenda 21, which included policies for Water Management (See Table 6.2) that have eventually found their way into national legislation relating to river channel management. In addition, internationally focused non-governmental organizations (NGOs) have been formed to promote public representation in water issues. Two such examples include the World Water Council (WWC), formed in June 1996 with the objectives of 'promoting awareness of global water issues and facilitating conservation, protection, development, planning and management of world water resources' and the International Rivers Network, formed in 1985 to campaign for social justice in water resources and especially to campaign against the construction of large dams. Although the remit of each NGO is broader than a concern solely with river channels, their activities reflect part of the global environmental consensus towards issues that affect the management of river channels.

Overall, the second-half of the twentieth century saw a dramatic increase in environmental concerns that has changed approaches to river channel management, especially in the legal requirement for environmental assessment as a tool for arbitrating competing management options.

7.4 New approaches: reducing the legacy through river restoration

In response to legislation demanding environmental assessment and sustainable management approaches, river management agencies began to re-formulate their operations to ensure that environmental values formed a core element both of new projects and also of the 're-management' of rivers previously subjected to engineering operations. Re-management operations were conceived to reverse the legacy of environmentally detrimental measures such as channel straightening, enlargement, constructed embankments and hard engineering structures in a process that became known as river restoration. While early restoration approaches focused almost exclusively upon pollution control, flow protection measures at planned dam sites, protected fish and wildlife habitat refuges and 'nature-like' channel designs (Petts *et al.*, 2000), the scope soon broadened as part of an ongoing evolution of water resources management (Chapter 6). This included, for example, the advent of increased stakeholder involvement in management decision-making and much closer scrutiny of management actions, especially in urban areas where the river corridor is often the only remaining 'green' space of local merit but has to be subject to appropriate levels of flood management and erosion control.

River restoration evolved from the field of *restoration ecology*, popularized following a symposium in Madison in 1984 (Jordan *et al.*, 1987). Restoration ecology was defined by the Society for Ecological Restoration as '... the process of repairing damage caused by humans to the diversity and dynamics of indigenous ecosystems ...' (Jackson *et al.*, 1995: 71) and was motivated by a desire to reduce losses in biodiversity and habitat diversity. The ultimate focus is on restoring badly damaged ecosystems to their pre-disturbance condition (Cairns, 1991) in terms of structure and function (Downs *et al.*, 2002a), using a contemporary undamaged site or knowledge of former conditions for guidance, but it has long been acknowledged that 'perfect' restoration is not possible (Berger, 1990) because all restorations are exercises in approximation (Figure 7.4). Restoration ecology projects encompass: judgment of need; an ecological approach; setting goals and evaluating success;

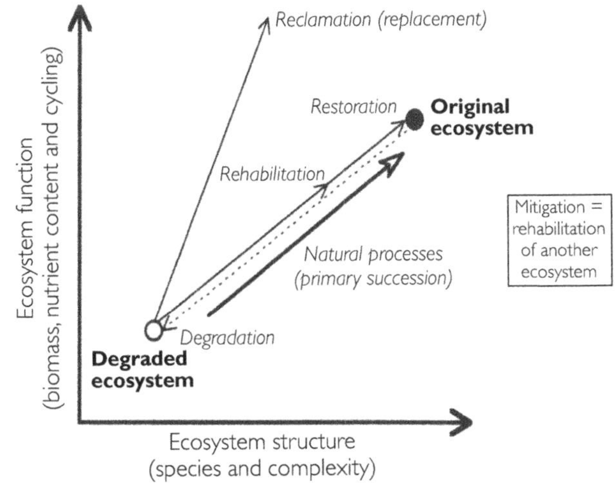

Figure 7.4 *Ecological restoration expressed as a combination of ecosystem structure and function. Ecosystem degradation usually implies a reduction in measures of ecosystem structure and function but not necessarily in equal proportion. Restoration, in its purest sense, implies returning structure and function measures to their original value but difficulties in achieving true restoration can be expressed using terms such as rehabilitation and replacement (Bradshaw, 2002). See Figure 10.12 for an empirical expression of this figure*

and evaluation of limitations of the approach (Jackson *et al.*, 1995). Therefore, river restoration is fundamentally about the re-creation of naturally functioning river environments in which the key steps are the identification of historic patterns of habitat development, the identification of developmental constraints and the relief of those constraints (Ebersole *et al.*, 1997) so that human actions no longer restrict the development of natural patterns of diversity (Frissell and Ralph, 1998). Ecological restoration can be viewed as having complementary emphases wherein the act of restoration is a technique for application and at the same time the ultimate test of research understanding (Jordan *et al.*, 1987). As such, it can never be undertaken in full certainty of the eventual outcomes. For river channel management, an explicit recognition of uncertainty marks an important distinction from management methods that have been traditionally based on (an apparent) certainty of outcome – this contrast remains a vexing dilemma for river managers used to 'guaranteed' operations. It also marks the break of channel management from being a 'modern' technological pursuit of the twentieth century into a potentially 'post-modern' entity.

Key landmarks in the evolution of river restoration commence with attempts to restore water quality from about the 1950s, following recognition of intense water quality problems resulting in human health hazards and near-zero fish life in many lowland rivers (Brookes and Shields, 1996b; Table 7.3). Physical restoration of river channels using geomorphological principles was attempted in small channels across the USA from around 1970 (Brookes and Shields, 1996), recognizing that '… the pressure to modify streams in the urban environment is not going to go away. Rather, that pressure will increase. It is imperative, therefore, that any stream modification be done in such a way that the maximum utility of the total stream environment is achieved' (Keller, 1978: 119). However, at this time, physical restoration projects focused on mitigation and enhancement works (Brookes, 1988) or works designed to benefit the physical habitat of a single species, often fish. Many river restoration schemes continue to be led by fisheries interests, especially to arrest the decline in habitat quality for commercial or sports fisheries such as for salmonids. Projects involving substantial modifications of channel morphology became popular from the mid-1980s, for instance, restoring the sinuosity of the Stensbaek (Denmark) to its historical, pre-channelized condition (Brookes, 1987b). Echoing the ideals of restoration ecology, river restoration was soon proposed as the process of recovery enhancement wherein an ecosystem is returned to a condition that closely resembles unstressed surrounding areas (Gore, 1985a).

Scientific demonstration projects, large river restoration projects and literature on integrated approaches began to appear in the 1990s. In Europe, opportunities for river restoration were first aided by the foundation of the River Restoration Project (RRP) in the UK, set up in 1991 as a nonprofit interdisciplinary group to promote river restoration projects. Following funding by an EU 'Life' grant and from South Jutland Community Council, three closely monitored demonstration restoration projects were established for three lowland rivers encompassing two rural sites (the River Cole in southern England, Figure 7.5, and the Brede Å in Jutland, Denmark) and one suburban site (the River Skerne in northern England; Holmes, 1998). In 1998, the RRP evolved into the River Restoration Centre (RRC) to provide a focal point for the dissemination of information concerning river restoration, following the creation, in 1995, of the European Centre for River

Table 7.3 Key landmarks in the evolution of river restoration

Focus of approach	Approximate year of initiation	Examples
Water quality restoration	1950	Ohio River, USA (Pearson, 1992); Thames estuary, UK (Gameson and Wheeler, 1977); various sites, USA (Patrick, 1982)
Biological rehabilitation of regulated river by compensation flows	1960	Compensation flows required since first reservoirs in UK, specific biological criteria later (e.g. Baxter, 1961)
Mitigation and enhancement of channels impacted by engineering works	1970	Various sites, USA (Shields, 1982)
Single species restoration/ rehabilitation of small rivers and streams	1975	Wisconsin, USA (White, 1975)
Geomorphological restoration/ rehabilitation of small rivers and streams	1975	North Carolina, USA (Keller, 1975; Nunnally and Keller, 1979); Jutland, Denmark (Brookes, 1987) Bavaria, Germany (Binder *et al.*, 1983)
Habitat rehabilitation of regulated rivers based on high flow releases	1980s	Various sites, including western USA (e.g. Reiser *et al.*, 1989)
Scientific demonstration projects of restoration	1985	Kissimmee River, Florida, USA (Toth *et al.*, 1993); River Brede, Denmark; Rivers Cole and Skerne, UK (Holmes and Nielsen, 1998; Vivash *et al.*, 1998)
Large river and floodplain restoration projects	Late 1990s	Kissimmee River, Florida, USA (Toth, 1996)
Integrated catchment approaches to restoration	Proposed, but no significant projects to date	See, for example, National Research Council (1999)

Source: Developed from Brookes and Shields (1996b).

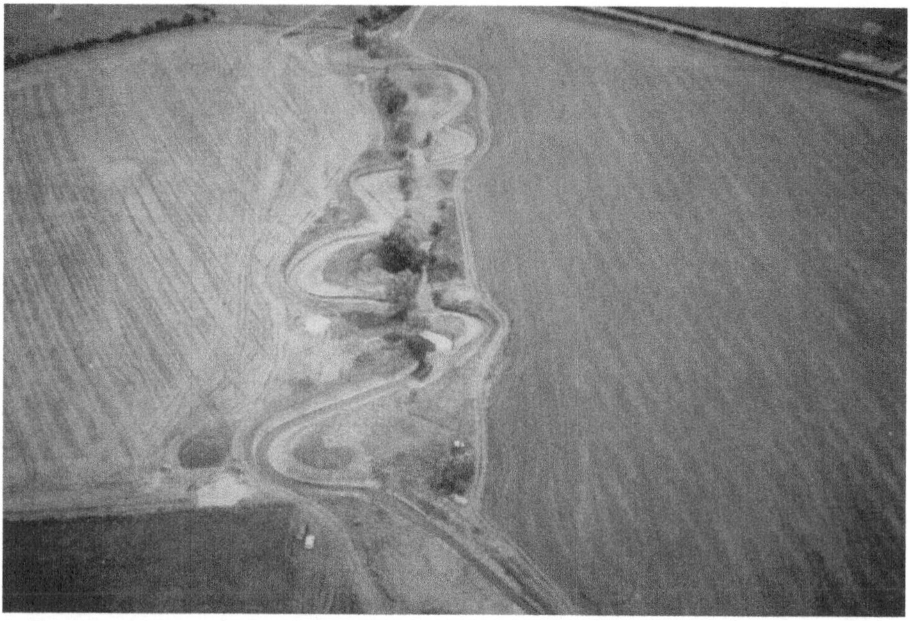

Figure 7.5 *The River Cole in southern England shortly after morphological reconstruction. The Cole was one of three scientific demonstration restoration projects funded by an EU 'Life' grant and facilitated by the River Restoration Project/Centre. (Photograph provided by A. Brookes and taken originally by C. Wheeler, formerly National Rivers Authority.) View looking upstream of the northern end of the project (planform in Figure 10.7). One of the other scientific demonstration restoration projects, the Brede Å in Jutland, Denmark, is shown on the front cover of this book*

Restoration (ECRR) with strategic aims in sharing experiences and information, acting to connect restoration practitioners and developing national networks on river restoration, and the organization of conferences and workshops (Nijland and Cals, 2001a). To date, three international conferences have been convened to develop aspects of river restoration, including 'theory and practice in lowland stream restoration' (Osborne et *al.*, 1993), 'the physical dimension' (Hansen et *al.*, 1998) and 'practical approaches' (Nijland and Cals, 2001b). In the USA, experiments have begun in restoring large rivers, such as the Kissimmee River in central Florida (Toth, 1996). Once a 166-km-long meandering channel with extensive wetland habitat and tributary sloughs, the Kissimmee was channelized for flood protection into a 90 km, 9-m-deep canal destroying 17 400 ha of river-wetland habitat and reducing wildfowl populations by 90% (Shen et *al.*, 1994). Restoration projects have begun aimed at restoring the ecological integrity of the river corridor (Toth et *al.*, 1993, 1998; Toth, 1996; Warne et *al.*, 2000). Literature on integrated approaches to river restoration, begun in 1992 with the comprehensive *Restoration of Aquatic Ecosystems* (National Research Council, 1992), was extended with volumes related to principles, process and practices for river restoration (Brookes and Shields, 1996a; Federal Interagency Stream Restoration Working Group, 1998), the rehabilitation of degraded habitats (De Waal et *al.*, 1998), the particular requirements of urban river restoration

(Riley, 1998) and those of watersheds as a whole (National Research Council, 1999). However, despite increasing interest in river restoration, the total length of restored channels worldwide remains very low.

Inevitably, the question is asked what is river restoration restoring? At a technical level, and derived from restoration ecology, the pursuit involves re-creating naturally functioning river environments using reference conditions from a neighbouring undisturbed river or a 'pre-disturbance' historical period (see Chapter 9). More broadly, the question raises a series of philosophical challenges for river restoration requiring the definition of 'an appropriately robust metaphysics and meta-ethics' (Attfield, 1994). Central to this issue is a core concern for whether nature is simply a social construct (i.e. do environments exist before the environed creature?; Attfield, 1999), and related questions such as:

- *does nature exist?*
- *can nature be reconstituted, recreated or rehabilitated?*
- *how sustainable is restoration?*

Nature is a human construct and its existence can be questioned because the imprint of humans across the entire planet means that 'we have built a greenhouse, a human creation, where there once bloomed a sweet and wild garden' (McKibben, 1989: 91). In this case, river restoration can only ever be 'artefactual' (Katz, 1992), as human technology is again imposed on natural river systems just as it was for so-called river improvements of the twentieth century (see Chapter 2), potentially representing a further 'unnatural' human imposition on natural systems. In defence, restoration offers at least the prospect of affording nature greater respect than was previously the case. With regard to the reconstitution of nature, related issues include whether restoration is possible physically and whether a restored system can ever have the same value as a natural system (Elliott, 1997). It is clear that, although better schemes may be expected with practice and better knowledge of river systems, an exact pre-disturbance environmental replication is unlikely and a 'rough and approximate restoration' is the most likely endpoint even with favourable political and economic conditions (Elliott, 1997: 77). Exact historical replication would in any case be undesirable because of changes in catchment populations that will have affected the fundamental catchment regimes of water and sediment (National Research Council, 1992). It is also argued that the value of 'faked nature' can never approach that of natural, undisturbed systems, whose values are defined, in part, by the longevity of their continuous existence and evolution, an argument not against restoration itself but a case for the preservation of natural areas (Elliott, 1997). This issue is of great importance to river restoration where, to date, restoration has often occurred as compensation for development activity in some other, largely undisturbed, location. Sustainable actions are a cornerstone of reconstructive post-modern environmental ethics (Oelschlaeger, 1995), but are largely unproven. However, considerable social and political momentum now favours restoration-based objectives for river channel management and includes factors such as the release of marginal lands (often near rivers), a focus on human quality of life, landscape quality and moves towards cost-effectiveness in flood defence operations (Sear, 1994). Faced with this momentum, the challenge is to develop practical approaches that allow sustainable solutions to be achieved within the context of the new objectives facing river channel management, acknowledging that individual approaches cannot be appropriate to all river environments (see Gregory and Chin, 2002, regarding urban river channels).

7.5 New methods: initiating design-with-nature

To achieve river channel management in its contemporary context requires that the technical operations employed are environmentally sympathetic and appropriate to the ecosystem under scrutiny: there is a need to implement a holistic, catchment-based, sustainable approach consistent with the ethos of 'design-with-nature'. Roots of the design-with-nature movement in rivers can be traced to the concerns of fishermen and hunters for loss of wildlife habitat in mid-nineteenth century Europe (Petts *et al.*, 2000) but such concerns were largely disregarded in favour of hard engineering design solutions until the mid-1960s when a movement began towards a softer management approach that mimicked the landscape's 'natural' characteristics. Rather than stemming from the earth and environmental sciences as might have been expected, it was the disciplines of landscape architecture and ecology that became most significantly involved. The ideas of the landscape architect Jens Jensen at the University of Chicago, of Aldo Leopold at the University of Wisconsin, who advocated a land ethic that 'changes the role of Homo Sapiens from conqueror of the land-community to plain member and citizen of it' (Leopold, 1949), and others in the 1930s were influential in a general sense (Berger, 1990), but it was a book *Design With Nature* by Ian McHarg, a landscape architect at the University of Pennsylvania (McHarg, 1969, 1992) that promoted ideas of wider importance than landscape architecture itself. The ideas were first developed in respect of the city where it was suggested in 1964 (McHarg and Steiner, 1998) that aspects of nature in cities would be improved through watershed planning. *Design With Nature* provided a method whereby environmental sciences, especially ecology, could be used to inform the planning process and the intent and language used subsequently appeared in the 1969 National Environmental Policy Act and other legislative instruments. Landscape architects such as McHarg were, therefore, prominent in envisioning an ecologically-based planning approach, to match ecological analogies already offered in the innovative building architecture of Frank Lloyd Wright (e.g. Smith, K., 1998) and those used in art and aesthetics (Tuan, 1993).

Design-with-nature type solutions for river channel management were first characterized as 'working with the river rather than against it' using a blend of experience and engineering judgement (Winkley, 1972), but there were also calls for alternatives to channelization where the designer should appreciate that streams are open systems and understand the implications of convergent-divergent flow processes, of geomorphic thresholds in stream behaviour and of the complex relations implicit in erosion, deposition and sediment concentration processes (Keller, 1975, 1976; Keller and Hoffman, 1977). A movement began centred on a reverence for rivers (Leopold, 1977), fuelled by a fuller understanding of impacts on channels of regulation and channelization (Petts, 1984; Brookes, 1988, respectively). Calls for environmentally sympathetic design procedures became commonplace (e.g. Binder *et al.*, 1983), usually beginning with mitigation measures designed to offset environmental impacts, then proceeding to enhancement devices for local environmental improvement and, finally, river restoration, a change in approach that took 20 years in the Thames River basin, UK (Gardiner, 1988a).

Achieving design-with-nature solutions for river channel management requires innovation in three ways. First, 'softer' methods of engineering the channel are required that eliminate the deleterious effects of previous methods (Chapter 4) and result in significant, sustainable environmental improvement. Second is a need for a greater role for environmental planning as a non-structural approach for underpinning 'cause and effect'-based management, that is, to offset the impact of human activities in the drainage basin and so reduce the need for symptomatic solutions. Third, the management process itself needs revision, not only in greater community involvement in management decision-making, but in ensuring that inventive solutions are applied and that *their impact is understood*. In this matter, the uncertainties implicit in pursuing river restoration have directed managers towards an adaptive management approach (e.g. Holling, 1978; Walters, 1986).

Developing softer engineering methods represents a significant challenge, not only for technical reasons. To begin, river managers have had predominantly a 'traditional' training in civil engineering, based on fluid mechanics and emphasizing structural measures as the solution rather than as part of the management system (Williams, 2001). In addition, methods have been largely predicated on engineering judgement (Hemphill and Bramley, 1989) with design drawings that emphasize permanence as a key characteristic (often to avoid future litigation), making the process inherently conservative and often unable to break from the practices of the past (Mintzberg, 1980). Despite this situation, a suite of 'softer' engineering alternatives have arisen based on two hybrid engineering principles. The first is *ecological engineering* (Odum, 1962), where natural energy sources and self-regulating processes are used to design, construct, operate and manage sustainable landscape and aquatic systems, consistent with ecological principles, to benefit both humanity and nature (Mitsch and Jorgensen, 1989; Barrett, 1999; Bergen *et al.*, 2001). Applications have concentrated on the restoration of disturbed ecosystems and on the creation of new ecosystems, with a goal of better performance, less cost, multiple benefits and acceptance by the public and regulators (Barrett, 1999), and underpinned by concepts of self-design and self-organization, a systems approach to biology, and the goal of sustainable ecosystems (Mitsch, 1998). The discipline demands knowledge both of engineering theory and core elements of quantitative ecology, systems ecology, restoration ecology, ecological engineering, ecological modelling and ecological engineering economics (Matlock et *al.*, 2001). For river channel management, the technique is most commonly associated with *biotechnical bank and bed stabilization*. Biotechnical solutions use living or dead vegetation (e.g. large woody debris) alone, or in combination with structural elements such as rock, to provide bank stabilization (p. 306). Stabilization using tree trunks provides a visually agreeable surrogate for rock or concrete options, using the slow decay of the trunks and the irregularity of their root boles to provide habitat value in addition to strengthening the bank or bed. Where living vegetation is used, the bank is strengthened mechanically by the binding properties of the root mass and hydrologically by the ability of roots to drain soils during water uptake (Simon and Collison, 2002). Numerous manuals now illustrate biotechnical solutions as part of environmentally acceptable stabilization solutions (e.g. Ward *et al.*, 1994; Gray and Sotir, 1996; Rosgen, 1996; Federal Interagency Stream Restoration Working Group, 1998; River Restoration Centre, 1999).

Alongside ecological engineering is *engineering geomorphology* or *geomorphic engineering*, wherein 'the geomorphic engineer is interested in maintaining (and working towards the accomplishment of) the maximum integrity and balance of the total land–water ecosystem as it relates to landforms, surface materials and processes' (Coates, 1976: 6). However, while engineers have been aware of the importance of site geomorphology for thousands of years (Fookes and Vaughan, 1986; Brunsden, 2002), geomorphology classes have never been part of compulsory engineering training, limiting the geomorphological appreciation of engineers despite marked improvements in the theoretical basis and technical capabilities of geomorphology over the past 30 years (Brunsden, 2002). As such, the term 'engineering geomorphology', implying the application of geomorphic theory and techniques to solve planning, conservation, resource evaluation, engineering or environmental problems (Brunsden *et al.*, 1978; Giardino and Marston, 1999), is probably more accurate than 'geomorphic engineering', which implies the existence of a hybrid discipline for which there is little evidence (cf. ecological engineering). Engineering geomorphology is now in its second phase of application in the UK (Hooke, 1999), providing information and design principles that raise awareness of geomorphic process dynamics in river projects, of the connectivity present in geomorphic systems and the long-term variability in geomorphological processes and landforms. Geomorphological inputs include environmental assessments of channel condition (see Chapter 8) and design principles for channel design, restoration schemes and bank protection. A third phase of engineering geomorphology application is predicted to integrate understanding of landform complexity, positive feedback, non-linear responses, event sequences, understanding of time-dependent processes, episodic change and inheritance, and a search for holistic solutions (Hooke, 1999; Brunsden, 2002). Under this scenario, the use of natural processes of water and sediment discharge to achieve management solutions becomes a truly central tenet and thus a 'geomorphic engineering' descriptor becomes fully consistent with the notion of design-with-nature.

Increasing realization that river channels are sensitive to water and sediment changes in the catchment has led to a greater appreciation of the need to manage land and water together (Newson, 1997) to achieve causally-based channel management solutions. This fact has promoted the use of non-structural solutions based on environmental planning as a design-with-nature measure (e.g. Dunne and Leopold, 1978). Approaches include catchment-level water and/or land management policies appropriate to ensure natural water and sediment regimes including the prospect of benign neglect that allows natural channel recovery, floodplain or river corridor policies designed to maximize the buffering capability of the riparian zone in separating the river channel from intensive land occupation, and in-stream flow allocations (National Research Council, 1992; Downs *et al.*, 2002a). Floodplain management measures have often centred on zoning flood-prone lands for flood control insurance purposes (e.g. US National Flood Insurance Program) but are increasingly being used to buy or lease flood-prone lands ('easements') for wildlife conservation purposes and to allow unconstrained river channel migration. Similarly, instream flow allocations that have traditionally focused on minimum environmental flows (Petts and Maddock, 1996) are increasingly turning towards high flow prescription to stimulate geomorphological processes that result in morphological changes in the channel and, from this, habitat benefits for instream and riparian flora and fauna.

In addition to novel engineering and planning measures designed to achieve design-with-nature, new management processes are also being applied. Foremost among these is *adaptive management* (Figure 7.6), which has a central theme of actively learning through experience (Holling, 1978; Walters, 1986) to deal effectively with systems characterized by uncertainty (McLain and Lee, 1996). This is important because of the difficulty in forecasting the responses of rivers to intervention (Chapter 5). Adaptive management can be defined as:

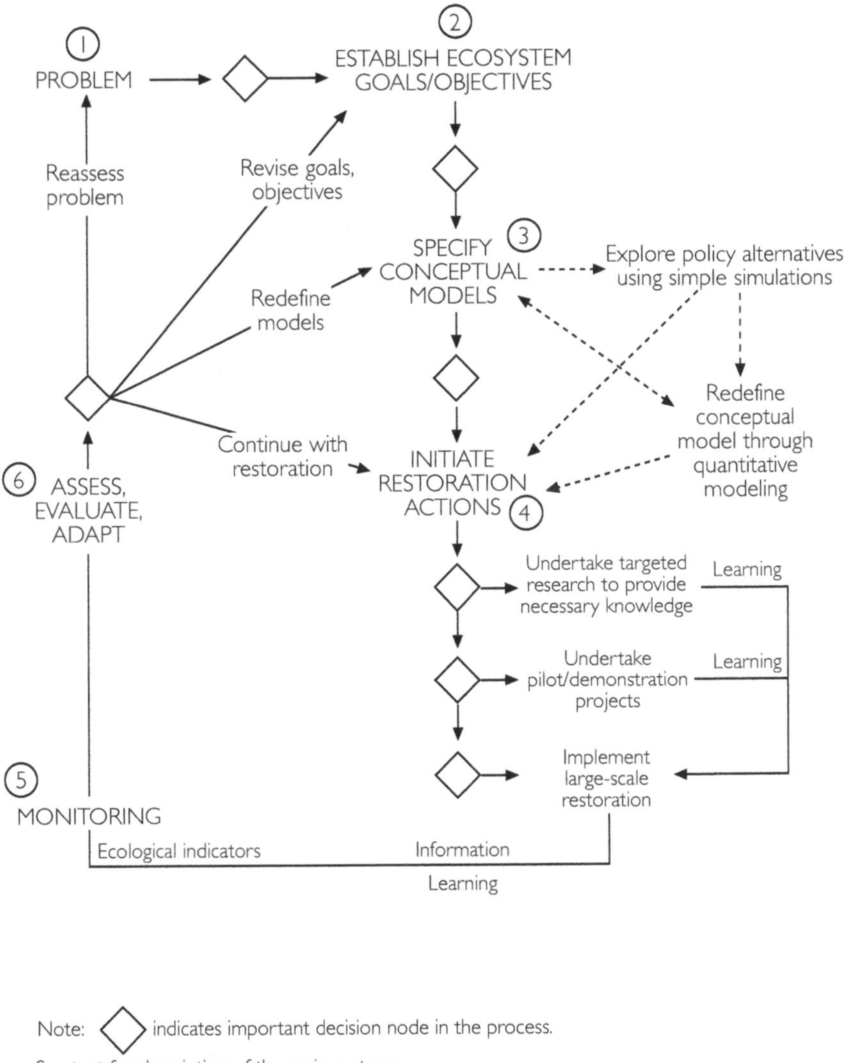

Figure 7.6 *Processes and feedback loops inherent to the process of adaptive management (Sacramento River Advisory Council, 2000): the figure is known informally as the 'Healey ladder' after its creator, M. Healey, University of British Columbia*

> *... an innovative technique that treats management programs as experiments. Rather than assuming that we understand the system that we are attempting to manage, adaptive management allows management to proceed in the face of uncertainty. Adaptive management uses each step of a management program as an information-gathering exercise whose results are then used to modify or design the next stage in the management program. In adaptive management, there is a direct feedback between science and management such that policy decisions can make use of the best available scientific information in all stages in its development. (Halbert and Lee, 1991: 138)*

This approach, using statistical and mathematical modelling to highlight uncertainties and optimize solutions under an experimental framework (Walters, 1986), marks a distinct break from river channel management based almost entirely on experience and intuition (Hemphill and Bramley, 1989) that has tended to result in conservative solutions and an unwarranted belief in design 'certainty'. Requirements of river channel management are inherently demanding of new techniques to achieve design-with-nature and thus an approach that can explicitly address uncertainty, using 'learning by doing' (Walters, 1986), a continuous process of planning, experimenting, monitoring and evaluating (Smith *et al.*, 1998), is highly appropriate. Adaptive management approaches to channel management should incorporate the application of all derived knowledge into designs so that cutting-edge, nonconservative management can be applied with minimum unnecessary risk (Downs and Kondolf, 2002). Two other important facets of adaptive management that break with twentieth-century engineering methods are, first, the importance of post-project appraisal in determining success, requiring careful setting of process-related 'success criteria' arising from the project objectives in advance of the project installation and, second, an integral requirement to adjust programmes and policies in light of experiences gained during the adaptive management programme (see Chapter 10).

7.6 Requirements emerging

Significant advances in the scientific understanding of river channel environments were achieved in the twentieth century, enabling evolution of suitably holistic approaches to management planning and engineering (p. 168). Management began to be driven by the realization that natural river channel forms, processes and adjustment, the history of human impacts and the legacy of previous management methods combine in such a way that every drainage basin presents a unique set of management challenges. These challenges need to be accommodated when setting management objectives, and resolved effectively in order to implement channel management approaches that achieve the sustained delivery of water resources, flood control and channel stability, and wildlife conservation.

Four central issues were distilled from the 'retrospect' and 'realization' of previous chapters (Table 7.1). First are the scientific advances related to multidisciplinary initiatives that will continue both to underpin advances in river channel management and to set the challenges for future research. The second issue relates to how new perceptions about rivers resulted in national and international legislation requiring environmental assessment as the basis for trying to understand, uphold and improve environmental value for river channel management. This translates as a need to 'know your river' through environmental assessment before attempting to manage it, and is reviewed in Chapter 8. The third issue followed from the fact that in long-settled regions of the world, subject to large-scale administration, river channels had long been managed as simple conduits for transferring water either towards or away from centres of population to support the extraction of wealth from rural areas for dispersal through cities (Petts *et al.*, 2000). However, consequent to legislation demanding environmentally sympathetic management approaches, a requirement evolved whereby river channels already subject to previous management operations needed to be re-managed using an explicitly historical approach: 'the age of river restoration had begun' (Petts *et al.*, 2000: 495). Aspects of river restoration are reviewed in Chapter 9 as part of a movement towards conservation-based management approaches. The fourth issue requires new, environmentally benign methods of management based on the design-with-nature approach developed by conservationists in ecology and landscape architecture. The new methods are derived from principles of ecological engineering, engineering geomorphology, environmental planning and adaptive management. However, for conservation-based management methods not to be cast as glorified civil engineering (Budiansky, 1995), these techniques need to be applied convincingly: Chapter 10 reviews this progress.

IV

Requirements

These three chapters focus on the present state of the art of river channel management, including environmental assessment and post-project appraisal; post-modern river management and river restoration; and environmentally aligned river engineering involving working with the river.

8 Environmental Assessment of River Channels

Synopsis

The need to 'know your river' prior to management (Chapter 7) resulted in a suite of environmental assessment techniques for river channels differentiated by their level of detail and purpose (8.1). The broadest inventories of the morphology and habitat condition of extensive lengths of channel are usually classified to indicate homogeneous river reaches that may require similar management activity (8.2). Analysis of the catchment historical context provides the framework for understanding the cause-and-effect basis of perceived management problems and for deriving appropriate objectives and strategies for management (8.3). Specific management responses should be planned cognisant of the potential for hazard associated with a particular reach; this can be estimated using a series of analyses accordant with the four main components of the channel's sensitivity to change (8.4). Future improvement in river channel management will depend upon evaluating the success of project implementation over the medium- to long-term perspective (8.5), and by linking the environmental assessment of river channels into the approaches and methods of post-modern river channel management (8.6).

8.1 The need to 'know your river'

Environmental assessment of river channels has become a core river channel management activity. In part, this stems from a realization that successful management requires a scientific appreciation of the potential for hazard posed by the particular channel (Chapters 4 and 5). Environmental assessment is also a response to competing water resources interests that now include legislated conservation concerns that demand an understanding of the 'state of the environment' and the potential for harm to conservation assets prior to management activity (Chapters 6 and 7). Each river presents a unique set of challenges for river channel management as a function of natural channel forms and processes conditioned by the catchment structure, the impact of past and present human activities on the channel including previous channel engineering, and the priorities of water resources management. History indicates that management 'solutions' imposed without consideration of conditions specific to the river channel provide only short-term operational benefits unless supplemented by regular maintenance and further capital works. They can also lead to environmental degradation including a loss in the functional integrity of the catchment hydrosystem (Chapters 2 and 3). Public desire for natural resources conservation now demands management with a concern for the probable future condition of a river channel reach, according to decisions based on its contemporary environmental state and within its catchment context and history. This is the basis for trying to ensure 'sustainability' in channel management operations. Every river must be assessed systematically in relation to its natural dynamics and degree of previous human impact: you must 'know your river' before attempting to manage it.

Environmental assessment (EA) can be defined generally as:

> *a process for identifying the likely consequences for the biogeophysical environment and for man's health and welfare of implementing particular activities and for conveying this information, at a stage where it can materially affect their decision, to those responsible for sanctioning the proposals. (Wathern, 1988: 6).*

While the growing use of EA in river channel management can be attributed to the various factors outlined above, the necessity for it has been ensured primarily by the increasing conservation component applied to water resources legislation at both national and international levels (pp. 181–4). In the USA, early test cases of EA under the National Environmental Policy Act of 1969 set the stage for ensuring that assessments were detailed, specific, did not pre-empt the conclusions and considered a full set of alternative project solutions (Brookes, 1988). In England and Wales, environmental assessment for 'significant' river and near-river projects was eventually formalized when 1985 European Union legislation became enacted at a national level in 1988 (Table 7.2). Shortly thereafter, the National Rivers Authority (now Environment Agency) began developing specific EA procedures for river channels and habitats (e.g. Brookes and Long, 1990; National Rivers Authority, 1992). Worldwide, there is now virtually no prospect that significant channel management activity could proceed without a comprehensive assessment of potential environmental effects. Indeed, a common reason for management delay often derives from arguments over the details associated with predicted potential effects. It therefore behoves the agency or individual responsible for management to undertake assessments that are perceived appropriate to the problem or issue.

There are generally acknowledged to be three levels of pre-project environmental assessment of river channels (Schumm, 1977a; Thorne *et al.*, 1998; Thorne, 2002) that include:

I inventories of river channel condition at the catchment scale ('state of the environment' surveys), designed to assess environmental values and identify management issues (Section 8.2);

II analyses that place a river reach in the context of its catchment history to understand the changing hydrological and geomorphological conditions that affect the current condition of the reach (8.3);

III sensitivity analyses related to channel dynamics and indicating the severity of channel hazards, the fragility of channel assets and the suitability of proposed management actions (8.4).

These assessments are nested. Level I surveys provide a network-wide assessment of geomorphological and habitat assets as the basis for identifying reaches of interest, usually based on 'reconnaissance' survey methods. These surveys inform planning decisions and are a logical investment for any hydrosystem with repeat calls on channel management. Level II focuses on a prioritized reach and encompasses a series of qualitative and quantitative assessments designed to understand cause and effect as the basis for determining the most appropriate management approaches. Level III assessments are more detailed site-specific analyses of the mechanics of change as the basis for applying appropriate management methods. Single projects generally require Level II and III assessments as the basis for sustainable management action. There is also a fourth and critical category of environmental assessment concerned with post-project appraisal:

IV monitoring and evaluation of project success at an appropriate timescale and in an adequate spatial framework to faithfully reflect channel response to the management action, as the basis for improving future actions (8.5).

Overall, EA represents the application of scientific methods to inform appropriate management approaches and methods (Chapters 9 and 10, respectively). A summary of the knowledge obtained and management purposes of each survey level is provided in Table 8.1.

8.2 Inventories of the channel environment

Catchment-wide, strategic assessments of river channel condition involve collecting baseline data on channel morphology and/or habitat to form a 'state of the environment' inventory as the basis for channel management planning. The information collected is likely to indicate (a) the linear extent of remaining 'high value' channel reaches within the network (i.e. channel assets); (b) the extent and type of direct channel modification by previous engineering and maintenance operations; and (c) the locations of channel instabilities (i.e. potential channel hazards) and/or recovery potential (Brookes and Long, 1990; Table 8.1). Such information can be used to assign management approaches and priorities such as the preservation of any remaining high value reaches, environmental enhancement and restoration of degraded reaches and, where necessary, remedial actions for previous management operations. An assigned 'protection value' also provides an indication to land planners and prospective developers of the likely strength of opposition or encouragement of proposed floodplain development actions (Brookes and Long, 1990). Identifying high value reaches is also one means of identifying 'reference reaches' (Hughes *et al.*, 1986; Kondolf and Downs, 1996) that can, in some instances, provide a valuable template on which to base designs for channel management in neighbouring areas or to set project environmental targets (p. 276).

Catchment morphological surveys

Catchment morphological surveys require an assessment limited in detail in order to cover many kilometres of survey within an acceptable time frame. Because geomorphological survey data is rarely collected routinely by management agencies, existing information is likely to be minimal so that the cost and time required for catchment-scale

Table 8.1 The purpose and application of four levels of river channel assessments

Assessment	**Channel knowledge obtained**	**Management purpose**
Level I Catchment baseline inventory	• Introductory channel understanding where no other information exists • Recognition and typing of erosion- and deposition-related channel hazards • Length and value of conservation-related channel assets, including habitats • Previous management actions	• Inventory of state-of-the-channel environment • Understanding likely compliance with statutory regulations • Determining necessity for further studies • Contribution to floodplain planning • Determining opportunities for environmental improvement/restoration, including setting environmental flows
Level II Catchment-historical analyses	• Cause and nature of channel instability and changes • Cause and nature of habitat losses • Conceptual model of hydrosystem operation before and after human disturbance	• Understanding whether proposed project will likely cause net environmental change, or the proposed environmental change (statutory compliance related) • Basis for setting management approach and objectives (success criteria) • Basis for setting management approach for sustainable actions • Basis for outlining conceptual project design
Level III Channel dynamics analyses	• Understanding of channel operation as (potentially): ratio of forces, proximity to thresholds, recovery potential, incremental sensitivity • Extent and risk of erosion and deposition	• Determining whether channel risk warrants remedial action • Determining prospect of management approach succeeding • Determining measurable success criteria • Determining specific

Assessment	Channel knowledge obtained	Management purpose
	• Susceptibility and fragility of channel conservation features • Likely impact of a proposed action • Likely stability and resilience of conservation measures	management methods • Basis for deriving detailed project design plans
Level IV Post-project appraisal	• Compliance of project with design intention • Performance of project relative to design intention • Sustainability of project relative to catchment hydrology and sediment transport	• Evaluating project success against success criteria • Evaluating project as a learning experience: the contribution to geomorphological, habitat and engineering knowledge • Evaluating unanticipated actions that require counterbalancing management • Evaluating efficacy of management method

data collection will be prohibitive (Downs and Thorne, 1996). Accordingly, catchment morphological surveys were developed (Kellerhals *et al.*, 1976) to facilitate rapid collection of applicable information using a reconnaissance survey in association with analysis of supporting maps and aerial photographs. Stream reconnaissance is derived from the technique of geomorphological mapping (see Cooke and Doornkamp, 1990) and is based on measuring key parameters of the channel morphology combined with the use of skilled, interpretative assessment of the channel conditions recorded in a systematic format (Thorne, 1998), often using a pro forma checklist to ensure standardization (Figure 8.1). The data inventory is likely to include a combination of quantitative measurements of channel geometry, qualitative observations of valley characteristics and riparian land uses, channel dynamics and evidence for previous management activity, and interpretation regarding channel recovery, estimated environmental value and recommendations for management (Downs and Brookes, 1994). Ideally, such surveys should indicate the balance of erosion forces and bed and bank resistance in the river channel, the relationship with upstream and downstream reaches, processes of channel change and a link to the character of the channel system as a whole (McEwan *et al.*, 1997). Information is usually collated

Outline

(1) Scope of survey and proposed management plan
- a. Details of survey
- b. Management proposal
- c. General comments

Region

(2) Valley characteristics
- a. Valley description
- b. Floodplain land uses
- c. Riparian/corridor land uses

Site

(3) Channel characteristics
- a. Position in floodplain
- b. Planform characteristics
- c. Bank characteristics
- d. Substrate characteristics

(4) Channel dynamics – as indicated by:
- a. Morphology
- b. Vegetation
- c. Structures

Catchment

(5) Catchment characteristics with a bearing on the outcome
- a. General information (input from desk study)
- b. Upstream
  - i. Channel character
  - ii. Potential influences (natural/land use/channel management)
  - iii. Evidence of previous/recent adjustment
- c. Downstream
  - i. Channel character
  - ii. Potential influences (natural/land use/channel management)
  - iii. Evidence of previous/recent adjustments
- d. Sediment budget

Evaluation

(6) Management interpretation
- a. Previous works
- b. Evidence of recovery
- c. Project appraisal
  - i. Conservation value of channel characteristics – reasons reference above
  - ii. Conservation value of channel dynamics – reasons reference above
  - iii. Summary conservation value (evaluation of potential for loss)
- d. Project recommendations
  - i. Strengths and grounds for opposition to scheme (if any)
  - ii. Minimum mitigation measures required
  - iii. Opportunities for local enhancement
  - iv. Potential for full restoration

Figure 8.1 *Skeletal framework for reconnaissance evaluations ahead of proposed river projects (Downs and Brookes, 1994)*

for channel reaches that are of different lengths possessing a homogeneous set of attributes and representing a distinct management challenge.

The utility of catchment morphological surveys can be evidenced by a proliferation of schemes developed around the world based on similar geomorphological principles but differing according to environmental factors critical to the region (e.g. Kellerhals *et al.*, 1976; Platts *et al.*, 1983; Water Victoria, 1990; Brookes and Long, 1990; Downs, 1992; Reid and McCammon, 1993; Rosgen, 1994; British Columbia Forest Service, 1995, 1996; National Rivers Authority, 1995; Brierley and Fryirs, 2000; Brierley *et al.*, 2002). One primary purpose of surveys is to provide suitable background material for more detailed, project-centred assessments (see 8.3 and 8.4). In addition, baseline surveys using reconnaissance approaches can provide a methodological basis for field studies of channel form and process, a format for collecting qualitative information and quantitative data on the fluvial system and a data supply for classifications and other analyses in support of sustainable river management (Thorne, 1998: 37). Quality assurance and consistency of interpretative observations is especially important and requires that the surveyor is trained in geomorphological principles, with a structured survey sheet used to minimize data omissions and subjectivity. In lowland rivers in temperate climates, reconnaissance surveys may be representative of channel conditions of the order of 10–100 years (Downs and Thorne, 1996), similar to the usual time frame for river management projects, whereas, in dryland rivers punctuated by large flood events, applicability may be limited to much shorter periods.

Baseline habitat surveys

Baseline habitat surveys provide additional information relating to the conservation aspects of channel management decisions. Such surveys explore the link between greater habitat heterogeneity and higher river biodiversity, avoiding some of the inherent problems of biological surveys (Harper and Everard, 1998) and the need for intensive laboratory analysis (Newson *et al.*, 1998b). Habitat surveys are used to indicate the functional 'health' of river system in terms of structure and connectivity (see Chapter 3) and as a tool in setting conservation targets through river restoration (p. 184–9).

Instream river habitats are a function of the interaction between channel morphology, channel bed and bank sediment and large woody debris, flow regime (including temperature) and hydraulics (primarily velocity–depth combinations), sediment transport characteristics and the presence and character of riparian and inchannel vegetation. Because of the nested link between channel processes, form, habitat and biota, habitat surveys can extend either 'top down' from morphological surveys or 'bottom up' by grouping data from biotic surveys. In the former category are approaches based on 'physical biotopes' (Rowntree and Wadeson, 1996; Newson *et al.*, 1998b) wherein physical features of the channel are combined with flow character to distinguish hydrogeomorphological units such as 'riffle', 'run', 'glide' and 'pool' (Table 3.3). Similarly, the River Styles geomorphological baseline survey (Brierley and Fryirs, 2000) has been extended to include 'hydraulic units' designated primarily by parameters representing flow hydraulics and substrate particle size classes, but also incorporating aquatic vegetation, organic matter, bank morphology and riparian vegetation (Thomson *et al.*, 2001; Figure 8.2). As a bottom-up approach, ecohydraulic analysis has been used to distinguish habitat building blocks or 'functional habitats' associated with distinct depth–velocity combinations. For example, functional habitats for lowland rivers include 'gravel', 'sand', 'filamentous algae', 'tree roots', etc., based initially on statistical analysis of invertebrate samples (Harper *et al.*, 1992, 1995; Kemp *et al.*, 1999; Table 8.2). Alternatively, the building block in the Instream Flow Incremental Methodology stems from expert understanding of fish habitat preferences defined by flow velocity, depth and bed material in its Physical HABitat SIMulation

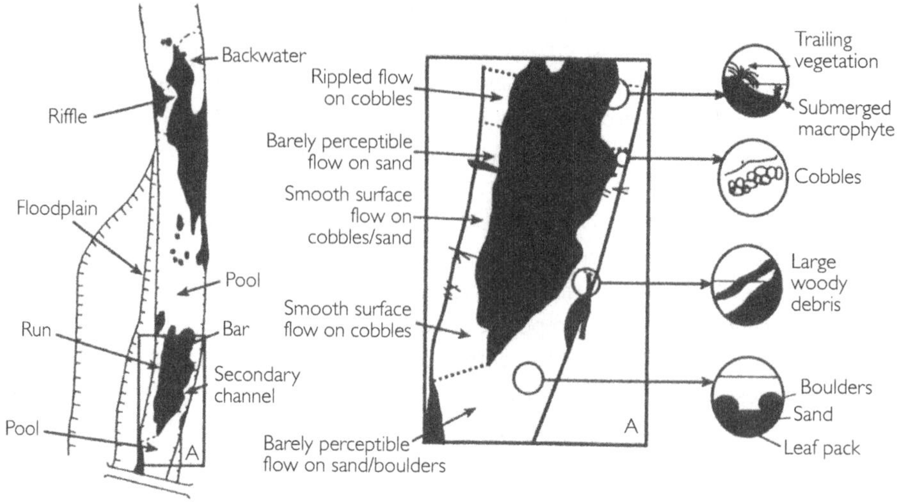

Figure 8.2 *Schematic representation of the hierarchical relationship between geomorphic units, hydraulic units and microhabitats based on the River styles procedure (Thomson et al., 2001)*

component (PHABSIM; Bovee, 1982, Figure 7.2) and one-dimensional hydraulic models to score the Weighted Usable Area of fish habitat. For practicality, habitat surveys usually involve reach-level rather than catchment-level survey.

Instream habitat assessments are now being integrated within more general assessments of the physical structure of river corridors that include assessment of channel modifications, riparian vegetation, floodplain land use and hydrological connectivity of river-floodplain environments. In Europe, investigations for a unified habitat assessment method are now underway following the EC Water Framework Directive requirement for 'hydromorphological' assessment of the ecological quality of rivers (European Community, 2000). Field tests of three such schemes, the UK 'River Habitat Survey' (Raven *et al.*, 1997), the French '*Système d'Evaluation de le Qualitié du Milieu Physique*' (Agence de l'Eau Rhin-Meuse, 1996) and the German '*Länderarbeitsgemeinschaft Wasser*' (LAWA, 2000) revealed difficulties in comparing between schemes owing to factors related to feature definition, survey strategy, assessment criteria and scale-dependency (Raven *et al.*, 2002). Scale-dependency may be the most challenging concern because habitat surveys are rarely linked into a hierarchical system of assessment (Newson and Newson, 2000), linked explicitly to physical processes or to the time required for habitat units to develop and be modified (an exception is the River Styles approach, e.g. Thomson *et al.*, 2001; Brierley *et al.*, 2002). These concerns reflect the fact that 'hydromorphological' surveys are still inventories of habitat at one point in time and are insufficient for sustainable river channel management for which more detailed environmental assessment is required.

Channel classification schemes

Baseline morphological and habitat surveys record large amounts of information that, for effective management applications, have to be condensed into a limited number of classes

Table 8.2 Functional habitats in lowland rivers

Functional habitat
Exposed rock and boulders
Cobbles and pebbles
Gravel
Sand
Silt
Marginal plants
Emergent plants
Floating-leaved plants
Submerged, broad-leaved plants
Submerged fine-leaved plants
Mosses
Filamentous algae
Leaf litter
Woody debris
Tree roots
Trailing (overhanging) vegetation

Source: From Newson *et al.* (1998b).

(or scores) with similar collected attributes and sub-divided into reaches displaying a similar overall character. As such, river channel classification schemes have become necessary for: recording existing river channel conditions and setting management priorities; assisting in defining the end state for river restoration schemes (i.e. to define the reference reach); and providing initial information about management measures that are likely to be successful (Kondolf, 1995b).

While classifications of natural river systems are generally hierarchical (Frissell *et al.*, 1986), channel management classifications have mostly addressed the meso-scale both spatially and temporally, classifying reaches of 10^1–10^3 m in length and having an applicability of 10^0–10^2 years for lowland geomorphic data (Downs, 1995b). The inherent difficulty is that the reach scale in river classification displays an interdependency of process and morphology requiring an understanding of both aspects for successful classification (Frissell *et al.*, 1986). For management purposes classifications must describe not only physical variability, but also inform decision making. As a result, their classes are partly a reflection of management concerns pertaining to a particular region, dictated by climate and land use. Some management classifications also stress *processes* of formation, rather than static morphology or habitat, as the basis for making interpretations about channel behaviour from its state. Where interpretation of channel processes is required, this adds a nonquantifiable aspect to the field survey that implies greater staff training and careful adherence to quality control if the scheme is not to lose its reproducibility (Kondolf, 1995b). One simplification in managed river systems is that channelization and bank protection schemes often relate directly to the value of the riparian land use, causing a distinct break in channel morphology and processes between land uses and making zoning of the channel easier than in natural fluvial systems where progressive changes are more common (Downs, 1995b).

Channel management classification schemes include those based on differences in:

- channel and floodplain morphology and/or habitat;
- contemporary channel adjustment processes;
- linked sequences of channel adjustment (from Downs, 1995b).

Morphology-based schemes usually have no specific temporal reference and are based either on classification, wherein the channel network is sub-divided and fitted to a defined scheme of geomorphic or habitat units (e.g. Brookes and Long, 1990; Rosgen, 1994), or characterization, wherein point samples of channel attributes are combined without *a priori* expectation to determine channel locations of similar character (e.g. Mosley, 1987). Characterization has the advantage of not requiring field interpretation of boundaries between different channel classes, and is more amenable to statistical analyses, but the lack of continuous coverage may cause critical locations of management importance to be missed. Classification of baseline morphology surveys can result in maps of homogeneous channel morphology/habitat reaches that are ideal for communicating planning-level information regarding channel conservation values and for determining locations for more detailed assessment (Figure 8.3). Classifications based on adjustment processes have been developed to indicate types of erosion, deposition and lateral migration found in engineered channels in comparison with natural channels (Brice, 1981; Brookes, 1987a; Downs, 1992; see Figure 8.4). They require that the surveyor has sufficient experience to make interpretative field judgements, including whether presumed adjustments are progressive, reflect a response to the last big flood event or are relicts of processes that are no longer operative. The advantage of the approach over the morphological inventory is that it encourages the experienced surveyor to reflect on cause-and-effect processes that underpin sustainable approaches to management, with classifications based on a location-for-time approach for predicting future channel adjustments. Some have arisen from channel geometry comparisons aimed directly at detecting predictable sequences of channel adjustment owing to knickpoint erosion (e.g. Schumm *et al.*, 1984; Simon, 1989) whereas others result from linking observed sequences of channel types (e.g. Nanson and Croke, 1992; Rosgen, 1996; Brierley and Fryirs, 2000). Their value lies in forecasting management challenges that will result from the upstream and downstream propagation of channel adjustment (p. 222).

Classifying baseline channel conditions is an effective and efficient means of communicating the results of environmental assessment to river managers, stakeholders and other specialists whose actions are conditioned by channel morphology and processes. However, a concern exists that management actions derived directly from a classification scheme without critically reviewing the contributing data or without further analysis can be flawed (Kondolf, 1995b), partly because the same channel morphology can result from different processes. This means that that same management response to a single channel type is unlikely to succeed in all cases. There are now several documented examples where channel restoration actions based primarily on an assumed reference channel type deter-

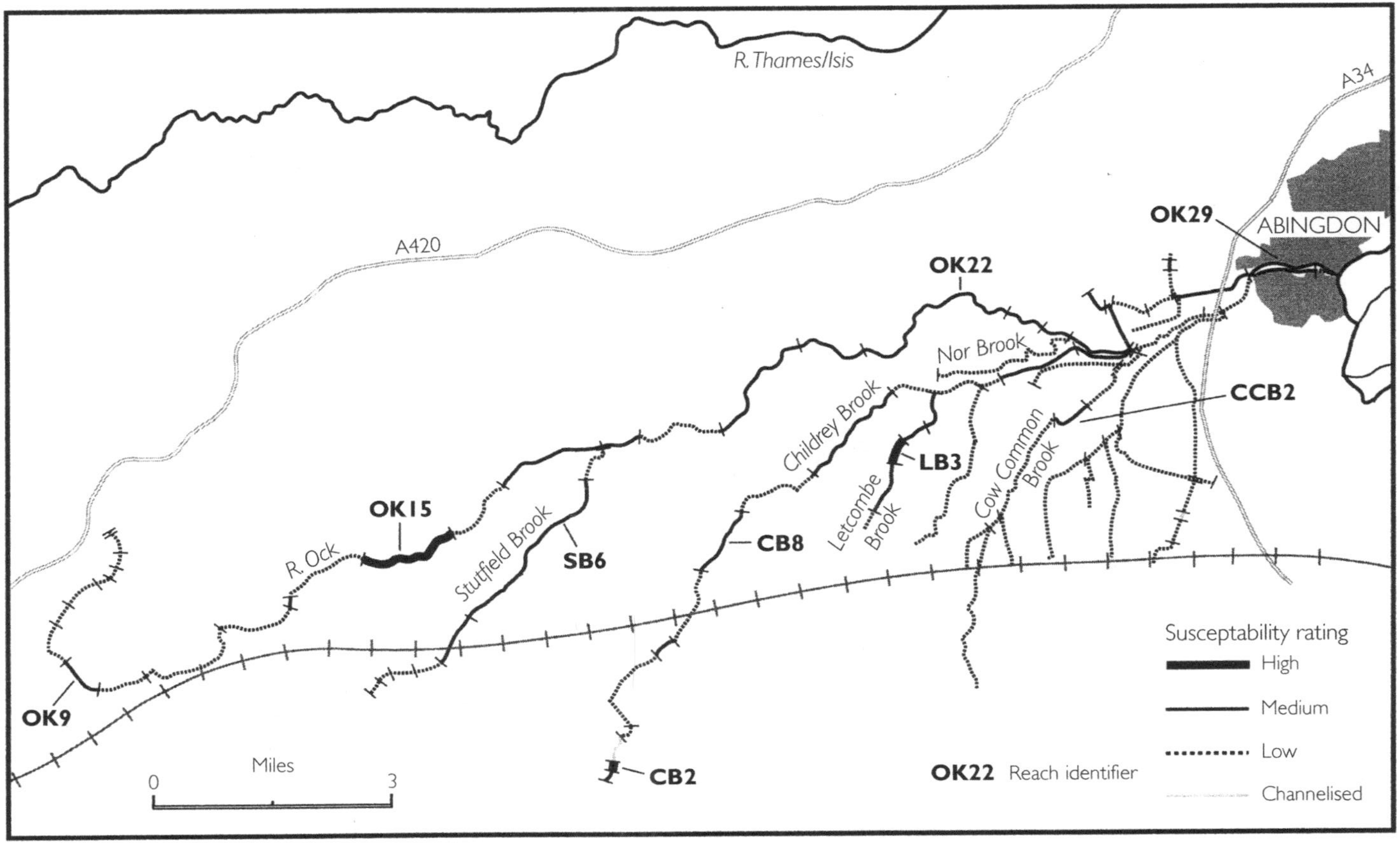

Figure 8.3 *Example output from a geomorphic baseline inventory of geomorphological conservation values along the River Ock and its tributaries, south-central England (based on the reconnaissance method of Brookes and Long, 1990; see also Downs, 1995b; Thorne et al., 1998)*

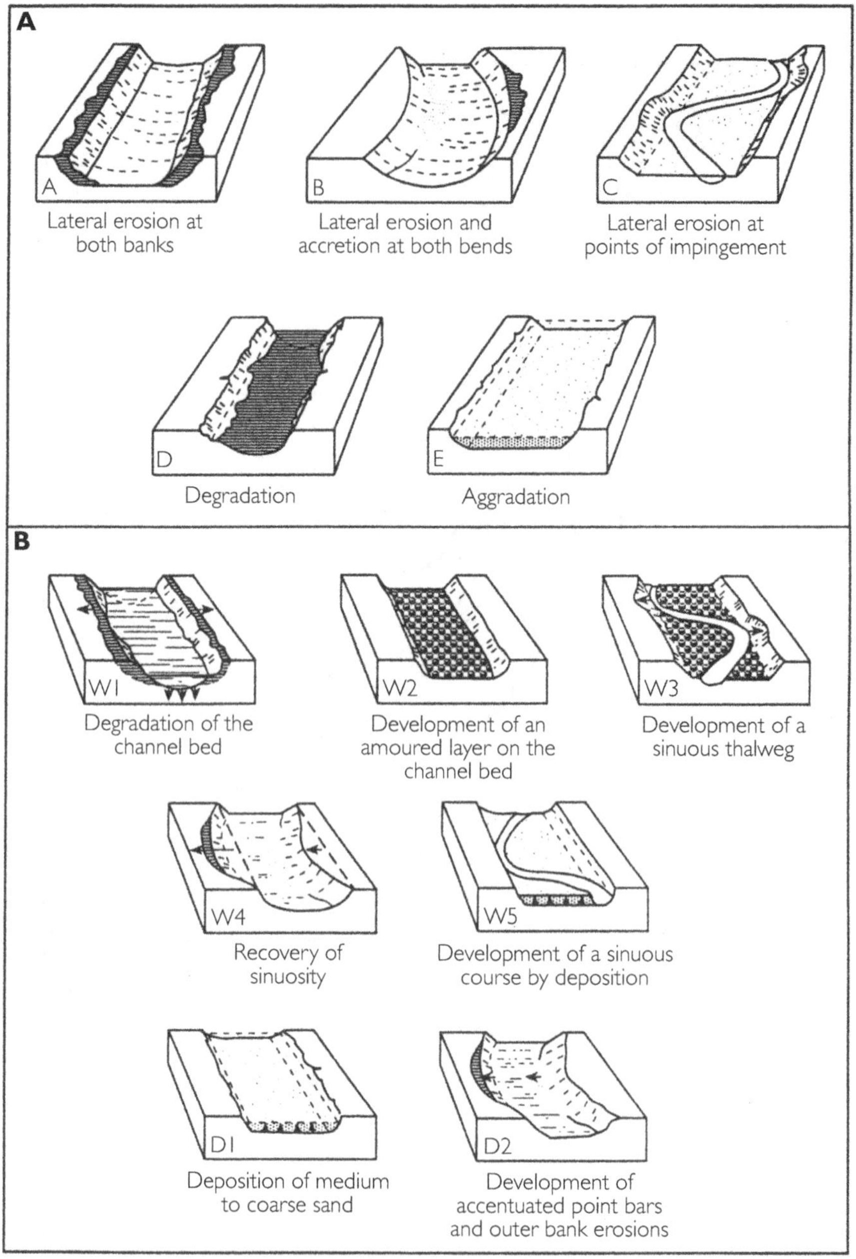
A
A
Lateral erosion at both banks
B
Lateral erosion and accretion at both bends
C
Lateral erosion at points of impingement
D
Degradation
E
Aggradation
B
W1
Degradation of the channel bed
W2
Development of an amoured layer on the channel bed
W3
Development of a sinuous thalweg
W4
Recovery of sinuosity
W5
Development of a sinuous course by deposition
D1
Deposition of medium to coarse sand
D2
Development of accentuated point bars and outer bank erosions

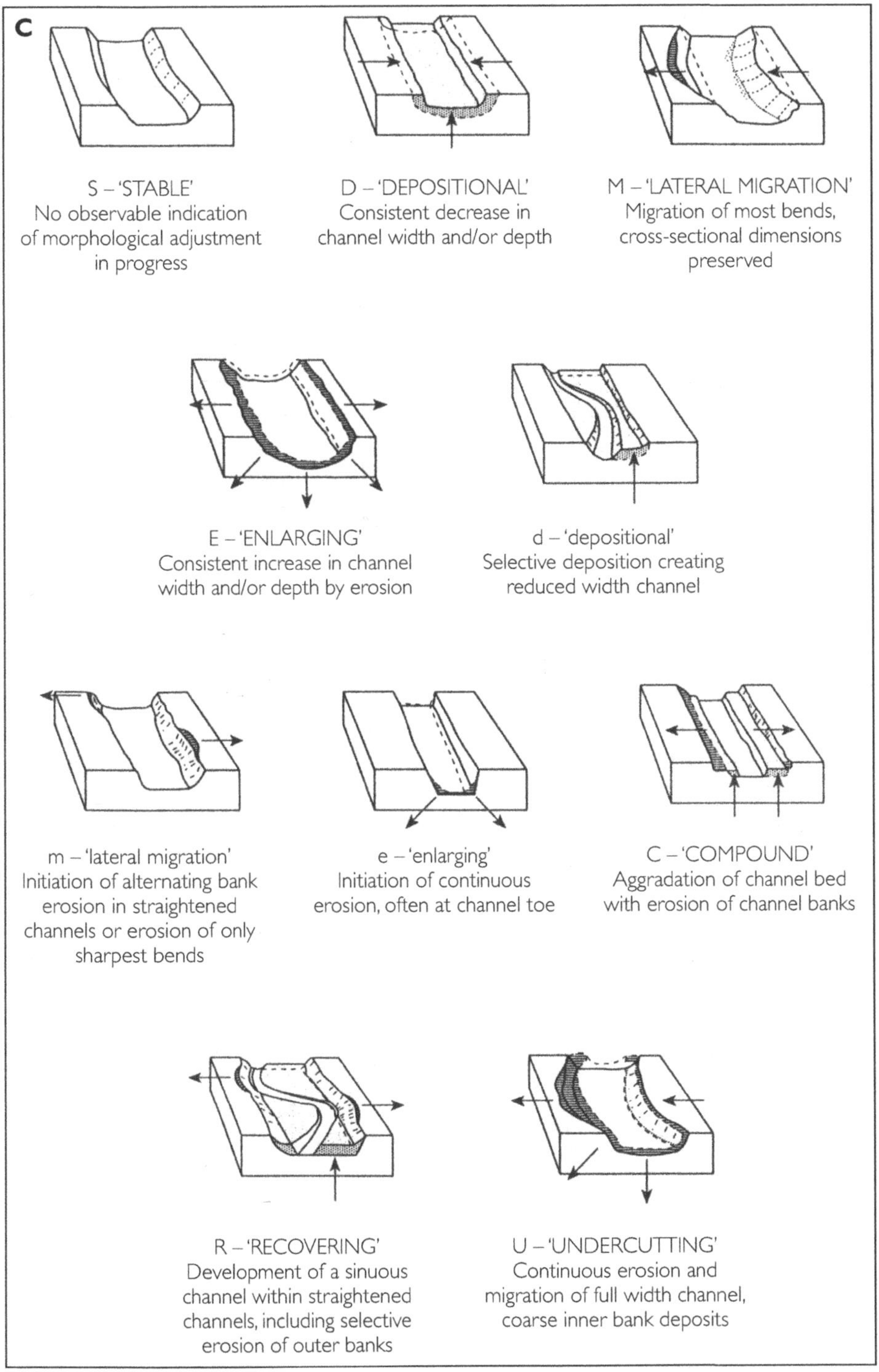
C
S – 'STABLE'
No observable indication of morphological adjustment in progress
D – 'DEPOSITIONAL'
Consistent decrease in channel width and/or depth
M – 'LATERAL MIGRATION'
Migration of most bends, cross-sectional dimensions preserved
E – 'ENLARGING'
Consistent increase in channel width and/or depth by erosion
d – 'depositional'
Selective deposition creating reduced width channel
m – 'lateral migration'
Initiation of alternating bank erosion in straightened channels or erosion of only sharpest bends
e – 'enlarging'
Initiation of continuous erosion, often at channel toe
C – 'COMPOUND'
Aggradation of channel bed with erosion of channel banks
R – 'RECOVERING'
Development of a sinuous channel within straightened channels, including selective erosion of outer banks
U – 'UNDERCUTTING'
Continuous erosion and migration of full width channel, coarse inner bank deposits

Figure 8.4 *Example classifications of river channel adjustment occurring in river channels subject to human disturbance processes and river engineering* ***A*** *Brice, 1981;* ***B*** *Brookes, 1987a;* ***C*** *Downs, 1992 (Downs, 1995b)*

mined by a channel inventory scheme failed in the first flood flow following implementation (Smith, 1997; Kondolf *et al.*, 2001; see pp. 313–4). Channel inventories of morphological condition and habitat are often necessary starting points for channel management planning but do provide *sufficient* information for designing a management solution. It is vital to supplement classification data with more detailed, project-centred channel analyses at restricted scales.

8.3 Analyzing the catchment historical context

The geomorphological sensitivity of river channels is complex (Chapter 5) and there may be significant time lags between changes in a driving variable and the channel's response. It may also involve 'system memory', whereby contemporary processes are directed by the long-term inheritance of floodplain morphology and sediments (see p. 146). 'Catchment historical' analysis is the means by which changing water and sediment regimes are associated with the environmental history of the catchment, allowing cause-and-effect processes to be isolated and providing a starting point for understanding channel sensitivity to change. The specific benefits of this understanding are three-fold (developed from Kondolf and Larson, 1995; see Table 8.1):

- understanding the nature of channel changes and habitat losses and linking them to management concerns: usually involving factors that have led to channel degradation such as land use changes, channelization or regulation;
- setting appropriate objectives for channel management and/or restoration: either with reference to the 'pre-disturbance' channel conditions or habitats, or according to an 'intermediate' disturbance channel condition that can be feasibly attained within contemporary flow and sediment transport regimes;
- setting appropriate strategies to achieve these objectives: choosing strategies that work within the framework provided by knowledge of historical fluvial processes and resilience of the channel system.

An associated outcome of catchment historical analysis is to guide the most appropriate analysis of geomorphological sensitivity at the reach scale as the basis for a channel management solution (Section 8.4).

Environmental assessment of the catchment historical context requires historical data sources to support a process-based understanding of channel adjustment. A variety of sources exist (Kondolf and Larson, 1995; Table 8.3). Many historical analyses extend back as far as the earliest reliable map sources (*c.* 150 years), unless reliable narrative accounts exist. The account is therefore only a partial record of disturbances except in those areas of the world subject to rapid European colonization in the mid-nineteenth century (e.g. western USA, Australia, New Zealand) where a period of 150 years may encompass the entire period of significant 'stresses' placed on the fluvial system. All historical data for reconstructing past channel environments should be used cognisant of its associated limitations (see Hooke and Kain, 1982; Petts, 1989; Downward, 1995; Kondolf and Larson, 1995; Gurnell *et al.*, 2003). Investigators should also be wary of considering evidence from the earliest reliable data source as representative of the pre-disturbance condition, especially in semi-arid environments or those subject to cyclical storm activity, such as El Niño:

Table 8.3 Potential components of a historical analysis of fluvial systems

Data source	Purpose
Flow records	Used to reconstruct a flood history of significant events, flood-frequency and flow-duration relationships for year-on-year flow indications and climate trends.
Repeat editions of large-scale topographical maps	Commonly used to reveal quantitative changes in channel width and rates of bank erosion, and to document changes in channel pattern and the approximate age of channel engineering works, bridges and other near-channel infrastructure.
Aerial photographs	Used for the same purpose as large-scale maps, but additionally allowing interpretative analysis of the 'textual' quality of the image related to features such as land use changes, riparian and inchannel vegetation, and channel sediments. Channel depth analysis may be possible.
Remote-sensing images	For similar purposes as aerial photographs but especially suited for discerning amounts of erosion and sedimentation in large rivers.
Ground photographs	Used for interpretative analysis of channel bank, cross-section and planform condition in time if the photographs can be referenced to a common point. Quantitative analyses through photogrammetry may be possible.
Repeat surveys of channel cross-sections	Allows quantitative and qualitative interpretations of channel changes, especially of channel depth, which may be difficult to achieve with maps and photographs.
Narrative accounts	Can provide indications of past channel environments, including vegetation communities, but must be seen within the prevailing cultural attitude towards the environment.
Floodplain stratigraphy	Uses floodplain and in-channel sedimentary evidence, probably in conjunction with techniques for dating organic or inorganic material, to reconstruct past geomorphological environments. For instance, absolute dates of floodplain stratigraphy may be indicated by carbon 14 dating of organic material while stratigraphic analysis of pollen records or heavy metal traces may be used if they can be correlated to a master sequence of ambient environmental conditions.
Vegetation composition and age	The expected progression in the composition of terrestrial vegetation on emergent bar and floodplain surfaces can be used to indicate the relative age of floodplain surfaces. Dendrochronology can be used to provide absolute dates of floodplain surfaces and rates of floodplain deposition.

Source: Based on Kondolf and Larsen (1995).

in these cases, channel form may simply reflect the last big storm, which may not represent 'typical' conditions.

A second major category of historical data involves using floodplain sediment analysis in conjunction with dating techniques (Smart and Francis, 1991) to determine the timing of phases of incision and aggradation in channel systems, and using the spatial characteristics of floodplain deposits to understand earlier channel conditions. These analyses can encompass longer time periods than other historical data sources (e.g. 10^2–10^8 years) but have not been used extensively in channel management assessments, partly owing to a perceived mismatch between the time periods they cover and the time frame of interest to channel management. There are also cost issues in sampling floodplains intensively enough to benefit site-specific management problems while being sufficiently extensive to indicate trends in the river channel's evolutionary development. However, important long-term dating techniques such as thermoluminescence dating (Page *et al.*, 1996) and radio carbon dating (Lowe, 1991) are now being supplemented by the improved precision and breadth in palaeoecology analysis using combined macrofossil, diatom and beetle indicators (Brown, 2002), and the prospect of yearly near-term date estimates obtained from dendrochronology (Nanson *et al.*, 1995) and isotopes of radioactive materials such as beryllium-7 (Blake *et al.*, 2002), caesium 237 (Walling, 1999; Owens *et al.*, 1999), and lead 210 (Byrne *et al.*, 2001). Evolving digital data technology and nonintrusive depth sounding techniques (e.g. sonar) may also reduce the future need for inherently destructive techniques such as floodplain trenching (Nanson *et al.*, 1995).The importance of such methods is demonstrated, for example, from palaeochannel evidence where recognition is mounting that lowland river channels in north west Europe were often multithread until approximately at least 2000 years ago with implications for understanding of channel change processes, the importance of large woody debris and the legacy of riverine and floodplain habitats (Brown, 2002). Likewise, the sedimentary architecture of palaeochannel fills can be interpreted to indicate the type of channel adjustment and provide evidence for the processes involved and type of historical channel adjustments (e.g. Collinson, 1986; Brown, 1995, 1996; Figure 8.5).

Merging numerous historical data sources across the catchment to indicate the 'story' of disturbance to a project reach in terms of changing water and sediment regimes requires significant conceptual and interpretative understanding of geomorphological systems. Essentially, the analysis is guided by four fundamental questions (Montgomery *et al.*, 1995), namely:

1 How does this landscape work?
2 What has its history been?
3 What is its current condition?
4 What are the possible/desirable future states for this landscape?

Some means is required of organizing the collected material as the basis for a cause-and-effect interpretation. Practical methods for assembling the information include watershed analysis (Montgomery *et al.*, 1995), cumulative impacts analysis (Reid and McCammon, 1993), fluvial audits (Sear *et al.*, 1995) and sediment budgets (e.g. Reid and Dunne, 1996). Sediment budgets often begin with an initial stratification of the watershed into dominant

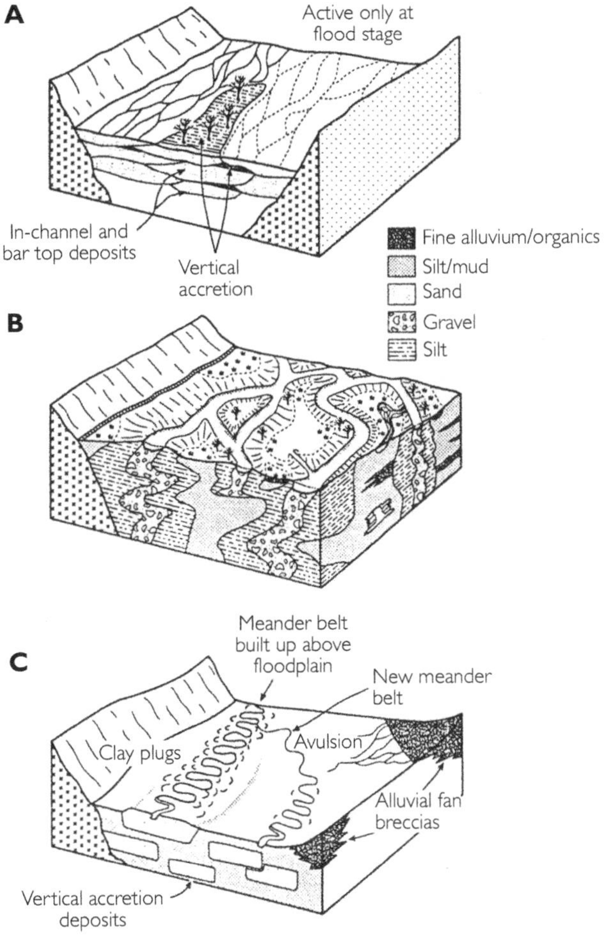

Figure 8.5 *Block diagrams illustrating differences in floodplain sedimentary architecture according to different river channel types:* ***A*** *low-sinuosity sandy braided floodplain;* ***B*** *intermediate sinuosity anastomosing floodplain;* ***C*** *sinuous avulsion-dominated floodplain (Brown, 1996)*

geomorphic process domains based on landscape-level structure and processes evident from analysis of geology, topography, channel pattern, slope and land cover. This may use a variety of historical sources combined with watershed models (Montgomery and Dietrich, 1994; Montgomery *et al.*, 1995, 1998). Quantifying the process of sediment production and downslope movement, storage, transfer and deposition (Dietrich *et al.*, 1982) is generally achieved through field estimates of sediment volumes (Reid and Dunne, 1996) combined with regionally derived process rates from experimentation. Alternatively, the 'fluvial audit' method integrates historical data with baseline survey data, using graphical displays to relate the prevailing sediment conditions and channel changes in the project reach to the generation and transport of sediment within the catchment (Sear *et al.*, 1995; Figure 8.6). The products include a summary time chart of catchment changes that may have impacted the fluvial geomorphology of the reach, a map showing the locations of catchment features of

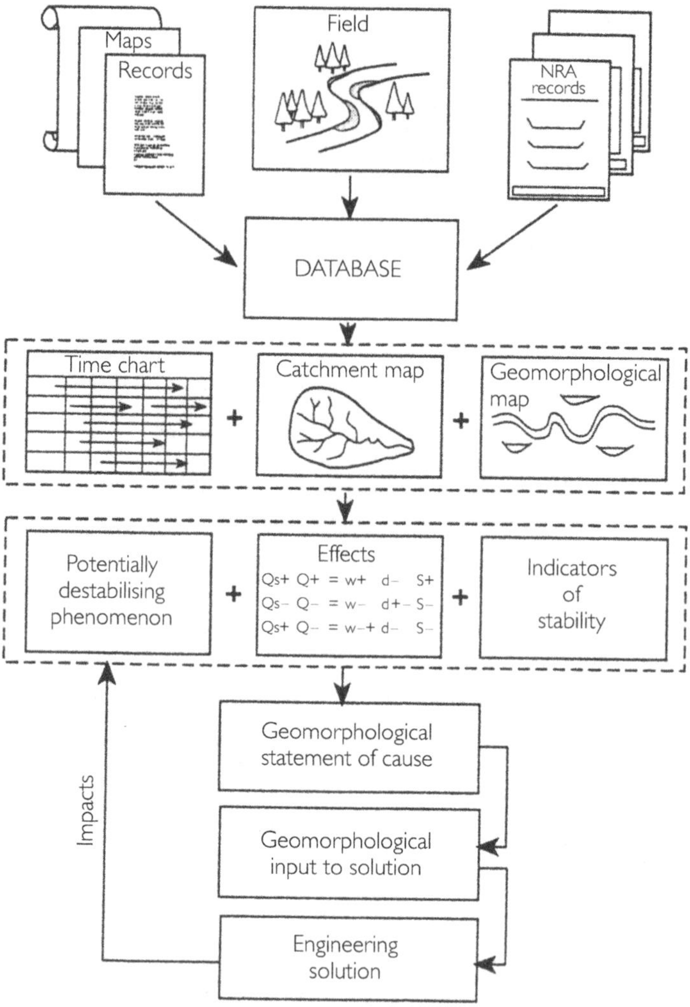

Figure 8.6 *An interpretative method for placing catchment geomorphological processes in their historical context for the purposes of river channel management: the 'fluvial audit' (Sear et al., 1995)*

apparent importance to the fluvial geomorphology of the reach and a detailed geomorphological map of the project reach and its immediate surroundings. These products are the basis for using expert geomorphological interpretation to derive a list of 'potentially destabilizing phenomena' that contribute to channel hazards of instability, erosion or deposition, and that may need to be tackled at source to derive a sustainable project solution. Whatever method is used, the fundamental challenge is to create an understanding that emphasizes process over the description of condition, change over stasis and, possibly, that extreme conditions may be more important than averages (Reid, 1998). This leads to the prospect that a thorough qualitative understanding of cause-and-effect relations may be of greater management use than partially quantified knowledge (Reid, 1998). Other desirable attributes of watershed analysis are outlined in Table 8.4.

Table 8.4 Desirable properties of a generally applicable watershed analysis

Property
Fits the particular needs of the agency or organization instituting it
Evaluates any potentially important impacts
Evaluates impacts at any point downstream
Evaluates impacts accumulating through both time and space
Evaluates the influence of any expected kind of land use activity
Evaluates any lands within the analysis area
Uses the best available analysis methods for each aspect of the analysis
Incorporates new information as understanding grows
Can be done for a reasonable cost over a reasonable length of time
Produces a readable and useable product. Is credible and widely accepted

Source: Reid (1998).

8.4 Estimating the sensitivity of channel hazards and assets

The third level of environmental assessment incorporates specific, reach-level, analyses of the dynamics of channel processes and follows an understanding of the catchment baseline and historical context. Management applications include gauging the risk associated with channel-related hazards and the susceptibility of channel-related assets (p. 120) and the potential stability and resilience of environmental restoration efforts (see Table 8.1). Such 'geomorphological dynamics assessments' (Thorne, 2002) may use a variety of analyses, some of which may still be under research development, and so the list of applicable analyses will change in time and include analysis specific to a particular geomorphological type of channel environment. Stemming from the notion of channel hazards being dimensions of the channel's sensitivity to change (p. 124), four categories of assessment are outlined below related to contemporary adjustment potential, proximity to adjustment thresholds, recovery potential and predicting incremental adjustment.

Contemporary adjustment potential: hazard locations

The starting point for detailed hazard assessment of river channels is to determine whether appreciable channel adjustment is likely under contemporary conditions, and where it will occur. In terms of geomorphological sensitivity, this centres on the ratio of disturbing to resisting forces. One method, adopting an historical approach, determines the likelihood of channel planform change as an extension of past changes. Sequential changes in the planform of the Lower Rillito Creek, Arizona, were grid mapped using aerial photographs spanning a 107-year period (Graf, 1984) (and 61 years on the Salt River, Arizona, Graf, 2000) indicating the proportion of time that the channel had occupied each cell in the grid. The resultant map longitudinally differentiated channel reaches that are more or less likely to migrate and indicated the future river corridor width in probability terms. The resultant map indicates lateral channel hazard potential and could be used to discourage floodplain development potential within the highly at-risk corridor.

Channel adjustment potential can also be gauged according to the ratio of disturbing to resisting stream power, that is, the ability of the river to entrain and transport sediment versus the resistance of the channel boundary to erosion (Bull, 1979, 1988). Stream power (p. 75) is a regional estimate of the channel's ability to do work based on the product of bankfull discharge and channel slope. In practice, there is no consistent, quantitative regional method for estimating the 'resisting power' of stream banks (Nanson and Croke, 1992) and so one application classified the likelihood of river channels in the UK to change based solely on their 'disturbing' stream power (Ferguson, 1981). The success of such regional analysis will depend on the trend in disturbing stream power being of appreciably greater magnitude than the environmental differences between the analysed rivers (i.e. resisting forces).

A further alternative is based on analysis of the channel dimensions. Assuming a log-linear trend in downstream hydraulic geometry (see Knighton, 1998), the impact of a disturbance such as river regulation or urbanization can be assessed by sampling downstream bankfull channel capacity dimensions and comparing the trend projected from the undisturbed areas to the actual channel dimensions measured in the disturbed areas. Deviations from the expected relationship indicate the type and severity of the impact, that is, the extent to which the channel has widened or deepened beyond its expected dimensions (e.g. Park, 1977a; Petts, 1977). However, the approach does not allow for downstream changes in the dominant processes of channel adjustment (Ebisemiju, 1991; Lawler, 1992) and, in searching for linear trends, locations of critical management importance may be overlooked (Gregory *et al.*, 1992). A more process-based analytical approach to hydraulic geometry assessment involves comparing calculations of flow-frequency, stage-discharge and sediment rating with discharge, cross-sectional dimensions, channel gradient, bed-material size and bankfull flow capacity (Thorne *et al.*, 1996a). The data are used to indicate the extent to which the channel has been altered from its natural condition by comparing the magnitude of the calculated dominant discharge with the channel's measured bankfull capacity. The approach assumes an alluvial river channel in which, under natural conditions, the dominant discharge and the bankfull discharge are similar (e.g. Andrews, 1980). Therefore, such analysis may not be appropriate in headwater streams (Newson, 1980), semi-arid or arid channels (Wolman and Gerson, 1978) or in areas dominated by alternating regimes of floods and drought (Nanson, 1986; Warner, 1987). In appropriate locations, the difference between the estimated dominant discharge and bankfull capacity should indicate the contemporary significance of human impacts such as channelization and, by implication, expected recovery processes such as sedimentation or erosion (p. 221). For the River Blackwater in lowland England it was concluded that channel maintenance operations have resulted in a bankfull channel capacity that is now four times that required to convey the dominant discharge but that low stream power and sediment availability restrict the channel's ability to recover by deposition processes (Thorne *et al.*, 1996a; Table 8.5).

Assessing the potential for river bank erosion is important for channel management because while erosion can represent an environmental asset associated with channel migration, it may also be deemed a hazard if agricultural land and buildings and structures close to the river channel are threatened. One prospect is to undertake a detailed recon-

Table 8.5 Quantitative analysis of bankfull channel dimensions for the River Blackwater, southern England, illustrating the impact of flood control measures on channel parameters including the bankfull discharge

Symbol	Explanation	Measured value from survey	Regime equation (Hey and Thorne, 1986)	Calculated value from regime equation	Percentage difference
w	width	12 m			
	vegetation I		$4.33\ Q_d^{0.5}$		
	vegetation II		$3.33\ Q_d^{0.5}$	6.36 m	+47
	vegetation III		$2.73\ Q_d^{0.5}$		
	vegetation IV		$2.34\ Q_d^{0.5}$		
d	Mean depth	1 m	$0.22\ Q_d^{0.37}\ D_{50}^{-0.11}$	0.58 m	+42
d_{max}	Maximum depth	1.4 m	$0.20\ Q_d^{0.36}\ D_{50}^{-0.56}\ D_{84}^{0.35}$	0.80 m	+43
R_w	Riffle width	11.5 m	$1.034\ w$	6.57 m	+43
R_d	Riffle mean depth	0.9 m	$0.9151\ d$	0.53 m	+41
$R_{d\ max}$	Riffle maximum depth	1.0 m	$0.912\ d_{max}$	0.73 m	+27
F	Form ratio	12	$w\ /\ d$	10.9	+9
A	Cross-section	12 m^2	wd	3.69 m^2	+70
L	Meander wavelength	90 m	$12.34\ w$	79 m	+12
Z	Meander arc length	50 m	$6.31\ w$	40 m	+20
p	Sinuosity	1.12	$3.5\ F^{-0.27}$	1.84	–64
S	Channel slope	0.001	$0.087\ Q_b^{-0.43}\ D_{50}^{-0.04}\ D_{84}^{0.84}\ Q_s^{0.10}$	Not known	Not known
ω	Stream power	13 W m^{-2}	$9810\ Q_d\ S\ /\ w$	5.6 W m^{-2}	+57
V	Velocity	0.3 m s^{-1}	$Q_d\ /\ A$	1 m s^{-1}	–233
R_v	Riffle velocity	0.35 m s^{-1}	$1.033\ V$	1.03 m s^{-1}	–194
Q_b–Q_d	Bankfull – dominant flow difference	16.4 $m^3\ s^{-1}$	$(Q_{bankfull} - Q_d)$	3.65 $m^3\ s^{-1}$	+350

Notes: Q, discharge; D, diameter of bed material size of specified percentile.
Source: Thorne *et al.* (1996a: 480).

naissance survey of bank conditions, including the bank material, its vegetation, erosion characteristics, geotechnical failure mechanisms and toe sediment accumulation (Thorne, 1998). The compiled material can be used to indicate whether erosion is on-going, whether it is aided or constrained by the bank vegetation and whether it is driven by fluvial processes or by mass-failure mechanisms. The assessment has been developed into an approach for 'appropriate bank management' (Thorne *et al.*, 1996b) according to the rate and type of erosion mechanism (p. 306). An alternative approach assesses bank erosion potential near bridges using a weighted index of diagnostic variables including the character of the channel bed and banks, the stage in the channel 'evolution' model (Simon, 1989), vegetation characteristics, the constriction to flow presented by the bridge and the angle of flow approaching the bridge. Each observation is allocated a score and the higher the overall score, the more probable it is that the channel instability is 'critical' with regard to the bridge structure. Field development of the technique on 1100 sites in west Tennessee,

USA suggested that a score of over 20.0 should be considered critical (Simon *et al.*, 1989; Simon and Downs, 1995).

Proximity to adjustment thresholds: hazard likelihood and type

Beyond analysing locations that are sensitive to channel adjustment, it is beneficial to know how likely the adjustment is, and of what type. This is achieved by assessing the channel's proximity to different thresholds in channel behaviour. Where channels are close to a threshold condition, the threshold is more likely to be crossed during a storm event or because of human modification of the channel hydrosystem. Activity thresholds are often related to human disturbance and may need to be disentangled from natural changes in flow, gradient and sediment character that cause threshold-based downstream transitions in channel morphology and habitat types (Church, 2002: 72). For restored rivers in the lowlands of England and Denmark, two thresholds in specific stream power (stream power per unit bed area) were identified (Brookes, 1990) between which restoration schemes were determined to be 'stable' in that they did not suffer excessive erosion or deposition. Above approximately 50 W m^{-2}, instream habitat devices or recreated morphological features often failed through erosion, whereas below 15 W m^{-2} schemes failed through deposition of sediment that smothered instream features including recreated pools and riffles. Recognition of these thresholds followed earlier work that identified a threshold of 35 W m^{-2} over which channelization schemes were often unstable (Brookes, 1987a, b; Figure 5.7). For locations in which these thresholds apply, the values provide bounding thresholds to guide management solutions. Extending this technique for distinguishing floodplain character, case studies were used to distinguish between high energy noncohesive floodplains in confined headwater areas (>300 W m^{-2} bankfull stream power), medium energy noncohesive floodplains in relatively unconfined valleys (10–300 W m^{-2}) and low energy cohesive floodplains in laterally stable low gradient environments (<10 W m^{-2}) (Nanson and Croke, 1992). Within the tripartite division were distinguished secondary floodplain types based on differences in stream power and geomorphic process factors (Figure 3.6). The data were extended to identify potential transformations between floodplain classes in the long-term, providing a perspective for guiding management applications to be truly sustainable in the context of evolving sedimentary conditions (Brown, 1996).

Many geomorphic assessments of proximity to threshold have used a discriminant function to distinguish between different channel patterns, particularly the meandering–braided pattern threshold. Using these discriminators during management planning can indicate whether the intended management action is likely to result in changes in channel pattern. Pattern is generally discriminated using an empirical relationship between channel slope (S) and bankfull discharge (Q_b) or mean annual flood (Q_m) in the form $S=aQ_b$, perhaps using an additional characteristic such as the median grain size (D_{50}). One determination of the meandering–braided transition in sand-bedded channels was $0.0041Q_m^{-0.25} < S < 0.007Q_m^{-0.25}$ (Lane, 1957) whereas in gravel bed rivers the threshold was determined at $S = 0.042Q_b^{-0.49}D_{50}^{0.09}$ (Ferguson, 1984). The identified thresholds have potential as guidance for channel management designs (e.g. to design the right *type* of channel) but the empirical nature of the equations will limit their general applicability (see

Thorne, 1997: 206–10). Future discriminators may be based more fully on the energy available for adjustment and thus achieve a greater range of applicability. For instance, in 228 rivers with a bankfull discharge greater than 10 $m^3 s^{-1}$, specific stream power (ω_{vt}) based on valley slope (S_v) and bankfull discharge was used to discriminate braided river channels in sand- and gravel-bedded channels from meandering channels with a sinuosity greater than 1.3, as $\omega_{vt} = 900D_{50}^{0.42}$ (van den Berg, 1995), where $\omega_v = 2.1S_vQ_b^{0.5}$ in sand-bedded rivers and $\omega_v = 3.3S_vQ_b^{0.5}$ in gravel-bedded rivers to accommodate for cross-sectional shape differences.

One problem in discriminant analysis is that channel stability or pattern is distinguished using just one function, involving relatively few variables, despite the complexity inherent to fluvial systems. An alternative is to use analysis such as logistic regression to indicate discriminant function as a probability of occurrence to accommodate for the variability inherent to the input data. Such analysis has been used to predict a probabilistic threshold between single and multithread channels (Miller, 1988) the probable occurrence of channel scouring (Tung, 1985), of tributaries being significant sediment sources (Rice, 1998) and of various types of channel adjustment occurring (Downs, 1994, 1995a). The latter discriminated numerous channel reaches in the Thames catchment, England, according to factors including channel bed materials, gradient, previous engineering operations and land use. Reaches showing signs of enlargement had strong positive relationships with high gradients and with previous straightening and, logically, were inversely related to confinement by bank protection. Channels exhibiting characteristics of lateral migration were negatively correlated with all engineering operations and favoured only by sand and gravel bed material whereas static channels were, not surprisingly, positively correlated with engineering operations, especially bank protection and flow regulation. Deposition in channels occurred in low gradient, fine-grained reaches and was not found in established urban areas. Discriminant analysis on the effects of urbanization on channel stability (Doyle *et al.*, 2000a) elaborates this finding, indicating that runoff changes following urbanization decrease the recurrence interval between flows competent to mobilize the median grain size on the channel bed, and increase the ratio of shear stress to the critical shear stress for sediment transport.

Recovery potential: hazard persistence

Gauging channel recovery potential involves understanding the temporal dimension of channel adjustment to estimate the persistence and probable end-point of the hazardous condition, for example, the changing nature of an erosion or deposition hazard through time. This understanding is central in ensuring that the most appropriate management solution is applied at the right time, avoiding costly errors associated with implementing unnecessary or ineffective measures. As so many channels have been subject to disturbance by human activities, this task often becomes one of gauging the ability and means by which the channel will recover from a previous disturbance such as channelization, which can be a one-time, 'pulse' disturbance (Underwood, 1994). Recovery does not imply that the channel will necessarily return to the pre-disturbance condition. Recovery may not occur at all if the channel has insufficient stream power and sediment supply to react to the imposed changes, or if operational maintenance by dredging periodically returns the channel to its modified condition or if the channel perimeter is armoured by bank protection that can resist erosion during high flow events. Where recovery does occur, it involves both one-dimensional and two-dimensional geomorphological feedback processes that are understood better conceptually than they are mechanically.

One option is to extrapolate forward monitored channel adjustments. In Big Pine Creek, Indiana, Barnard and Melhorn (1982) used sequential aerial photographs to plot the return of channel sinuosity in time since the channel was straightened in 1932 (p = 1.0). They projected a log-linear rate of recovery that indicated that the original sinuosity (p = 1.42) would return by 2080 (Figure 5.8B). Alternatively, proximity to thresholds analysis using discriminant functions, such as the sinuosity–slope product or stream power, can be used to determine whether a channel seems likely to retain its imposed planform (e.g. where a meandering channel has been straightened or where a braided channel has been made single-thread) or whether some form of recovery seems probable.

Channel recovery rates are rarely uniform over time. They have often been conceptualized as a log-linear decay curve that possesses a reaction time (e.g. time until the first effective flood) and a relaxation time (time to return to former conditions or to achieve new conditions) (Figure 4.8). An example for eroding gullies in the Colorado Piedmont determined rates of change using space-for-time substitution, dating the ages of ponderosa pine trees upstream to estimate the minimum age of establishment of the gully floor surface (Graf, 1977). The headward extension of the gully network following disturbance, probably caused by land use changes in the early twentieth century, was suggested to have a relaxation half-life of 17 years to reach the new steady state gully length (i.e. three-quarters of the length will be achieved in 34 years, etc.). Space-for-time substitution is also used in versions of the 'Channel Evolution Model' that documents channel recovery from straightening (Schumm *et al.*, 1984; Harvey and Watson, 1986; Simon, 1989; Figure 5.9) using the channel width–depth ratio (w:d). In Oaklimiter Creek, Colorado (Harvey and Watson, 1986), the original channel (w:d ≈ 4.0–7.0) was straightened, causing an increase in channel gradient that triggered erosion. First the channel deepened as the bed incised (w:d ≈ 3.0–4.0), then the channel widened as the oversteepened banks suffered mass failure (w:d ≈ 5.0). Subsequently deposition occurred as material eroded from upstream came to rest in the eroded reach (w:d ≈ 6.0). Eventually, equilibrium conditions re-formed in the incised channel, but the 'bankfull' channel is now inset within the eroded section (w:d > 8.0) so that the upper bank surface becomes a terrace. Conceptually, this is an example of complex response (Schumm, 1973) where a single impact causes multiple channel responses over time at particular locations as a consequence of the propagation of the original disturbance upstream and downstream. In long profile, the changes conform to a log-linear decay curve of migrating degradation upstream of the 'area of maximum disturbance' (AMD, the straightened reach), a reach of aggradation downstream of the AMD and a later wave of secondary aggradation in the AMD reach as material from upstream erosion deposits in the reach (Simon, 1989). Knowing the temporal and spatial sequence of these changes can be used to ensure that grade control measures are sited effectively (pp. 303–8).

Processes of channel recovery following land use changes, wherein channels respond to changing water and sediment regimes, vegetation cover and composition, are less well understood. In the Bega catchment, New South Wales, Australia, different recovery pathways are being plotted using space-for-time linkages according to channel types under the river styles method (Fryirs and Brierley, 2000). 'Cut-and-Fill', 'Transfer' and 'Floodplain Accumulation' river styles are seen as being either largely consistent with the pre-disturbance condition, in an unstable state far removed from the pre-disturbance condition, at a transitional stage showing signs of initial recovery, recovering in a resilient mode back towards a near-to-pre-disturbance condition or recovering but towards a newly created condition (Figure 8.7). The essential control on the recovery processes is the spatial relationship between sediment supply, storage and transfer. Cut-and-fill reaches, found at the base of the escarpment in the upper catchment, for example, were formerly sediment storage locations but became significant sediment sources following entrenchment, with the potential for recovery restricted by the limited sediment supply from upstream.

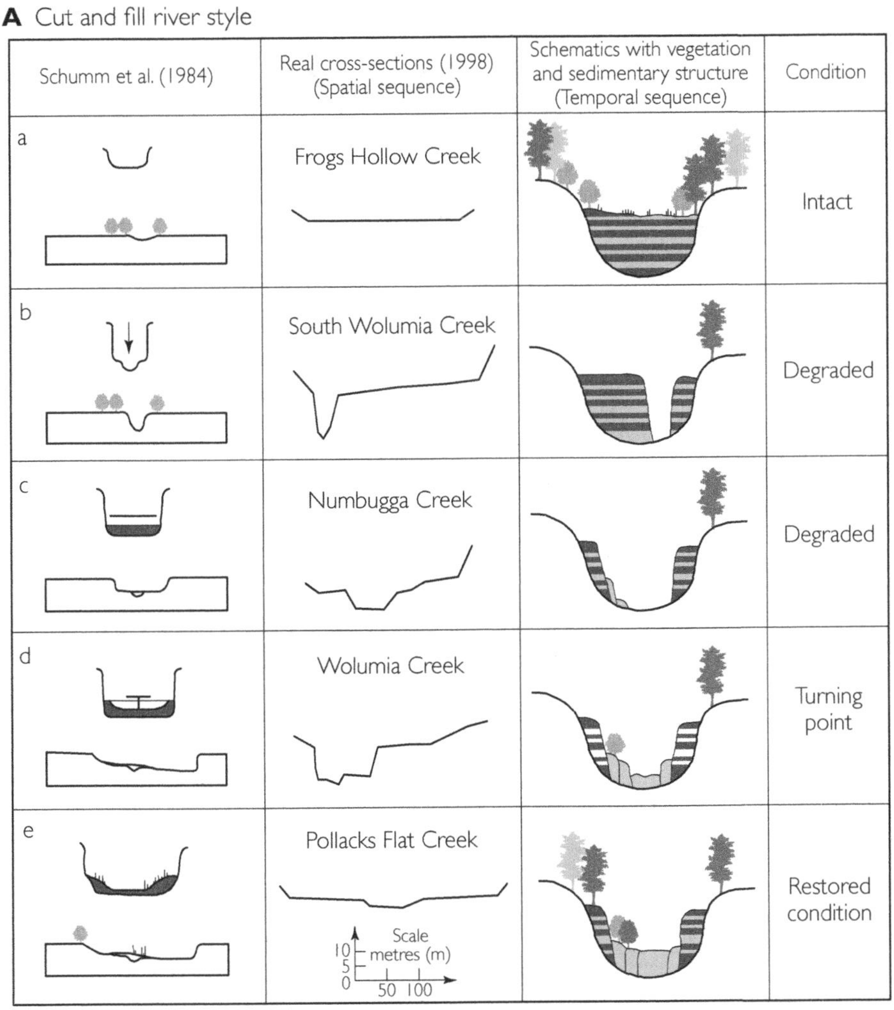

Figure 8.7 *Schematic representation of the evolutionary stages in river adjustment following land use changes for different 'river styles'* (continued)

B Transfer river style

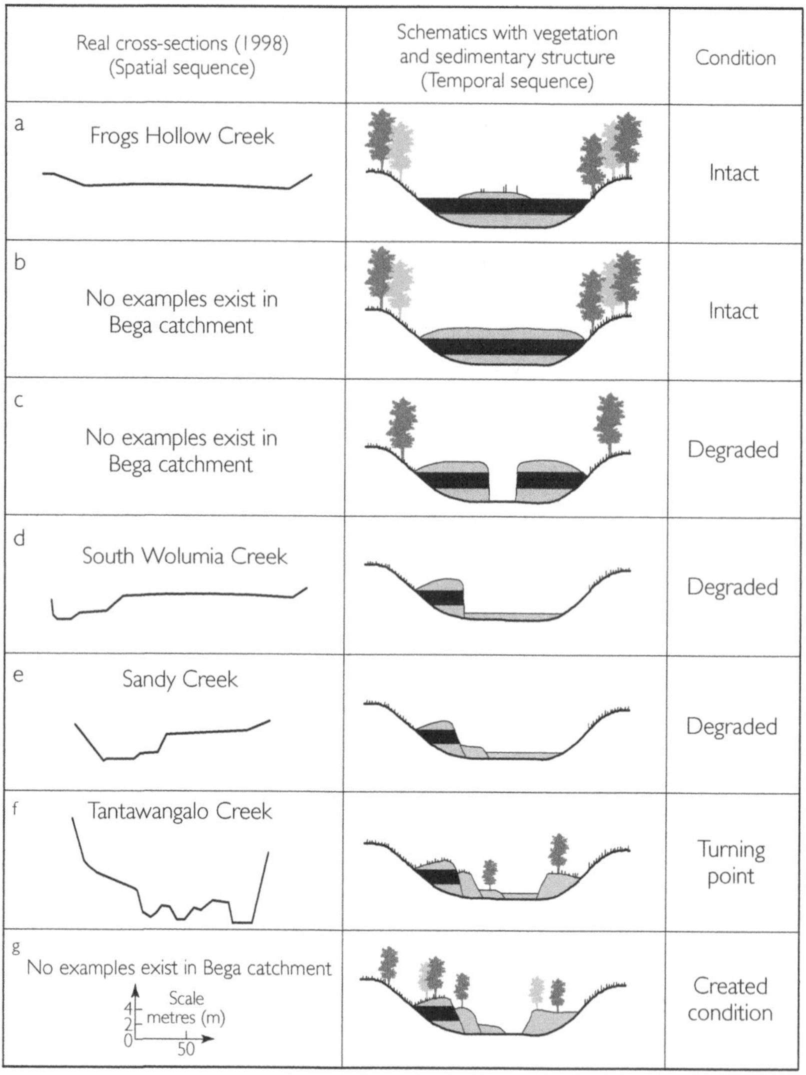

Figure 8.7 (continued) *Schematic representation of the evolutionary stages in river adjustment following land use changes for different 'river styles'* (continued)

C Floodplain accumulation river style

Real cross-sections (1998) (Spatial sequence)	Schematics with vegetation and sedimentary structure (Temporal sequence)	Condition
a No examples exist in Bega catchment		Intact
b Lower Bega River d/s of Bega Bridge 		Degraded
c Lower Bega River at Groses Creek 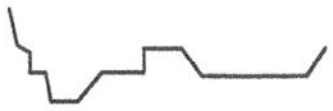	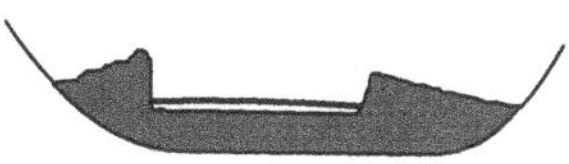	Degraded
d Lower Bega River at Grevilles 	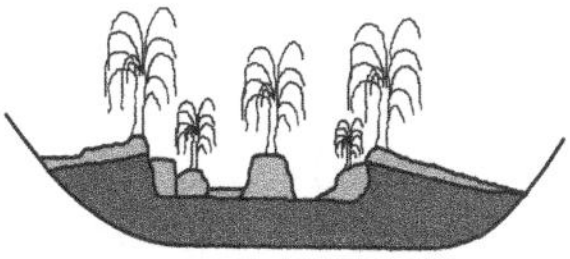	Turning point
e No examples exist in Bega catchment 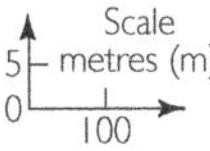 Scale metres (m) 5 0 100	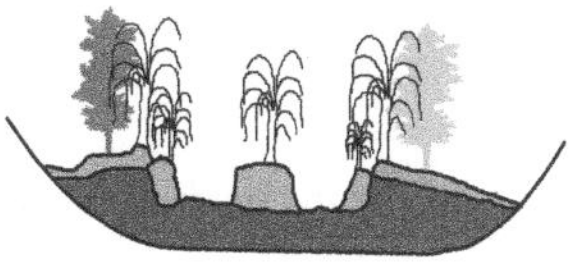	Created condition

Figure 8.7 (continued) *Schematic representation of the evolutionary stages in river adjustment following land use changes for different 'river styles'* (continued)

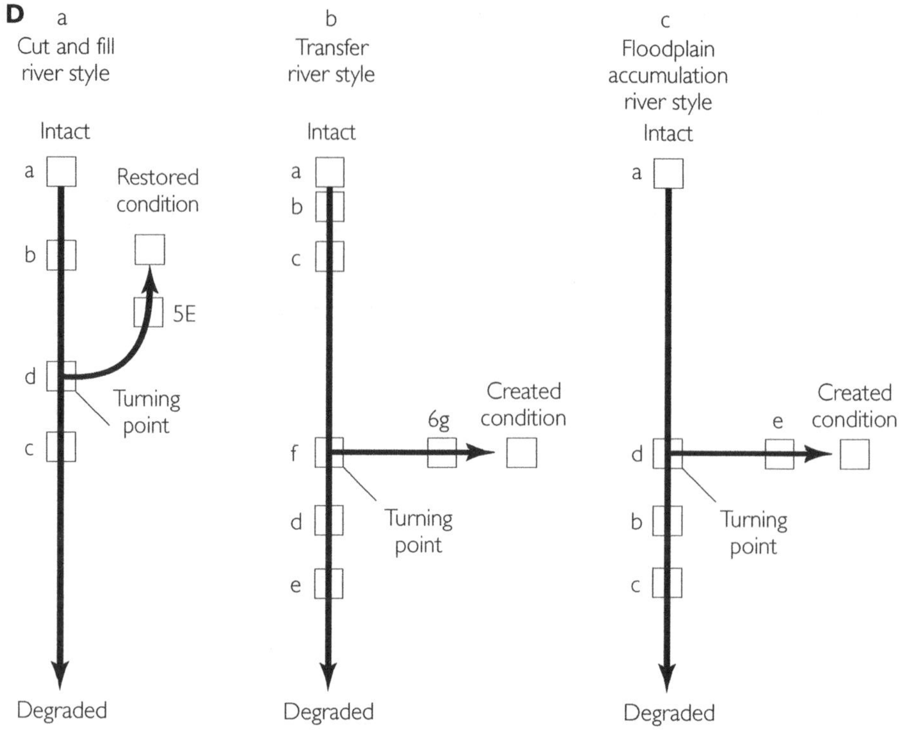

Figure 8.7 (continued) ***D*** *illustrates the evolution within the context of river recovery to a previous ('restored'), new ('created') or 'degraded' channel condition (Fryirs and Brierley, 2000)*

Predicting incremental adjustment: hazard magnitude

Predicting incremental adjustments of river channels using analysis to simulate the magnitude of adjustment according to the proportional magnitude of change in input variables (Downs and Gregory, 1993, 1995: 124) is perhaps the ultimate measure of channel hazard sensitivity and has the potential for use in estimating the probable impact of proposed water resources measures, channel engineering operations or land use changes. However, the challenge of cause-and-effect simulation modelling of channel adjustment ideally requires catchment-based physical models that can accommodate the lag effects inherent to geomorphological systems (i.e. 'system memory') and this is beyond the current scientific capacity. Current efforts towards incremental modelling of channel adjustment are based largely upon short-term changes at the cross-section or reach scale. Catchment-scale, process-based sediment transport models are under development (Downs and Priestnall, 2003) and have, to date, been led by concerns for soil erosion: such as in the Water Erosion Prediction Project (Ascough *et al.*, 1997) or the Mediterranean Desertification and Land Use programme (Brandt and Thornes, 1996; Kirkby *et al.*, 1998).

At the cross-section scale, research into processes that drive mass failure of river banks according to the shear strength of the bank and the mechanical and hydrological effects of vegetation on stream banks (e.g. Osman and Thorne, 1988; Simon *et al.*, 1999; Simon and Collison, 2002) have been distilled into a model for channel bank stability. The model predicts whether the factor of safety for wedge-shaped, planar failures is exceeded according to properties of the bank materials, pore-water pressure and the type of vegetation on the bank top. While the model does not incorporate drawdown effects, the prospect of bank toe erosion or the hydrological effect of vegetation, it allows the user to undertake sensitivity analysis of the bank using 'what if' scenarios. For instance, it is possible to simulate whether erosion of the bed or the bank toe will make the channel banks less stable (by making the banks higher and by altering the bank geometry, respectively).

Assessments of reach-scale channel changes are difficult to simulate because morphological changes are tertiary responses that follow from the dynamics of hydrological and sediment processes, the primary and secondary drivers, respectively. Therefore, reach-scale models are dependent on accurate estimations of sediment supply and transport, and on the ability of the sediment processes to interact with the topography and grain size distribution on the channel bed to determine the resultant channel form (Figure 8.8). Because bed topography does not respond to sediment changes in a uniform manner, there are fundamental difficulties in using one-dimensional mathematical models to indicate channel bed changes. Not only do they rarely encompass the effects of channel banks on sediment supply, but sediment is assumed to distribute evenly over the channel bed of the receiving reach thus altering the bed gradient in the next iteration of the model. In reality, it is quite

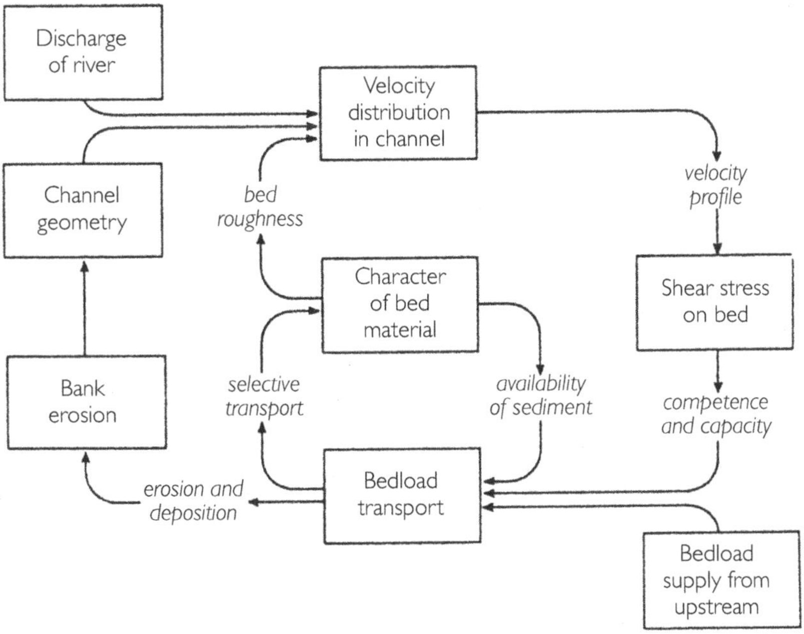

Figure 8.8 *Conceptual representation of feedback loops and process–response mechanisms associated with dynamic alluvial rivers, illustrating the difficulty in making accurate predictions of incremental channel adjustments (Richards and Lane, 1997, following Ashworth and Ferguson, 1987)*

possible that any additional sediment is stored temporarily on point or side bars leaving the gradient of the bed thalweg unchanged (Richards and Lane, 1997). Assessments of changing bed topography, based on two-dimensional hydraulic models linked to bedload transport as a function of shear stress, are in their infancy in gravel-bedded channels (Richards and Lane, 1997) but more developed in sand-bedded channels (van Rijn, 1987) where predicting the initiation of sediment transport is less complex. However, in terms of planform adjustment, it has been possible to test the application of a meander migration model (Johannesson and Parker, 1989) to the problem of bank protection on the Sacramento River, California (Larsen and Greco, 2002). The model predicts bank erosion rate at a series of nodes along the channel thalweg proportional to a dimensionless bank erodibility coefficient and to a steady state velocity factor (at the two-year recurrence interval) representing the difference of the near-bank velocity to the reach-average velocity. The model is used to simulate meander migration for a period of 50 years under various management scenarios (i.e. no rip-rap, existing rip-rap, extended rip-rap and managed avulsion; Figure 8.9).

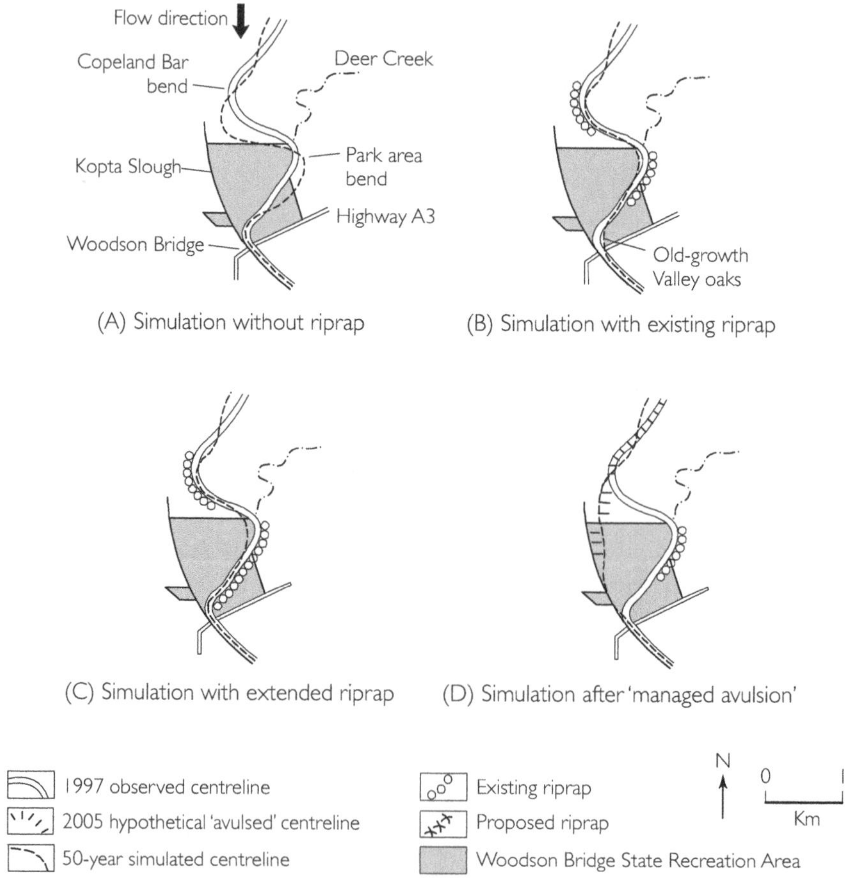

Figure 8.9 *Model simulation of planform adjustments of the Sacramento River, California, for a period of 50 years under various management scenarios (Larsen and Greco, 2002)*

Detailed environmental assessments of river channels are tools for understanding the sensitivity of channel dynamics and have great potential use in planning management responses appropriate to the specifics of channel hazards and the fragility of channel assets that are sustainable in the context of expected future changes (see Table 8.1). The preceding sections considered whether channel change is likely in particular locations, the type of change relative to thresholds in channel behaviour, how changes will develop in time and space and the magnitude of change relative to changes in driving variables, with the predictive challenge becoming more demanding in each section. The latter sections in particular represent frontiers for geomorphological research and because geomorphological processes are generally nonlinear, discontinuous and indeterminate, physically deterministic models may never be capable of fully predicting morphological change in river systems, not least because multiple equilibrium situations (i.e. channel end points) are possible for any combination of hydraulic variables (Richards and Lane, 1997). Therefore, it is important that model results are used in combination with field-based interpretation as the basis for directing management responses to perceived channel hazards.

8.5 Evaluating project success

A post-project appraisal (PPA) of monitoring and evaluation is an integral part of environmental assessment but was rarely applied to 'traditional' management schemes despite the long history of channel instability arising from channel engineering operations, such as bank erosion, knickpoint headcutting, structure collapse and accelerated rates of sedimentation (p. 98). Management agency commitments to maintenance operations (e.g. dredging channels of fine sediment and in-channel vegetation) or to remedial engineering work (e.g. repairs to bank protection or additional protection flanking the original scheme) are signs of the project's failure to sustain channel stability. Evaluations, if any, have tended to be *construction-centred* and *short term* (Downs *et al.*, 1999), for instance, conducting a survey of as-built conditions to ensure the engineering scheme was implemented as designed. Conversely, for projects where implementation is intended to initiate environmental improvement that is progressive over time as flood events re-shape the channel, evaluation is necessarily *performance-centred* and *long term*, and may involve conservation and/or channel stability criteria.

Post-project evaluation provides five generic benefits (Wathern, 1988; Brookes and Shields, 1996a; Skinner *et al.*, 1999; Downs and Kondolf, 2002):

1 judging whether the project conforms with intended design;
2 judging project performance in relation to the design intentions;
3 assessing whether management of unanticipated impacts is required;
4 learning from projects so that future practices and procedures can be improved;
5 assessing the efficacy of resource investment.

It is implicit that evaluation requires testable criteria to be stated during the design phase of the scheme (Brookes, 1996a; Watts and Fargher, 1999) and that the key to judging project effectiveness lies with periodic monitoring to compare with a baseline condition (Downs and Kondolf, 2002). In judging project performance, PPAs require a 'compliance audit' of the scheme's installation details relative to the design intentions, a 'performance audit' of the short-term functioning of the scheme relative to received flows and a 'geomorphological evaluation' of longer-term prospects of the scheme to meet its objectives (Downs *et al.*, 1997; Figure 8.10), especially the degree to which the medium- to long-term physical processes are in harmony with the prevailing and projected future catchment hydrology and sediment budget (National Research Council, 1992; Sear, 1994).

The effective use of PPA is increasingly viewed within the context of an 'adaptive management' project approach (p. 193), wherein empirical evaluations from progressive schemes are communicated back to the design process. Clearly, this also requires individuals (project managers, engineers, environmental scientists, etc.) to report back fairly on schemes that have unanticipated effects, and a cultural shift is required in channel management to enable individuals to report their 'failures'. Kondolf (1995a) suggests that, overall, an effective PPA requires:

- clear objectives – to provide a framework for evaluation and to identify any potential incompatibilities among the project objectives;

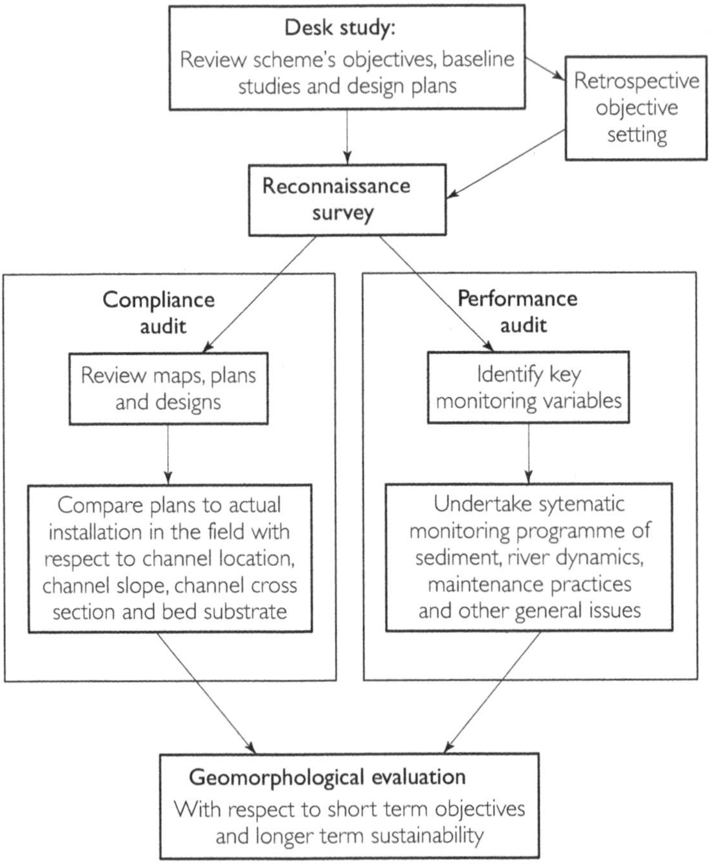

Figure 8.10 *Representation of the post-project appraisal process centring on three key parameters, namely a 'compliance audit', 'performance audit' and 'geomorphological evaluation' (Skinner, 1999, following Downs et al., 1997)*

- baseline data – as an objective basis against which later assessments are judged;
- good study design – to ensure clear evaluation even in complex river environments;
- commitment to the long term – because the impacts of a scheme may not be fully apparent for many years;
- willingness to acknowledge failures – that is, to learn from experiences.

The potential 'effectiveness' of the PPA in terms of its ability to benefit future design practice will vary largely according to the availability and quality of pertinent pre-project and post-project data (Downs and Kondolf, 2002). A summary of beneficial pre- and post-project information is given in Table 8.6. Of prime importance is the availability of testable objectives for

Table 8.6 Five levels of investment in geomorphological post-project information: data requirements for each level of PPA and typical level of understanding achieved

Components of PPA	**Category of post-project appraisal**				
	'Full'	**'Medium-term'**	**'Short-term'**	**'One-shot'**	**'Remains'**
Pre-project	*Level of commitment*				
Success criteria	explicit	explicit	explicit	implicit or explicit	implicit or nonexistent
Baseline surveys	thorough	thorough	thorough	partial or thorough	none
Documented design rationale	explicit	explicit	explicit	implicit or explicit	implicit or nonexistent
Design drawings	thorough	thorough	thorough	thorough or conceptual	conceptual or none
'As built' (record) drawings	exist	exist	beneficial	beneficial	none
Post-project / follow-up	*Level of commitment*				
Periodic or event-driven monitoring	>10 years	5–10 years	<5 years	single survey	single survey
Supplementary historical data	beneficial	beneficial	beneficial	necessary	necessary
Secondary analytical procedures	probably unnecessary	possibly unnecessary	beneficial	highly	beneficial necessary
Opportunities for learning					
Understanding compliance with design intentions	yes	yes	yes	possible	speculation
Indication of short-term geo-hydraulic performance	yes	probable	possible	possible	speculation or none
Longer-term evaluation of geomorphological sustainability	probable	possible	speculation	speculation	speculation or none

Source: Adapted from Downs and Kondolf (2002).

the scheme, baseline survey data and the extent of post-project monitoring relative to the actual installation (which may differ from the design intentions). Data availability and quality is likely to vary widely according to available funding, the methods chosen for analysis and the specific environmental context of schemes. Consequently, investment levels in post-project appraisal vary from an ideal scenario, where all pre-project information is available and a long period of monitoring has ensued (a 'full' PPA), to the worst case in which only the sketchiest design intentions for the scheme are understood, no baseline survey data exists and no periodic or event-driven monitoring has occurred (a 'remains' PPA – Table 8.6). Not surprisingly, the potential for insight from appraisal varies with the level of data investment. The 'full' PPA after 10 years or more of monitoring should allow for the management scheme to be seen in terms of project sustainability, because the monitoring period should have included a number of geomorphologically effective flow events (p. 61) allowing the channel to interact with the management 'solution'. As the period of monitoring is reduced, secondary analyses become important to draw out trends from limited data (Downs and Kondolf, 2002).

Methods for judging conservation performance will vary with the specific success criteria of the scheme. Where biotic improvement is the key concern, species surveys (e.g. fish counts) may be the ultimate test of the implemented scheme but, especially with faunal surveys, high inter-year (and seasonal) variability as a result of biological limiting factors beyond the scope of the scheme (e.g. upstream or downstream conditions, accidental pollution, predation) may mask the scheme's true value, requiring long monitoring periods to detect population trends (Clayton, 2002). More robust are performance measures that indicate habitat improvement or geomorphological sustainability as a measure of *potential* biotic gain, involving a re-survey and comparison of habitat variables using a variety of top-down or bottom-up approaches (p. 234). Post-project evaluation should also place the channel in its temporal context, indicating whether the channel evolution since project installation has occurred during 'average' flow conditions or whether flows have been of greater or lesser magnitude than would be expected for the period. This is important because projects are often expected to evolve morphologically during high flow events and thus the speed at which the river evolves towards the intended final form and function depends upon the magnitude of effective geomorphological events both before and after the installation of the scheme. Such an evaluation in a lowland river in Nottinghamshire, UK concluded that flow deflectors installed to increase channel bed diversity increased the long-term potential for sediment transport at the deflector sites by 25% (prior to scour occurring) but in the 2.5-year period since installation, low flows caused sediment transport potential to be 40% below the long-term average expectation (Downs *et al.*, 1999). This indicates that making general conclusions about deflector performance in the elapsed period would be premature.

Post-project evaluations are now frequently demanded by funding agencies and constitute a fundamentally new challenge for channel management, creating an upsurge of interest in PPA in parallel with river restoration (e.g. Everest and Sedell, 1984; Frissell and Nawa, 1992; Kondolf, 1994c; Beschta *et al.*, 1994; Kondolf and Micheli, 1995). Dissemination of evaluation outcomes is critical to ensure that managers learn from mistakes (Kondolf, 1994c), that is, to inform future projects about 'good' and 'bad' practice (Downs and Kondolf, 2002): results from a selection of schemes are reviewed in Section 10.3.

8.6 Environmental assessment in context

During just 35 years since the inception of environmental assessment for river channel management projects, assessments have become probably the single most important aspect of the management process. Most countries have legislation that demands assessment to a reasonable standard, nearly all river projects concerned with flood control or curtailing river movement require assessment of the extent of hazard and require reach or off-reach mitigation, and all conservation and restoration-oriented projects require answers to the value-laden questions 'what are the environmental assets of the present channel?' and 'what assets do we wish the channel to have in future?'. Furthermore, as a corollary of the search for sustainable solutions, the assessment questions are becoming more specific and more challenging, especially with regard to time and space: questions such as 'how long-lived will be the downstream impacts to actions taken here?' present fundamentally challenging questions to environmental scientists that in turn are shaping research agendas. Poor quality assessment data and poor quality interpretations of data lead to increasingly frequent litigation.

In summary, environmental assessment in river channel management can provide the basis for:

- determining elements of importance in determining project success (section 8.5);
- setting appropriate and attainable goals for management projects (section 9.2);
- developing management approaches based on cause-and-effect (section 9.3);
- specifying targets for the scientific and engineering learning experience (section 10.1);
- choosing appropriate methods that work with the river to provide project success (section 10.2).

The primary threat to the utility of environment assessments for river channel management is insufficient appropriate long-term monitoring data. In post-project appraisals, the existence of long-term monitoring data reduces the need for 'creative' secondary analysis (Downs and Kondolf, 2002) and the meaning of such monitoring data is less likely to be questioned. The lack of suitable monitoring data is also partly responsible for the development of reconnaissance styles surveys (Downs and Thorne, 1996; Thorne, 1998). Despite this, there is still no requirement for standard measurement networks or long-term monitoring as part of the UK Environmental Change Network, even with the emerging legislative requirement for European Union nations to evaluate spatial and temporal compliance of channel hydromorphological condition (the physical outcome of the inter-relationship between the flow regime and the channel perimeter) as part of water quality objectives (Sear and Newson, 2003). Suitable long-term data would be obtained from flow gauging, sediment transport gauging, periodic repeat surveys of monumented cross-sections and channel long profile, and repeat aerial and oblique photography from monumented positions. Additional data would include analysis of floodplain stratigraphy and conservation of historical records regarding management of the channel, including undertaking post-implementation record ('as-built') drawings. Of these requirements, the likelihood is that only flow gauging records may exist close to any particular site of concern. The data are pivotal for determining *change* in the river channel environment and for providing an appropriate spatial and temporal context to evaluation (p. 229). River channel changes are argued to encompass at least three modes including large-scale synchronous changes, long-term persistent change and socially significant (risk-defined) change (Sear and Newson, 2003); all are dimensions of channel sensitivity to change are integral to a risk-based strategy for management (Chapter 5). The utility of being able to detect changes using various long-term monitoring records, together with historical reconstruction and remote sensing, includes providing an 'early warning' system of environmental change, better determining the vulnerability of nature and society to change and better informing and guiding the public and policymakers (Parr *et al.*, 2003).

With suitable data, future environmental assessments should improve the understanding of connectivity between process, form, habitat and biota whether using top-down or bottom-up approaches. Top-down, increasing efforts will be made to achieve catchment-scale assessments to link sediment transport processes to channel form using sediment budgets. The impetus for this action is generally related to assessing the impact of land use changes on river channels, in particular the role of human activity in releasing increased amounts of fine sediment that impact on water quality and cause channel aggradation (and thence increased flooding) downstream. Frameworks for 'watershed analysis' have been drafted (Montgomery *et al.*, 1995, 1998), technological advances are allowing catchment-scale models with relevance to fluvial geomorphology to be developed (Downs and Priestnall, 2003) and the philosophical and practical issues related to real cumulative impacts analysis are being addressed (Reid, 1998). Bottom-up, biological and ecological scientists are increasing their understanding of links between flora and fauna and their habitat, allowing analyses with increasing certainty of the factors that limit biological and ecological populations, thus leading to more specific and practically attainable targets for conservation management. In between these extremes, at the reach-scale, improvement in understanding the sensitivity of geomorphological systems is required as the basis for better definition of the cause-and-effect links between process, form and channel habitat and especially the temporal dynamics of longitudinal (one-dimensional) and spatially distributed (two-dimensional) feedbacks in system response. Such improvements will increase the prospect of developing sustainable management approaches, especially where river restoration is required.

9 Conservation-based Management Approaches

Synopsis

New challenges for river managers (Chapter 7) have resulted in the development of conservation-based management approaches based on 'river restoration' and related strategies. These approaches are based on undoing river stress to result in partial system recovery, and are usually intended to resolve competing management concerns regarding water resources, channel instability and conservation (9.1). The goals for conservation-based approaches include statutory obligations of the management agency or agencies (e.g. related to conservation, flood defence, maintenance, channel instability, water quality) and aspirations of the catchment community (e.g. aesthetic, recreational, community action, development) that together form the basis of management schemes (9.2). Priorities for individual schemes should be specified according to the environmental condition of the channel and translated into approaches based on principles of: preservation and natural recovery, restoring flow and sediment transport, prompted recovery, morphological reconstruction and instability management (9.3). The evolution and future success of river restoration is dependent upon addressing a series of requirements, opportunities and limitations that challenge the philosophy and methods used by the channel management community (9.4).

9.1 New challenges for river channel management

In the late twentieth century, greater understanding of river channel systems (Chapters 3–5) combined with a shift towards concern for the environment (Chapters 6 and 7) and the results from riverine environmental assessments (Chapter 8) initiated a renaissance in approaches to river channel management. Essentially, management priorities expanded from facilitating water resources management (e.g. flood defence, land drainage, irrigation and regulation) and managing river channel hazards (e.g. controlling erosion and deposition) to include an integral concern for managing river channel assets (valued environments that depend upon natural river channel functions and dynamics). As such, management based on 'taming nature' (Chapter 2) was succeeded by approaches involving conservation-based strategies and an emerging awareness that 'invoking nature' could actually reduce the costs and improve the effectiveness of these longer-standing concerns.

In summary, the new management challenges are to:

1 *Manage rivers as 'fluvial hydrosystems' (Petts and Amoros, 1996a), incorporating knowledge of past, present and future conditions* (Chapters 3, 4 and 5). This requires an approach that acknowledges that river channel systems are highly interconnected and must be viewed within their historical catchment context in order to produce sustainable management solutions. Management actions may also be determined, at least in part, by the habitat requirements of various charismatic flora and fauna such as anadromous salmonids, otter, beaver, neotropical songbirds, black-faced spoonbills, cottonwoods, willows, black poplar and riparian and floodplain forest assemblages. The chosen species are usually rare and have demanding or particular habitat requirements that make them good indicators of the overall 'health' of the river ecosystem.
2 *Integrate conservation actions with water resources and hazard management* (Chapters 1, 2, 6 and 7). Conservation issues were long viewed as secondary, or directly opposed to, more traditional concerns such as flood defence, drainage, irrigation, bank erosion and regulation. Inventive management approaches are required to reconcile conservation with these traditional functions. One such prospect is the role of restoring meanders and floodplains to act, in part, as flood retention for a flood-prone settlement downstream (e.g. Sear, 1994).
3 *Protect naturally functioning river systems* (Chapter 8). Evidence from environmental assessments has indicated just how few 'natural' river channels remain: more than 98% of Danish streams have been modified (Brookes, 1987a), an estimated 96% of lowland rivers in the UK have been channelized (Brookes and Long, 1990), 96% degraded in The Netherlands (Jasperse, 1998) and only 2% of rivers in the USA are not fragmented by dams (Graf, 2001). Therefore, as one of a spectrum of management actions (Boon, 1992), preserving intact natural habitats (i.e. channel assets, Chapter 5) has been argued as a critical responsibility for contemporary management (e.g. National Research Council, 1992; Brierley and Fryirs, 2000; Graf, 2001).
4 *Re-manage degraded systems* (Chapters 2 and 7). The extent to which previous management operations caused environmental degradation of river systems means that contemporary management plans must do more than avert further system degradation in reaches previously subject to flow regulation, straightening, re-sectioning, bank revetment or detrimental land use changes. They must actually improve environmental conditions and undo the degradation legacy (see Section 7.4). This action is commonly referred to as river restoration.

Responding to these challenges requires actions that accord to the environmental 'state' of the river reach and its catchment historical context. In remaining semi-natural or scenic rivers, *preservation* initiatives are appropriate to prevent environmental damage (National Research Council, 1992; Graf, 1996; Gregory, 1997). In lightly degraded river reaches, management responses that *limit* further damage are required whereas *mitigation* strategies may compensate for locally deleterious impacts (Brookes, 1988). In highly degraded channels, strategies are required that reverse the extent of degradation, promoting environmental enhancement through the act of river *restoration*. Figure 9.1 depicts these management options as responses to a continuum in contemporary environmental degradation of the river environment, caused by increasing 'river stress' from land use changes

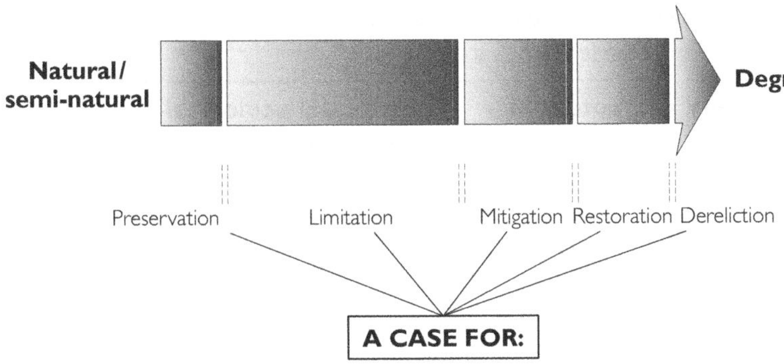

Figure 9.1 *Management options according to the environmental state of the channel (Boon, 1992)*

or previous river management operations. Contemporary first-order physical and biological stresses typical of river environments are likely to include (National Research Council, 1992: 188–206):

- problems of water quantity and flow mis-timing;
- morphological modifications of the channel and riparian zone;
- excessive erosion and sedimentation;
- deterioration of sediment quality;
- deterioration of water quality;
- introduction and invasive spread of alien species.

The cumulative impact of these stresses is to significantly degrade river habitats. Table 9.1 lists examples of specific ecological impacts arising from human activities in and beside river channels. This degradation is indicative of a breakdown in the physical and ecological integrity of the river system. Physical integrity refers to the active connection of river processes and morphology under the present hydrological regime to maintain a dynamic equilibrium of channel activity, with adjustments not exceeding limits of change defined by societal values (Graf, 2001: 6); ecological integrity requires a river system to be sustained (i) under current hydrological, water quality and sediment transport regimes, (ii) in the context of balancing the social, economic and environmental needs, and (iii) with a minimum level of maintenance (Petts, 1996: 546).

Conservation-based approaches are therefore directed towards avoiding or rectifying accumulated river stress to retain or restore the physical and ecological integrity of river channels while also maintaining channel hazard reduction and the sustainable delivery of water resources. Actions centred on undoing river stress are generally described as 'river restoration', which was defined as 'the complete structural and functional return (of a river) to a pre-disturbance state' (Cairns, 1991) (see Chapter 7, pp. 184–9). However, as full river restoration is rarely possible as a result largely of continuing water resources commitments and the need to protect floodplain infrastructure from erosion, deposition and flooding hazards, it is generally impossible to reverse all river stresses. The degree to which restoration is constrained under different conditions is reflected by the various terms used to describe the management actions (Table 9.2). Full restoration is also difficult to achieve

Table 9.1 Examples of the potential ecological consequences arising from selected human activity in and beside river channels

Human activity and potential physical effect	Potential ecological consequences
Dams	
Reduced flood flows leading to reduction in channel migration	Reduced diversity of riparian habitats
Reduced flood flows eliminate frequent scour of active channel	Riparian vegetation encroaches into channel
Increased base flows and raised alluvial water table upstream of dam	Waterlogging of vegetation
Reservoirs drown existing vegetation	Longitudinal connectivity of riparian corridor interrupted
Base flows reduced or eliminated	Riparian and instream vegetation severely stressed or dies
Trapping of bed-load sediments behind dam, release of sediment-starved water downstream, channel incision	Reduction in water table, stressing or killing bankside and riparian vegetation, loss of connectivity in instream habitat, loss of upstream spawning areas for anadromous fish
Channelization	
Channel excavation/dredging	Physical removal of fauna and flora, immediate reduction in number of fish present and species diversity, limited recovery in time, increase in non-native species suitability for fauna and flora
Silt released and re-deposited through excavation/dredging	Increased turbidity kills macroinvertebrates and is unfavourable to game fish species, reduction in suitable nest sites for salmonid species
Smothering of channel bed gravels with sand/silt	Changes in standing crop numbers and species diversity, reduction in oxygen to buried fish eggs
Reduction in low-velocity habitat and slack-water channel edge habitat	Increased drift of macroinvertebrates, lack of shelter for fish fry, reduction in primary productivity
Elimination of pools, riffles, point bars	Reduction in macroinvertebrate diversity, reduction in fish number and biomass
Reduction in deep water habitat	Reduction in numbers of large fish, implications for rate of production

Human activity and potential physical effect	Potential ecological consequences
Increase in water temperate, and in temperature extremes	Direct impact on fish survival and production, reduction in macroinvertebrates
Reduction in volume of large woody debris	Direct loss of habitat, reduced cover for fish and reduction in nutrient status and trapping ability
Weed cutting	Removal of fauna in weeds, increased macroinvertebrate drift, reduction in overhanging food source, reduction in nursery habitat
Creation of unstable, silty banks	Reduction in number of fish through loss of habitat
Reduction in total bank length where straightened	Reduction in overall area of available habitat
Forestry and timber harvest	
Removal of trees in riparian areas	Direct loss of trees, reduction in structural complexity for bird, mammal and insect habitat, partial elimination of large woody debris supply to the channel
Log transport on rivers leading to erosion of banks and simplification of channel geometry	Reduction in habitat complexity and consequent impact on fish, etc.
Removal of trees on hill slopes, resulting in increases in peak runoff and erosion	Bank erosion, conversion of vegetated bottomland and open gravel-bed channel, changes to species composition
Urbanization	
Settlement along riverbanks and on bottomlands	Riparian habitat replaced by urban infrastructure
Increased impervious surface upstream increases peak runoff, induced channel widening and incision	Increased velocities cause additional stress for fish and macroinvertebrates, water table may lower stressing riparian vegetation
Land drainage to make land suitable for development	Flow quantity alteration from natural situation and impacts on instream fauna, desiccation of riparian vegetation

Source: From Downs *et al.* (2002a); developed from Brookes (1988); Kondolf *et al.* (1996).

Table 9.2 A selection of terms used to describe river restoration and closely associated management strategies

Term	Definition
Restoration	*The act of restoring (a river) to a former or original condition.* River restoration is the process of recovery enhancement. Recovery enhancement enables the river or stream ecosystem to stabilize at a much faster rate through the natural physical and biological processes of habitat development and colonization. Recovery enhancement should establish a return to an ecosystem that closely resembles unstressed surrounding areas (Gore, 1985c). The complete structural and functional return (of a river) to a pre-disturbance state (Cairns, 1991). The complete structural and functional return of a biophysical system to a pre-disturbance state (National Research Council, 1992). Return of an ecosystem to a close approximation of its condition prior to disturbance (National Research Council, 1992). The totality of measures that change man-induced alterations to rivers in such a manner that the ecological functioning of the new state resembles a more natural river (Muhar *et al.*, 1995).
Full restoration	The complete structural and functional return to a pre-disturbance state (Brookes and Shields, 1996b).
Re-establishment	*To make (a river) secure in a former condition.*
Recovery	*The act of restoration (of a river) to an improved/former condition.*
Prompted recovery	Morphological diversity is improved using geomorphological and hydraulic manipulation of natural processes (Downs and Thorne, 1998a).
Rehabilitation	*To help (a river) adapt to a new environment.* A partial structural or functional return to the pre-disturbance state (Cairns, 1991; National Research Council, 1992). Putting (a river) back into good condition or working order (National Research Council, 1992). (An approach to management having) due regard to the catchment geomorphological system (i.e. recreating form *with* function) whilst acknowledging the constraints to design and assessment existing in a multi-functional river management environment (Downs and Thorne, 1998a).
Enhancement	Any improvement of a structural or functional attribute (National Research Council, 1992).

Term	Definition
	Any improvement in environmental quality (Brookes and Shields, 1996b).
Creation	Bringing into being a new ecosystem that previously did not exist at the site (National Research Council, 1992). Development of a resource that did not previously exist at the site. Includes the term 'naturalization', which determines morphological and ecological configuration with contemporary magnitudes and rates of fluvial processes (Brookes and Shields, 1996b).
Naturalization	Recognizes that the concept of 'natural' is defined by the community relative to the modified state of the system and that the goal of naturalization is to drive the system as a whole toward a state of increasing morphological, hydraulic and ecological diversity, but to do so in a manner that is acceptable to the local community and sustainable by natural processes, including human intervention (Rhoads *et al.*, 1999).
Preservation	The maintenance of an aquatic ecosystem ... Preservation involves more than preventing explicit alterations ... Preservation also implies management ... of the aquatic ecosystems to maintain its natural functions and characteristics. Preservation is distinct from restoration and creation in that the functions and characteristics of the preserved ecosystem are presumed to exist, more or less, in their desired states. (National Research Council, 1992).

Source: Developed from Gregory (2000: 265) following a table of dictionary definitions (in italics) by Sear (1994: 170); Gregory and Chin (2002); Gregory (2002).

in practical terms partly because of problems in defining the pre-disturbance structural and functional details required as pre-cursors to designing the restored channel, and partly because of population increases, land use and climate changes that cause the historical pattern of runoff and sediment yield to be an inappropriate baseline for contemporary conditions (Downs and Thorne, 2000). Full restoration is also difficult to achieve philosophically, because time is an integral component of environmental value (Elliott, 1997; see Chapter 7). Overall, the National Research Council (1992) definition of restoration as 'the return of an ecosystem to a close approximation of its condition prior to disturbance' (p. 18) is more realistic although less romantic than Cairns' definition, and succinctly underlines the practical challenge of managing rivers as ecosystems to retain their physical and ecological integrity.

9.2 Defining the goals

The specific conservation, water resources and hazard management obligations of a river management agency are normally determined by local, national and international legislation. These obligations have often broadened in scope considerably in recent years as channel management has been set in the context of integrated river basin management (see Chapter 6), but also reflect the increasing role of participants from the catchment community in providing the initiative for management actions (see Chapter 7). Consequently, the re-management of river channels now includes goals that relate to community aspirations in addition to the statutory objectives of the management agency.

Management agency responsibilities

It is usually the responsibility of multipurpose agencies to integrate community management objectives with their own regulatory concerns for conservation management, channel hazard reduction and the facilitation of water resources management.

Conservation management objectives are dictated by the nature of river environment and the history and causes of river degradation in the catchment including those instigated by previous channel engineering operations. Actions at the *in-channel* scale are often designed to improve topographical and functional diversity to provide suitable hydraulic habitats for aquatic fauna. Actions may range from morphological reconstruction of a 'natural-looking' stream channel, often by reinstating a sinuous planform, to efforts involving substrate reinstatement to improve riffle habitats for invertebrates and fish. Substrate reinstatement may involve adding coarse bed material in channelized lowland streams now subject to sedimentation, such as Harper's Brook in eastern England (Harper *et al.*, 1998) or replacing cobbles with finer gravel thought to be suitable for spawning habitats such as in the Merced River, California (Haltiner *et al.*, 1996). Efforts to improve the diversity of *marginal and riparian* habitats are intended to either facilitate a more dynamic aquatic–terrestrial transition zone (Naiman and Décamps, 1990) or may also be seen as a means of encouraging channel bank stability. Actions can range from providing sheltered, slow water zones for fish fry and for establishing aquatic emergent plants in lowland rivers (Downs and Thorne, 1998b) to encouraging channel bank erosion that promotes the entry of large woody debris into the river and that can act as shelter habitat as it moves downstream (Piégay *et al.*, 2000). For riparian flora, flow re-regulation may provide a chance to re-establish threatened native tree species that rely on seasonal attributes of the pre-regulation hydrograph for successful growth. *Floodplain* habitat restoration may involve re-establishing floodplain forest species (e.g. Consumnes River, California; Haltiner *et al.*, 1996), reconnecting backwater channels intended to improve fish rearing habitat (e.g. the Rhône, Henry *et al.*, 1995) and the re-assignment of floodplain land uses to those that are more amenable to periodic flooding, such as on the River Ijssel in The Netherlands (Cals, 1994; Cals *et al.*, 1998).

Management goals centred on river restoration are also being applied to more 'traditional' management concerns such as hazard reduction. For instance, in higher energy channel systems, river restoration is often also intended to reduce systemic *instability and erosion* in incising rivers. Many river systems have become destabilized through channel straightening and will progress through a sequence of adjustment processes (see Chapter 5).

Stabilization measures include the use of grade control structures to halt the upstream erosion of knickpoints and thus prevent upstream channel bed incision, and the planting of trees as a form of revetment to provide long-term additional bank strength. Grade control structures may be constructed from native materials such as boulders (e.g. Rosgen, 1996), and the outside of vulnerable bends protected from erosion using stone spur dykes as deflectors, such as on Goodwin Creek, Mississippi, USA (Shields *et al.*, 1995a, b, 1998a, b). Using these measures, erosion control may provide both site habitat diversity and prevent the impoverishment of upstream habitats. Conversely, channel *maintenance* required to offset the sediment deposition hazard in channelized lowland rivers can be re-appraised through analysis of catchment sediment sources and transport. Tracing the origin of sediment sources may allow at-source control to eliminate 'excess' supply while channel alterations within the affected reach can be used to reduce its ability to act as a sediment trap. Habitat benefits should result from the reinstatement of more 'natural' system functioning. For instance, in the Mimmshall Brook, UK, more sustainable maintenance measures were required owing to progressive sedimentation of a rare river feature. Sediment sources and transport characteristics were identified prior to recommending a solution based on meander reinstatement, wetland creation and knickpoint stabilization to reduce flooding, sediment loads and bed erosion (Sear *et al.*, 1994; Newson *et al.*, 1997).

Restoration can also benefit water resources management. Reinstating floodplain wetlands by allowing natural flooding can provide cost effective *flood defence* for downstream settlements, for example, the Dutch 'Space for Rivers' initiative proposes to physically lower floodplain elevations and create set-back embankments that provide greater flood storage capacity and improved floodplain habitat diversity in the light of climate change and projected sea-level rises (Cals *et al.*, 1998; Cals and van Drimmelen, 2001). Alternatively, hydraulic modelling of flood defence provision for rural areas in England and Wales has shown many areas to be over-serviced (Ramsbottom *et al.*, 1992), permitting maintenance commitments to be relaxed and allowing natural recovery of some in-channel and bank margin habitats without conflicting with the stated level of service. One prospect in lowlands is to create a compound, two-stage channel that has ample flood conveyance capabilities but constrains low flows within a channel of a geomorphologically appropriate low-flow width. The lower reaches of Wildcat Creek, California, are notable in this regard, not least because local citizen groups campaigned for a more integrated project that included recreational space (Riley, 1989). A catchment perspective on river restoration can also benefit problems of *physical water quality*, especially where associated with accelerated rates of fine sediment production and transport. At-source measures for reducing fine sediment production have the potential to reduce intragravel fine sediment that reduces the quality of spawning habitats for salmonids and well as reducing the occurrence of chemicals bound to the sediments. Likewise, measures to reduce systemic channel instability can also reduce the re-excavation of contaminated floodplain sediments such as those deposited following uncontrolled metal mining upstream (e.g. Lewin and Macklin, 1987), while restoration projects that involve re-meandering a channel implicitly result in a longer channel with a better nutrient exchange and absorption capacity. The integration of restoration objectives with established concerns for urban water supply and irrigation requirements is also causing *flow regulation* strategies to be reconsidered. The pre-eminent example in this regard occurs in the Sacramento and San Joaquin catchments of California, USA, where parallel ecosystem restoration programmes (Anadromous Fish Restoration Program and the California Bay-Delta Authority) are examining the prospects for reversing the historical decline in native salmon species within the context of a rapidly enlarging population and irrigation-based agricultural practices. Together, these programmes spent US$678 million in the late 1990s (Kondolf, 2001) on a variety of initiatives including re-setting flow releases downstream of dams to be more beneficial to fish, the use of fish screens to reduce fish mortality at flow abstraction pumps and research into prospects for sustainable conjunctive use of a limited water supply.

Community aspirations

Community aspiration objectives in river restoration are commonly driven by 'quality of life' concerns of those who live by or use the river and its riparian zone. They often involve issues of riverine aesthetics, recreational opportunities, community involvement and development pressures.

Improving the *aesthetics* of a visually degraded riparian landscape is often related to the objective of restoring (or preserving) local elements of cultural heritage and may involve issues beyond the remit of the river management agency. Management actions may result indirectly in increased land values as the location is perceived as a more pleasant place to live, while the very existence of the restored environment may provide a 'non-use' value that can be part of the integral economic justification for the project (e.g. Turner, 1993b). There may be some conflict with conservation management goals where the valued cultural features involve rare but artificially created habitats or near-river buildings and other infrastructure. It is also notable that the aesthetic river values perceived by the catchment community (Woolhouse, 1994) and the conservation values placed on rivers by scientists do not necessarily correspond. In the UK, this is expressed by public concerns for access and safety (Green and Tunstall, 1992) and the desire for 'neatness' in the river environment (Gregory and Davis, 1993), although public surveys carried out following restoration schemes have been extremely positive and suggest that the restored sites are highly valued, especially in urban environments (Tunstall *et al.*, 2000). Management to provide additional *recreational* opportunity often occurs because the river corridor possesses the only remaining low intensity, publicly accessible, land use within the catchment. Consequently, efforts are directed at improving public access, although this may bring tension with conservation management efforts. A related theme is management to improve commercial or sports fisheries and has been one of the fundamental drivers of river restoration to date (Petts *et al.*, 2000). Fisheries improvement can require management actions to improve in-channel habitat diversity and to restore flow in regulated rivers.

Restoration-based approaches may also involve efforts at increasing *community involvement* in caring for their river, facilitating social cohesion by raising environmental awareness and pride in the local river. On the River Medway in Kent, England, 'hands on' community involvement in improving the river corridor landscape has been ongoing since 1988 and includes such actions as building soft bank revetments (Smith, 1994). In California, public participation has been typified by the activism of conservation-minded citizens (e.g. Griffin,

1998) and by organizations such as the 'Friends of the River', which was founded in 1973 as a counterpoint to the objectives of 'single purpose' management agencies (Section 7.3). They have a tradition of campaigning against development that threatens natural environments and for promoting inclusive stewardship of rivers based on stakeholder involvement. Almost all river channel management projects in California now include an active role for a local stakeholder group in the decision-making process (Haltiner *et al.*, 1996: 296–97). A good example is found in the lower 84 km of the Merced River in California's Central Valley where the Merced River stakeholder group composed of landowners, scientists and representatives from management agencies and nongovernmental organizations meet regularly to discuss and promote future management initiatives (http://www.mercedriverstakeholders.org/, last accessed 21 October 2003). Furthermore it should not be forgotten that restoration is often the 'mitigation' that allows *development* to proceed (Downs *et al.*, 2002a). This includes restoration as mitigation for floodplain building activity, where dedicated riparian conservation initiatives may be ordered as compensation for building beyond the immediate riparian zone (e.g. outside of the 100-year flood inundation envelope). It also includes restoration as mitigation for environmentally unfriendly channel management actions, for instance, where bank protection to protect floodplain infrastructure is allowed only if conservation actions are taken elsewhere along the river.

9.3 Approaches

Priorities and principles

River channel management actions can be prioritized in several ways. From a conservation perspective, prioritization stems from the need to manage river channels according to their environmental state (Figure 9.1) to achieve various management agency and community objectives while undoing accumulated river stresses to maintain or improve the physical and ecological functioning of the fluvial hydrosystem. Under this scenario, the highest priority is to preserve remnant lengths of near-pristine river channel and their supporting processes; essentially, to protect existing high quality habitats (Ebersole *et al.*, 1997; Brierley and Fryirs, 2000; Sacramento River Advisory Council, 2000; Graf, 2001; Roni *et al.*, 2002). Thereafter, management priorities are generally linked to cost and pragmatism, and are related to the hydrological connectivity and geomorphological recovery potential of the channel (e.g. Fryirs and Brierley, 2000; Clark, 2002; see Chapters 5 and 8). Reaches that are highly connected (or that can be re-connected easily) should be targeted before isolated reaches, and reaches with a high natural recovery potential should be tackled before those reaches that require significant assistance to regain natural process functions. Environmentally, the most highly degraded channels with the least prospect of stress reduction would be tackled last. However, in practice, various concerns related to priorities for water resources and channel hazard management, political pressures and the remit of the dominant governing agency may determine an alternative priority order. For example, from the perspective of social justice and inclusion, it may be highly desirable to tackle highly degraded channels in poorer urban areas despite the cost and modest potential for environmental enhancement (Clark, 2002).

Table 9.3 A hierarchy of management principles for sustaining or restoring the physical and ecological integrity of river systems. Principles are arranged according to order in the hierarchy 1 = high, 6 = low

Management principle	Description
1 *Preserve natural processes where they continue to function*	Protect naturally variable flow regimes and geomorphological processes associated with unconstrained flooding by nonstructural solutions designed to permit the river to continue to function dynamically.
2 *Limit process changes*	In reaches where natural processes continue to function, but where the threat of significant change exists, protect the natural processes using nonstructural solutions potentially in combination with measures designed to prevent systemic instability from reaching the valued reach.
3 *Restore processes where possible*	In regulated rivers, return the catchment flow and sediment transport regimes towards the unregulated condition (e.g. restore the flood pulse) to the extent possible based on daily-to-seasonal and annual-to-decadal flow regimes. When planning the flows, consideration should be given to the influence of contemporary and future land uses on catchment hydrology and sediment generation. In river systems where catchment flow cannot be restored, such as where flow disruption is the result of catchment land use changes or extensive channelization, accommodate the disturbance and pursue restoration by modifying local flow hydraulics and sediment transport processes using small-scale structures designed in their catchment context.
4 *Restore the natural channel geometry*	Achieve reach-scale restoration through direct morphological modification in rivers with low natural recovery potential. The process will initiate changes to local flow hydraulics and sediment transport processes and the implications of these changes should be fully understood and accommodated in the context of the prevailing catchment flow and sediment transport regime if the approach is to be sustainable.

Management principle	Description
5 *Restore the riparian plant community*	The riparian plant community can become a functioning part of the channel and floodplain geometry but the option is unlikely to succeed unless restored processes and/or morphology have created suitable habitats: in modified rivers there are often severe constraints on the extent of channel marginal habitats and variations in the water table depth away from natural conditions that may cause riparian plant communities to fail.
6 *Restore native aquatic plants and animals*	May be required in situations where the native aquatic flora and fauna have been eliminated in the past but is unlikely to succeed unless other restoration efforts have created the habitat niches required by the species, have restored the processes critical to the survival of native species, and have eliminated or placed non-native species at a competitive disadvantage.

Source: Developed from National Research Council, 1992; Sacramento River Advisory Council (SRAC), 2000.

Once prioritized, principles for management actions can be derived with the aim of managing the river as a hydrosystem to retain or restore physical and ecological integrity (Heiler *et al.*, 1995; Petts, 1996; Graf, 2001; Jungwirth *et al.*, 2002). The primary aim in reducing river stress is to re-create the natural variety of physical processes that promote structural and functional habitat improvements in the river channel, its riparian zone and floodplain, and thus create conditions suitable for the renaturalization of plant and animal species indigenous to the area. Related to this aim, a hierarchy of six management principles can be derived, organized by the desirability of preservation before re-creation, restoring process before form, and habitat before re-introducing biota (National Research Council, 1992; Sacramento River Advisory Council, 2000; Table 9.3). The hierarchy reflects the fact that restoration based on management principles related to the lower levels in the hierarchy are unlikely to be sustained without attention to upper order. For instance, attempts to restore a native riparian plant community are unlikely to succeed in a highly channelized river channel because of the reduced extent of aquatic–terrestrial transitional bank habitat and the consequent disruptions to the near-surface water table in the bank zone (Downs *et al.*, 2002a). Likewise, restoring a 'naturalized' channel morphology is unlikely to be sustainable without attention to the appropriate water and sediment regimes that drive the geomorphological processes. Conversely, restoration beginning with the attention to channel processes is expected to lead, over time, to improvements in the native plant and animal habitats.

The other key element in achieving the challenge of managing a river as an ecosystem is to understand the channel's sensitivity to change and thus its recovery potential (Chapter 5). This involves environmental assessment to appreciate longer-term geomorphological changes, to understand the dynamics of sediment supply, storage and transport within a catchment, to recognize whether a river channel is dynamically stable or unstable and to understand the spatial controls on channel adjustments in order to gauge the level of energy available for recovery and the type of recovery expected (Brookes, 1995; Fryirs

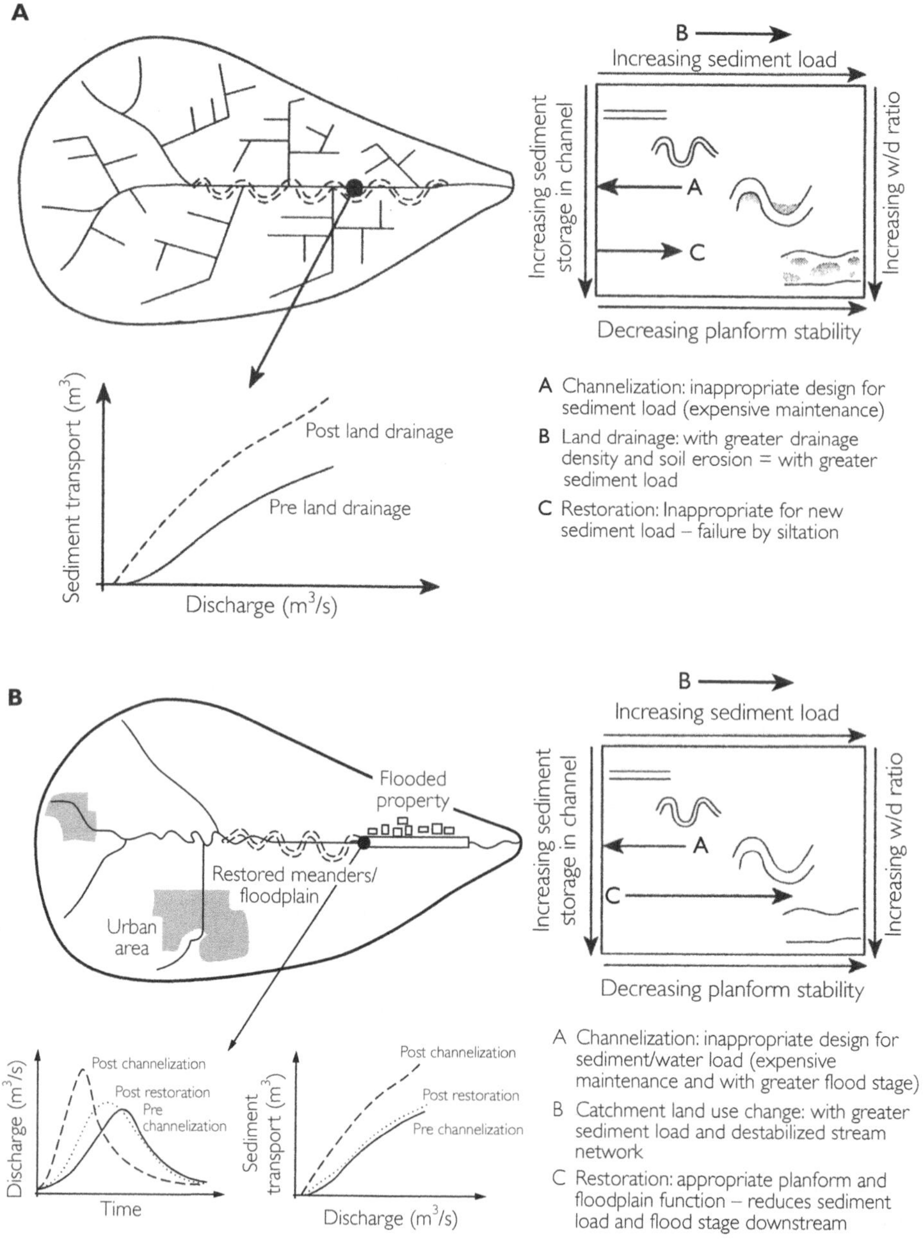

and Brierley, 2000; Newson *et al.*, 2002; Chapter 8). Conceptual examples illustrating management approaches based upon an understanding of the contemporary catchment sediment system and the direction of likely morphological change are provided in Figure 9.2A and B. The consequences of insufficient understanding of the hierarchy of principles for restoration actions and the characteristics of channel sensitivity and recovery potential is illustrated by the following fallacies in channel restoration design (Downs *et al.*, 2002a: 276–79):

- that an inherently stable, static channel geometry exists for every stream and will be sustainable;
- that a static channel is ecologically preferable to one that adjusts progressively in response to high flows;
- that restoration of form without consideration of catchment processes will be sustainable;
- that restoration projects can be designed solely from a channel classification system.

The diversity and varied histories of river channel environments mean that there cannot be one single approach to conservation-based channel management that is suitable in all instances. This is in direct contrast to approaches to management during much of the twentieth century wherein calculations of one-dimensional flow conveyance led to the imposition of a straight, trapezoidal (and possibly embanked) channel form irrespective of the hydrology, geomorphology and ecology of the host channel. Instead, assessments of the channel condition (including its catchment-historical context and recovery potential)

Figure 9.2 *Two hypothetical examples illustrating the importance of understanding the contemporary catchment sediment system and direction of likely future morphological changes in the channel prior to undertaking restoration (figure and text from Sear, 1994).* **A** *A lowland channel subject to 40 years of land drainage is restored to channel dimensions based on the pre-improvement channel and with a reinstated pool–riffle sequence, spaced according to regime relationships with channel dimensions. The scheme is expected to increase habitat diversity and improve aesthetics but in five years it fails from siltation of the channel bed and excessive weed growth that compromises land drainage. The problems occur because the agricultural activity developed on the drained soils produces far more fine sediment than in the pre-disturbance channel so the restoration channel dimensions are inappropriate to the contemporary catchment sediment regime. Solution of the problem requires that either the sediment is controlled at source by regulation of agricultural activity or creating dedicated sediment traps, or that a narrower and more sinuous channel is designed that is more appropriate to the contemporary sediment regime.* **B** *Restoration is used successfully to improve environmental value and to improve flood defence to an urban area. Historical analysis shows that river network changes have increased sediment loads and discharge peaks leading towards an urban area. The channelized river through the urban area was initially successful at conveying flood flows but, over time, sediment deposition has occurred in the overwidened channel leading to a progressive increase in the frequency of overbank flows. The solution involves restoring a meandering channel upstream at near to pre-disturbance channel dimensions. The channel floods regularly and the floodplain storage acts to delay and slightly to attenuate the flood peak, allowing sediment to be deposited on the floodplain rather than in the channelized reach. In addition to the river channel improvements upstream, the urban area now floods less frequently and the channelized reach requires less maintenance dredging*

provide the basis for selecting the most appropriate management principle described in Table 9.3, and for choosing management approaches that may be suitable in many instances. Table 9.4 illustrates how this procedure could be used to choose one of five generic management approaches that, in order of application preference, are:

- nonstructural policies: planning and land use regulation measures designed to address the causes of river stress that are at the root of management issues;
- improved network connectivity: to reinstate more natural flow and sediment transport regimes and thus promote natural hydrological and geomorphological processes;

Table 9.4 Guidelines for appropriate approaches to conservation-based river channel management based on existing environmental condition and channel recovery potential

Existing condition	Recovery potential	Typical catchment locations	Principle	Typical management approaches (see Section 9.3)
Near pristine/intact/ unimpaired	N/A	Uplands and foothills, remote from human disturbances	**Preserve process**	• Nonstructural measures
Near pristine, highly susceptible to disturbance/under threat of instability	N/A	Uplands and foothills, close to human disturbances	**Limit process changes**	• Nonstructural measures • Structural reinforcement
Mild–moderate degradation – connected to high value channels	High	Channelized in agricultural area: cobble-gravel-sand bedded channels	**Restore process**	• Nonstructural measures • Restore flow and sediment transport • Prompted recovery
'Petrified' in regulated system	Impeded	Downstream of dam in foothills/lowlands	**Restore process**	• Improved network connectivity • Prompted recovery • Morphological reconstruction
Mild–moderate degradation – isolated from high value reaches	Impeded/ moderate	Channelized in agricultural or urban lowland area: gravel-sand bedded channel	**Restore process**	• Nonstructural measures • Restore flow and sediment transport • Prompted recovery
Moderate–high degradation	Low	Channelized in urban lowland area: sand-silt bedded channel	**Restore geometry**	• Restore flow and sediment transport • Prompted recovery • Morphological reconstruction • Structural reinforcement
'Moribund' – highly degraded	Low/none	Channelized in lowland clay bedded channel	**Restore geometry and riparian plant community**	• Prompted recovery • Morphological reconstruction • Structural reinforcement

Source: Based on Boon (1992); National Research Council (1992); SRAC (2000); Brierley and Fryirs (2000).

- prompted recovery: instream measures designed to accommodate disturbance stresses and act locally to create habitat diversity through manipulating flow hydraulics and sediment transport;
- morphological reconstruction: direct re-creation of a naturalized channel geometry using construction techniques;
- instability management: structural techniques such as bank revetment used to prevent channel change in situations where infrastructure or the conservation value of neighbouring channel reaches is threatened and other approaches do not provide sufficient protection.

The measures associated with these generic approaches represent the fundamental tools available to river managers striving to achieve the objectives of conservation-based river channel management and are introduced below. A brief introduction to each approach is provided below, and methods typical to each approach are outlined in section 10.2.

Non-structural approaches to preservation and natural recovery

Catchment- and corridor-level planning approaches are suitable as the basis for preserving intact, valued or rarely found river landforms, for protecting relatively undisturbed channels against disturbance and for bringing about progressive improvements in disturbed channels with recovery potential. A variety of measures are associated with the approach and they are likely to be most effective if undertaken as part of a scheme of integrated river basin management (see Chapter 6). *Benign neglect* is achieved by planning policies that restrict development in the catchment (e.g. designation of the catchment as part of a preservation zone or national park) or that allow previously 'productive' land to fallow back to its natural vegetation type, allowing runoff and sediment production to the channel to revert over time to rates and volumes that are close to the pre-disturbance values. However, depending on the nature of the disturbance, recovery processes may not return the channel to its pre-disturbance condition but instead to a new, 'created' condition (Fryirs and Brierley, 2000; Figure 8.7).

A second category includes measures designed to minimize the impact of on-going human activities. It includes best practice source control policies for *land management* and *water resources management* designed to reduce the impact of land uses and water resources exploitation on rates of runoff and sediment production. It also includes 'semi-source' measures based on *floodplain management* that are designed to permit the river's spontaneous natural regulation functions (e.g. flooding, erosion and deposition) to occur irrespective of land and water uses beyond the bounds of the floodplain. Land management measures usually involve protocols for reducing the generation of additional runoff or sediment production from areas of agriculture, commercial forestry or urban development. Best practice water resource management policies will involve measures derived to balance water resources needs with measures to protect or improve channel hydrosystem functions and processes. Floodplain management focuses on zoning permissible land uses near to the channel to reduce the economic impact of flooding and thus reduce the need for structural flood controls that impact on the channel (see Chapter 6).

The third category of measures attempts to protect the river margin environment irrespective of land and water uses. *Buffer strips* retain moisture and sediment, offsetting enhanced runoff peakedness or volume and allowing sediment deposition prior to flows reaching the channel. Apart from sediment control, buffer strips contribute to corridor restoration through protecting wildlife and fisheries, and by facilitating water-quality and river bank stability improvements (Large and Petts, 1994). *Bankside fencing* (Figure 9.3A) is suitable where the primary cause of channel degradation is livestock access to the channel and in such situations restricted livestock access has been consistently linked with riparian vegetation recovery (e.g. Kauffman *et al.*, 1993) and the return of a narrower and deeper channel with better instream habitat (National Research Council, 1992). *Bankside tree planting* (Figure 9.3B) strategies have been suggested as the single most beneficial option for instream habitats (Ward *et al.*, 1994) by assisting in the creation of a stable environment in terms of light, shelter and temperature, by shading out excessive growth of aquatic macrophytes, by providing a source of large woody debris input and by providing links to terrestrial habitats beyond the channel edge.

Restoring flow and sediment transport processes to facilitate recovery

Process-based management approaches to restore flow and sediment transport are appropriate for channel hydrosystems regulated by dams, weirs, embankments and water transfers to or from the river. The approach involves altering the degree of flow regulation to partially re-naturalize flow and sediment regimes and thus stimulate hydrological, geomorphological and ecological processes to cause a gradual improvement in the channel's morphology and habitats. The approach is centred on improving network connectivity (see Chapter 3) and is suitable for projects that require a significant improvement to the environmental state of the channel and where the geomorphological recovery potential exists but is not realized under the existing flow and sediment regimes. A range of methods is available.

A

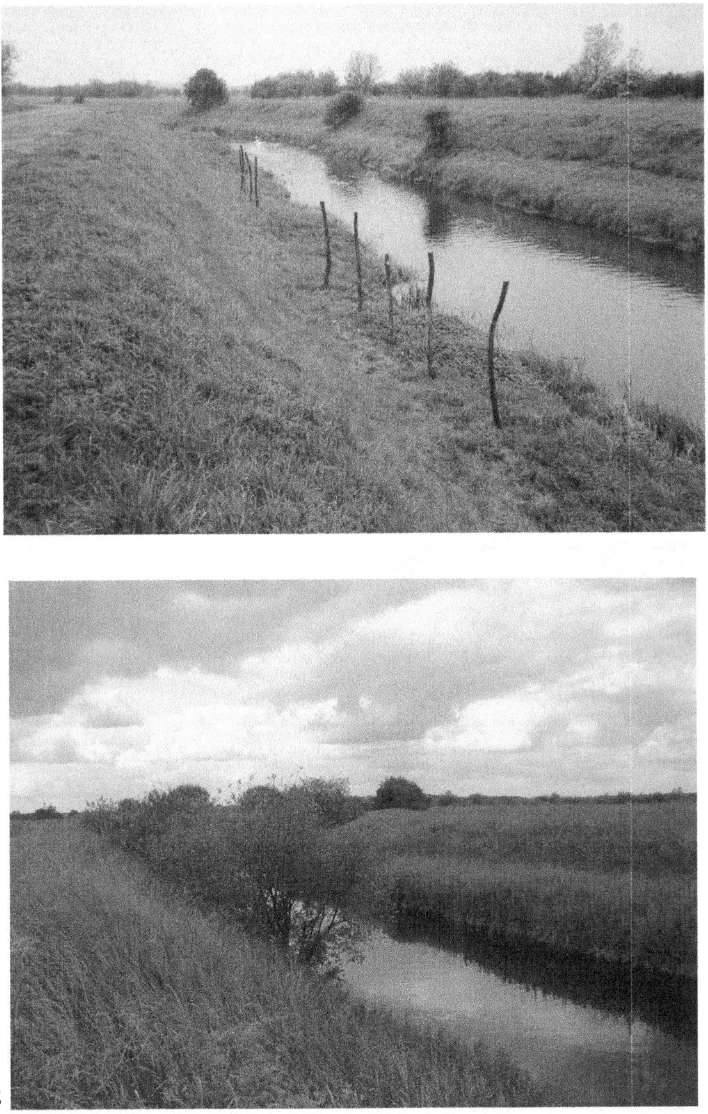
B1
B2

Figure 9.3 ***A*** *Bankside fencing used to prevent cattle ingress to a recently restored river reach, East Fork San Juan River, Colorado (photograph P.W. Downs, 1995). Flow is towards the camera.* ***B*** *Bankside tree planting to create shaded instream habitat on the southern banks of the River Idle, UK: (1) less than one year after planting in summer 1996 (photograph P.W. Downs, 1997); and (2) six years after planting (photograph K. Skinner, 2002). Flow is towards the camera. Design details in Downs and Thorne (1998a, 2000)*

Prescribed *environmental high flows* (Figure 9.4A), or flushing flows, involve the re-regulation of flows downstream of a control point (e.g. large dam or water abstraction offtake) in order to integrate conservation objectives with hazard management and water resources concerns. Whereas environmental flows have historically involved setting minimum flow standards, flushing flows involve the periodic release of a moderate flood flow

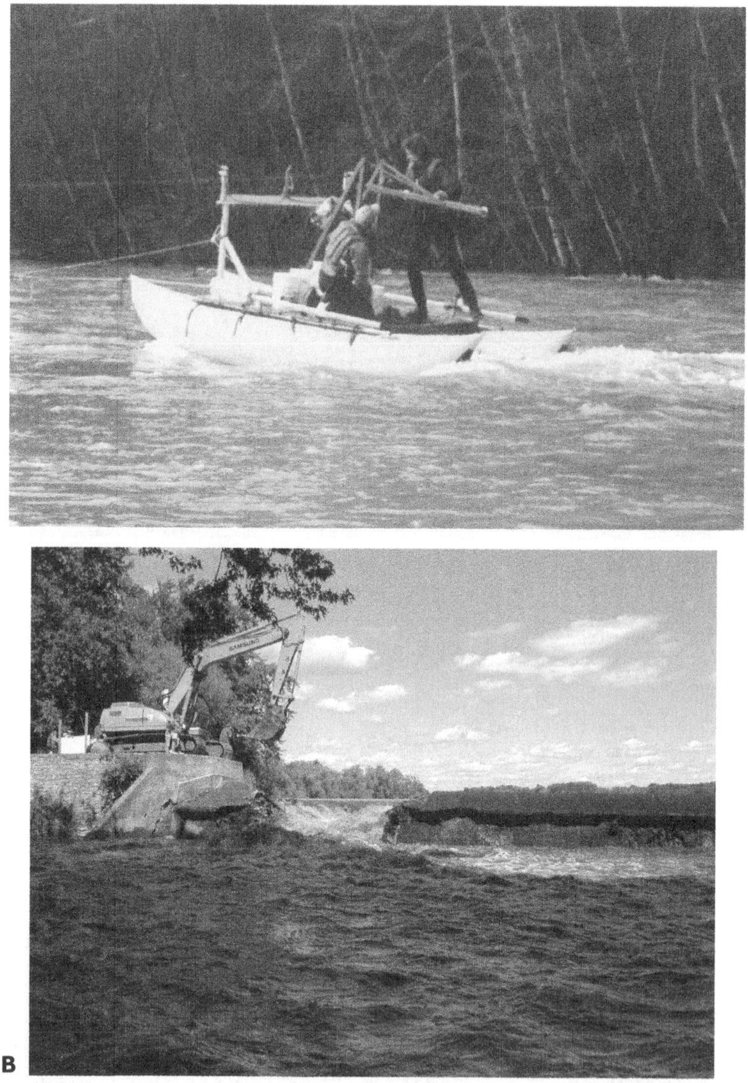

A

B

Figure 9.4 ***A*** *Sediment sampling during a prescribed environmental high flow release on the Trinity River, northern California (photograph S. McBain).* ***B*** *Restoring flow and sediment processes by dam removal: removal of the Rockdale dam on the Koshkonong River, Wisconsin, September 2000 (photograph M. Doyle). Details in Doyle et al. (2003b)*

to partially restore the 'flood pulse advantage' to channel ecosystems (see Chapter 3). The required high flows may be defined on the basis of their hydrologic return period or on their impact on sediment transport processes or the channel morphology (Reiser *et al.*, 1989; Kondolf and Wilcock, 1996). The flows released should be linked to target ecosystem objectives including geomorphological processes and habitat for invertebrates, fish and riparian vegetation (Downs *et al.*, 2002b). One concern inherent to flushing flows is that while they stimulate sediment transport, they do not address sediment supply issues.

Network connectivity can also be improved by *weir and obstruction removal* or the *set-back of embankments*. The removal of superfluous weirs and other instream flow obstructions, including structures that disconnect channel side arms from the main channel flow, provides benefit to the continuity of downstream and lateral instream flow and sediment transport processes. While the action may involve only local process modifications, the strategic benefit, particularly to fish, may be significant where the obstruction prevented upstream or lateral access to critical locations such as spawning or rearing grounds. Likewise, in relation to flood flows, setting back embankments to allow a greater area of regular floodplain inundation may facilitate sediment deposition processes that maintain floodplain microtopography and habitat, reduce the volume of instream sedimentation downstream, and be of significant benefit in providing seasonal floodplain rearing grounds for fish and foodweb improvements for floodplain fauna. It is important to determine by environmental assessment whether the potential habitat benefits provided by local actions such as reconnecting backwater or distributary channels will be sustainable in relation to the prevailing regional trend in sediment transport, that is, whether the measure be prone to significant sedimentation and thus only of temporary benefit. It is also important to determine whether former floodplain land uses will offset the benefit of their reconnection. For instance, the removal or breaching of flood embankments to permit more frequent floodplain inundation is often intended to benefit fish by allowing them regular access to floodplain food sources that are beneficial to their survival (Somner *et al.*, 2001). However, in former agricultural fields that were laser-levelled, this measure also carries with it a risk that fish will become stranded in shallow depressions because the floodplain lacks its natural microtopography and drainage that would route fish back towards the main channel as flows recede.

At a catchment-level, the ultimate expedient for improving network connectivity involves *large dam removal* (Figure 9.4B). This addresses flow and sediment transport processes as well as sediment supply issues and the role of the dam as a physical barrier to upstream and downstream movement of fish and other instream fauna. Dam removal provides the prospect of allowing natural recovery processes to initiate rehabilitation of channels and floodplains upstream and downstream of the former dam. Options for dam removal include complete removal, partial removal and staged breaching. While the option of dam removal is now actively being considered for river restoration purposes (e.g. The Aspen Institute, 2002; The Heinz Center, 2002; Bushaw-Newton *et al.*, 2002), dam removal has historically been prompted by the cost of rehabilitating privately owned dams that are now deemed unsafe (Shuman, 1995) and has thus been primarily a public safety issue. The main environmental challenge for dam removal is to provide a sufficient assessment of the complex effects that are probable following removal. Of particular concern is the fate of the 'sediment wedge' of deposited material stored behind the dam in terms of downstream sedimentation and impacts on water quality. Public concerns over downstream flood hazard and the loss of recreational amenity provided by the impounded lake may also be significant.

Prompting recovery to improve habitat diversity

Prompted recovery (Downs and Thorne, 1998a, 2000) or 'assisted natural recovery' (Newson *et al.*, 2002) generally consists of instream measures designed to modify energy

A

B

C

D1

D2

Figure 9.5 ***A*** *Log flow deflectors to stabilize restoration through morphological reconstruction, anchored by large boulders and assisted by tree planting, East Fork San Juan River, Colorado (photograph P.W. Downs, 1995). Flow is towards the camera.* ***B*** *Triangular flow deflector to create channel bed diversity and more varied instream habitat, shortly after installation, River Idle, UK. Constructed using gabion baskets, back- and top-filled to encourage vegetation establishment (photograph P.W. Downs, 1996). Flow is away from the camera.* ***C*** *Low weirs and pinned logs to create complex instream habitat, Huckleberry Creek, Washington (photograph P.W. Downs, 2000). Flow is towards the camera.* ***D*** *Construction of shallow water berm to create suitable conditions for reed planting to assist channel narrowing and creation of slow-flow habitat, River Idle, UK: (1) upstream end of berm shortly after construction (photograph P.W. Downs, 1996: flow is away from the camera); and (2) looking upstream from downstream end of berm six years later (photograph K. Skinner, 2002)*

conditions within the reach (Hey, 1994a) and thus locally influence flow hydraulics and sediment transport patterns. The approach is suitable for prompting or accelerating recovery processes to increase habitat diversity in disturbed channels, especially in locations where there is little prospect that non-structural measures will achieve significant changes. Essentially the approach is one of accommodating disturbance and is often applied in channelized rivers and in catchments significantly disrupted by urban or agricultural developments. A precursor for prompted recovery is to understand processes prevailing in the project reach in their catchment context and subsequently to implement measures that are appropriate to ambient environmental conditions including the altered catchment flow and sediment transport regime caused by human disturbances. Therefore, although the approach is often structural, it has several commendable aspects. First, where the measures are designed according to their catchment-scale process context, they can achieve sustained improvements. Second, the approach facilitates restoration while still accommodating land use changes and other river management functions such as flood defence, allowing it to be feasible even in channels where the prospects for conservation management is limited. Third, because the approach is largely based on 're-directing' natural forces, it can be cheap to install requiring only small channel structures rather than large-scale earth-moving works. Fourth, the approach can be applied incrementally over quite short reaches of river that are typical of those generally available for river restoration (although this strategy is unlikely to maximize habitat improvement potential – see Section 9.4). The approach requires patience from the funding authority because it may require several floods for habitat improvements to begin to take effect and may be criticized for being too conservative and less effective in terms of cost and environmental improvement than other approaches.

There are numerous measures available for prompted recovery. One is the placement of structures such as *deflectors* (Figure 9.5A,B), *low weirs, sills and vanes* to improve habitat by altering flow and sediment transport patterns to create conditions that are more akin to the optimal ecohydraulic configuration for native biota. Low-velocity habitats for fish shelter can be created using dedicated *cover devices* built into the bank, mid-channel or marginal *boulder clusters* for rest habitat, *floating and pinned logs* and installed *large woody debris* (Figure 9.5C) such as root wads and trunks. It is always preferable that the materials used match the type of material that would be found naturally in that channel environment. In lowland rivers, *reed planting* (Figure 9.5D) at the channel toe can also provide shelter for fish but may also trap nutrients and fine sediments and thus assist in reducing fine sediment loads and in the creation of a meandering low flow thalweg and greater channel bed diversity (Downs and Thorne, 1998a). For 'excess' coarse sediment, *bedload traps* may also be used to create areas of preferred deposition to offset sedimentation problems downstream.

Reconstructing channel morphology

Direct engineering of channel morphology to provide a geometry that is similar to the pre-disturbance condition can provide the foundation for ecological recovery. This measure is often pursued in lowland channels that have been straightened and resectioned into uniform, trapezoidal canals (see Chapter 2) and that have insufficient available energy or upstream sediment supply to be capable of significant natural recovery. Channels that possess sufficient fine sediment supply to re-establish a sinuous 'inner channel' within the channelized cross-section by means of patterned deposition (Nixon, 1966; Brookes, 1992), but not to re-establish a fully sinuous form, may also be candidate reaches. In these cases, engineering a new channel with a cross-section, planform and longitudinal profile that replicates a natural channel avoids the long time period required to achieve significant environmental benefit by natural or prompted recovery. However, reconstruction is generally a less preferable management approach than nonstructural measures, improved network connectivity and prompted recovery because historically it has encouraged the continuation of a form-led, symptomatic basis to management rather than one that is process-led and deals with cause-and-effect. In addition, morphological reconstruction is highly disturbing to channel habitats and biota in the short-term (e.g. Biggs *et al.*, 1998), construction has often been based on a highly homogeneous sinuous planform template irrespective of the channel's natural tendency (e.g. Smith, 1997; Kondolf *et al.*, 2001), and schemes have often employed unnecessary (and expensive) bank revetment techniques that perpetuate a view of channel stability as a static rather than a dynamic property. There is, therefore, the risk that schemes of morphological reconstruction may result in a sinuous version of a channelized river that does not lead to significant environmental enhancement and can be inherently unstable (e.g. Kondolf *et al.*, 2001).

Despite the potential drawbacks, morphological reconstruction may be the only realistic management approach in lowland channels incapable of recovery or in channels where continued severe instability is predicted. Additionally, it has the political advantage that engineering the channel morphology provides a clear and rapid demonstration of action to funding agencies, irrespective of real, sustained, environmental improvements. In Europe, where the majority of river restoration schemes, beginning in the 1980s in Germany and Denmark (Brookes, 1996a), have involved small streams with low stream power, most projects have involved morphological reconstruction. The approach commonly centres on *restoring sinuosity and cross-sectional asymmetry* (Figure 9.6A) of straightened channels to a form closer to the pre-disturbance morphology. Other reconstruction measures include *resizing flood channels* (Figure 9.6B) often by narrowing to offset inherent problems of sedimentation in 'oversized' lowland rivers, and constructing a *two-stage channel* (Figure 9.6C) utilizing the floodplain and the channel together to contain a prescribed return period of flood, but with a 'low-flow' channel designed to accommodate the natural bankfull flood event (see Brookes, 1988). Recent advances in channel design have included step-wise procedures intended to ensure that the reconstructed channel is designed within its prevailing hydrological and sediment transport regimes so that there is no net erosion or sedimentation within the reach (see Chapter 10).

Reinforcement to manage channel instability

Traditional management approaches were often accompanied by structural reinforcement of the channel banks and/or bed designed to ensure a permanently static condition of the river channel and thus permit maximum use of the floodplain. However, these measures were almost universally detrimental to instream habitat and conservation values, did not always succeed in preventing erosion and could exacerbate conditions of instability downstream or upstream (see Chapter 4). Keeping the river channel static 'at all costs' also promoted the precautionary use of structural revetments when there was no proven need for them, increasing management costs and further degrading habitats. Given this history, structural reinforcement of the channel perimeter for conservation-based management approaches is a contentious issue, and yet will continue to be needed in projects charged with protecting near-river infrastructure and agricultural land. Two trends in revetment use may reduce the future habitat impact of these measures.

A

B

C

The first is a change from 'hard' to 'soft' revetment techniques to reduce the environmental loss incurred while still reinforcing channels. For protection of inclined banks, it has become increasingly commonplace to use a combination of *live vegetation* (trees and shrubs; Figure 9.7A), *biodegradable materials* (buried rootwads, willow spilings, brush mats, coconut fibre rolls, etc. (Figure 9.7B)) and *geotechnical fabrics* for protection and, where harder protection is required, to use *loose rip-rap* that can be planted with vegetation, rather than mortared rip-rap, paving slabs or sprayed concrete. Vertical and near-vertical bank protection is often achieved using *gabion mattresses*, and soil-filled *cellular grids* lend themselves to the structural or decorative use of vegetation, rather than concrete walls or sheet piling, except in cases where development extends fully to the channel edge. For bed protection, *notched weirs* (Figure 9.7C) made from native stone or tree trunks and *mortared rocks* graded to the bed level to mimic riffles are alternatives to traditional concrete structures.

The second trend is towards a far more strategic application of reinforcement: this should take account of the channel condition and equilibrium trajectory, applying the right measure at the right time to minimize the negative environmental impact on the channel (Figure 9.7.D). It should use the minimum necessary extent of reinforcement to counter the tendency for overprotection while ensuring the scheme is properly designed, and it should use the minimum reinforcement required to constrain movement by using the 'softest' available solution based on ecological engineering principles (see Chapter 7). Achieving such a strategic approach requires that applications are based on environmental assessment of the channel conditions. For example, 'appropriate bank management' (Thorne *et al.*, 1996b) begins with a regional environmental assessment of the cause, severity and extent of the bank erosion problem as the basis for guiding the necessity of bank protection and, where a structural solution is required, the scope, strength and length of reinforcement. Similarly, in channels destabilized by eroding knickpoints, regional environmental assessment of the channel bed and banks is used to classify channel reaches according to their expected evolutionary sequence (Schumm *et al.*, 1984; Simon, 1989; see Chapter 8) as the basis for siting grade control structures where they will be most effective. In these ways, bank and bed reinforcement can be minimized and targeted to specific locations with a proven requirement, with the aim of reducing both management costs and environmental degradation.

Figure 9.6 *Three examples of morphological channel reconstruction in English lowland rivers:* **A** *detail of asymmetric meander created on the River Cole (photograph P.W. Downs, 1996). Flow is away from the camera. Compare with aerial view of restored reach (Figure 7.5) and planform design for project (Figure 10.7).* **B** *Resizing of a flood control channel in a constrained floodplain setting: Wraysbury River. Outer channel widened to provide flood capacity while limestone blocks have been used to encourage sediment deposition and formation of a narrower low flow channel where riparian vegetation can establish (photograph A. Brookes, c. 1991). Flow is towards the camera.* **C** *Two-stage channel created on the River Roding by lowering the right bank floodplain (photograph P.W. Downs, 1990). Flow is away from the camera. Design details in Sellin et al. (1990)*

A

B

C

D

Figure 9.7 ***A*** *Bank revetment using interlocking tree log structures ('engineered log jams', Abbe et al., 1997) to protect road to the right of the photograph, Cispus River, Washington (photograph P.W. Downs, 2000). Flow is away from the camera.* ***B*** *Bank revetment using tree logs driven into the bank to stabilize channel after morphological reconstruction, and with their root boles protruding to create pool habitat diversity: Trout Creek, northeastern California (photograph P.W. Downs, 1997). Flow is towards the camera.* ***C*** *Bed protection in an incising channel with an extremely constrained residential floodplain setting (note bridge and utility pipes): Kitswell Brook, north of London, UK. Protection is achieved using notched, blockstone, low-flow weirs with buried gabion protection to prevent enlargement of plunge pool. Bank protection is achieved using driven and bound larch-poles (photograph P.W. Downs, 1995). Flow is towards the camera.* ***D*** *Strategic application of structural measures to prevent channel incision; drop structure with rip-rapped plunge pool and baffle plate: Long Creek, Mississippi (photograph P.W. Downs, 1995). Flow is away from the camera. Conceptual design details in Figure 10.9B.*

9.4 Progress and requirements

As the pre-eminent icon for conservation-based river channel management, the 1990s saw a remarkably rapid growth in the popularity and enthusiasm for 'river restoration' from management agencies and members of the public. However, river restoration is still in its formative years and many of the collected works and conference proceedings on the subject have dwelt on the future prospects for river restoration (e.g. National Research Council, 1992; Osborne *et al.*, 1993; Brookes and Shields, 1996a; Boon, 1998; Nijland and Cals, 2001b). Table 9.5 summarizes a selection of the suggested requirements, opportunities and limitations as they relate both to the practical and scientific challenges posed by river restoration.

One of the most pressing practical issues is for riparian land availability: adopting a hydrosystem approach to river channel management necessarily focuses attention on managing both water and land together (Newson, 1997) and providing permanent 'space for rivers' (Cals *et al.*, 1998). Under this scenario, land acquisition becomes critical to ensuring meaningful environmental enhancement (Brookes, 1995). This is in direct contrast to twentieth-century channel management methods that often had as prime objectives flood defence and land drainage that permitted or protected intensive floodplain development.

Table 9.5 Examples of practical and scientific needs in river restoration

Practical needs	Scientific needs
Land: 'space for the river' including permanent floodplain acquisition or cooperative floodplain 'easements' and set-aside protected lands.	Enlarged multidisciplinary database through scientifically transferable baseline environmental assessment and post-project appraisal, to reduce project uncertainty.
Supportive legislation at regional, national and international levels, whereby restoration principles are integrated into water management strategies as a component of 'water quality' (and see Chapter 7).	Improvements in ecosystem understanding through funding for scientific investigation, including cross-disciplinary approaches (see Chapter 7), and explicit statements of project risk.
Supportive institutional arrangements especially in countries where ecosystem management has traditionally been under fragmented jurisdiction (see Chapter 6).	Adoption of a catchment historical perspective in projects to ensure that projects are developed conscious of the influence of past processes on the contemporary river environment, and an understanding of future environmental trajectories with and without further management activity (see Chapter 5).
Appropriate 'teams' to facilitate restoration being representative of the 'catchment community', including riparian stakeholders, community activists, planners, environmental scientists, engineers, economists and politicians (see Chapters 6/7).	Clear specification of physical and functional objectives, specifying the target flow regime (where regulated), geomorphological functions, habitat diversity and biota, and acceptable levels of flood control and water delivery (where appropriate).
Appropriate funding basis and political will, especially to facilitate restoration as an integral and justified goal for management activity, rather than as a secondary concern or as a by-product of development activity, flood control or water resources management.	Integration of efforts of disciplinary specialists, requiring an appropriate forum for integration, and including the integration of science into the decision-making process.
Improved communication of the benefits of restoration from practitioners to politicians and the public, and of the practical problems stemming from degraded rivers.	Improved communication of scientific information to managers and the public, also requiring a suitable forum, such as a stakeholder group (see Chapter 7).

Practical needs	Scientific needs
Better ecosystem understanding by managers to understand the consequences of certain actions, by engineers responsible for channel design and construction oversight.	Recognition of the need for different solutions for different environments, based on baseline investigations and the catchment historical perspective: no general approach can exist (see Chapter 8).
Action in the face of uncertainty: acknowledgement that restoration-based management actions have an inherent experimental component but that this uncertainty is an opportunity rather than a threat. Recognition that projects may have to be re-visited and supplementary actions taken.	Adoption of adaptive environmental management approaches to deal with uncertainty through learning by surprise using rational and transferable science, rather than learning by trial and error (see Chapter 7).

Source: Developed from National Research Council (1992); Osborne *et al.* (1993); Brookes (1996a); Haltiner *et al.* (1996); Brookes and Shields (1996c); Boon (1998); Downs *et al.* (2002a); Newson *et al.* (2002).

In densely populated countries, restoration may only be feasible on near-river lands owned by the local authority, creating a focal interest in the use of 'green corridors' in urban river settings (e.g. Barker, 1997) or on short lengths of river creating an 'island' of environmental improvement where the opportunities for off-site impacts to harm the restoration efforts are high (Brookes, 1996a). One alternative is to pursue planning policies that allow or dictate that intensive agriculture is not practiced to the river's edge, creating a river corridor. In the European Union countries, a range of policies have been developed that may promote this practice, including permanent set-aside, countryside stewardship schemes, the designation of environmentally sensitive areas and grants to establish new woodland. In the USA, riparian lands may be bought outright or conservation 'easements' created, wherein the land stays under private ownership. In The Netherlands, the 'Space for Rivers' programme strives to persuade landowners that it is in the common interest to move agriculture back from the river edge for reasons of flood defence as much as for river restoration (Cals *et al.*, 1998; Cals and van Drimmelen, 2001). However, the fact remains that many floodplain lands are in parcelled ownership, meaning that lengthy and complicated negotiations may be necessary for their release (Brookes, 1996). An alternative prospect on private lands is that river restoration may become 'development-led' (Brookes, 1995, 1996) whereby opportunities for restoration are greatest when land use is changing and development permission can be tied to enacting restoration initiatives. In the UK, the overall result has been a financially opportunistic rather than strategic development of restoration schemes (Newson *et al.*, 2002).

A critical element in introducing conservation issues to channel management is supportive legislation (see Chapter 7). In this regard, Denmark became a world-leader in river restoration initiatives because the Danish Watercourse Act of 1982 included special provision allowing stream restoration (Iversen *et al.*, 1993; Nielsen, 1996). Likewise, in Germany, where main river management is the responsibility of the states, Baden-Württemburg has been responsible for promoting rehabilitation of small streams since 1985 (Kern, 1992). Turning points in management activity can be traced to legislative changes: for instance, in England and Wales, the Water Resources Act 1991 was a breakthrough because it allowed environmental enhancement initiatives to be pursued that were not tied to flood defence projects (Gardiner, 1994; see Table 7.2). Across Europe, the next major changes in river channel management are likely to arise as a consequence of the search for suitable protocols for robust 'hydromorphological' assessments of fresh water status required by the European Union Water Framework Directive, and the requirement that all inland and coastal waters reach 'good status' by 2015 (European Community, 2000). In the USA, 17 measures have been proposed to underpin a National Restoration Strategy (National Research Council, 1992). The measures, aimed at promoting policies and cooperation at state and federal levels are based on four key elements, namely: goal setting, principles for priority setting and decision-making, the redesign of federal policies, and programmes and innovation in financing and use of markets (National Research Council, 1992; 375). Their impact, if incorporated in legislation, could be considerable.

As river restoration activity should be based on catchment-level analysis and actions, supportive institutional arrangements are critical and would allow river restoration to build upon experiences related to integrated river basin management (Mitchell, 1990; Newson, 1997; see Chapter 6). International restoration projects, for instance, require trans-boundary cooperation that has historically been very difficult to achieve in water resources management. The signing of accords such as the Lower Danube Corridor Declaration by the governments of Bulgaria, Romania, Moldova and the Ukraine (Bachmann and Wurzer, 2001) is necessary in cases where international boundaries run along river centrelines. Even nationally, river restoration may be seen to imply additional risks to riparian landowners, which can worry river management authorities. In setting up the EU-funded River Restoration Project demonstration schemes, the necessary clearances and finances were far easier to obtain in Denmark, where restoration was already practised and managed and funded primarily by one agency, than in England, where the Environment Agency (then NRA) would not support the project without detailed project management and legal agreements being signed, and where financial guarantees from the contributing partners were much harder to obtain (Holmes and Nielsen, 1998; 187–88). In many countries there is also a fragmentation of ecosystem management between numerous agencies such that the National Research Council (1992: 343) suggest that '(t)he politics and consensus building required for integrated management of the resource are often as complex as the ecosystem itself'. In the Pacific north west of the USA, there are typically 17 federal, state, tribal management and regulatory agencies involved in management of water as it passes through a project watershed (Shannon, 1998). This may lead to sub-optimal institutional arrangements such as in California where, by default, the flood control agencies are increasingly taking the role of the lead river restoration managers (Haltiner *et al.*, 1996: 322).

A related concern is the need to organize a team with the appropriate scientific and managerial skills to undertake a project successfully (Gardiner, 1991a; Federal Interagency Stream Restoration Working Group, 1998; see Chapter 10). Channel management was often led by engineers, reflecting the legacy from construction-oriented projects. However, this is no longer a necessity and, as channel management becomes more broad-based, so newly trained and existing managers will require additional training in order to understand the dynamics of river ecosystems that lie within their responsibility, possibly blurring the distinction between scientists and managers (National Research Council, 1999). Frameworks are also required for committing adequate (and long-term) funds to permit restoration-based management programmes to be carried through to their completion. This requires a significant shift away from a construction-centred and short-term financial perspective on channel management to one that funds stakeholder involvement, scientific data gathering and analysis, solutions that work with nature and are driven by weather events, and project evaluation following a significant period of effects monitoring.

Scientific constraints on river restoration centre on limits to understanding (Table 9.5). This can be conceptualized as the extent of safe, usable knowledge available for the project design. Information is either useless (unnecessary, irrelevant) or useful, and is either safe (basic principles with a low probability of error) or unsafe (specific knowledge with a high probability of error) (Statzner and Sperling, 1993). Additionally, some ecosystem driving forces are manageable (e.g. human action) while some are not (e.g. precipitation) so that uncertainties in system response will always exist, but management solutions should be based on the best available information. For management to progress, there is a need to make the useful 'unsafe' information 'safe' through experimentation, justifying the need for an enlarged, multidisciplinary database through environmental assessment, improvements in ecosystem understanding and the adoption of a catchment historical perspective to understand river channel sensitivity to change. In practical terms, this will require better pre- and post-project assessments and a greater ability in target setting (Downs *et al.*, 2002a; Newson *et al.*, 2002). Specific areas of scientific deficiency include the understanding of sediment transport processes during overbank flows, interactions between surface waters and the alluvial aquifer and safe methods for extrapolating information from neighbouring reaches (Brookes, 1996a). Generically, three concerns are suggested (National Research Council, 1992, 230–244), namely: conceptual limitations in ecosystem understanding that often involve a failure by ecologists to consider the river and its floodplain as one ecosystem; inadequate information databases for all disciplines; and a lack of appropriate expertise. The latter category includes a failure to see the project reach as part of the catchment, (despite numerous recommendations, e.g. Sear, 1994; Kondolf and Downs, 1996; Newson *et al.*, 2002), a failure to incorporate larger-scale habitat characteristics such as sinuosity and gradient and a failure to see rivers as an integral part of a dynamic physical environment.

The search for safe, usable knowledge also demands the integration of disciplinary expertise at the level of basic scientific research (see Chapter 7), particularly with regard to the cascading linkage between process, form, habitat and biota that is central to river restoration issues (e.g. National Research Council, 1992; Downs *et al.*, 2002a). Best use of this information will require a forum that integrates science into the decision-making framework: typically, decisions regarding river channel management have been made at a political or economic level without significant concern for the scientific implications of the decisions made. It also requires a clear translation of scientific details into the implementation process, especially related to site drawings (Brookes, 1996) where the precision required in river restoration may demand new skills of construction staff and requires scientific as well as engineering oversight during project implementation.

The recent history of river restoration is another impediment to the best use of scientific information. For instance, because all projects must be set in their environmental context, there will need to be a wide variety of project approaches undertaken at highly variable physical settings, spatial and temporal scales (Haltiner *et al.*, 1996) to provide transferable experiences that increase the prospect of applying the right solution at the right time. Also, the inability of scientists to define targeted habitat functions (and necessary supporting processes) clearly in quantitative terms puts conservation issues at a relative disadvantage to flood control and water resources issues when setting project objectives (Downs *et al.*, 2002a). Habitat objectives to date have usually relied on defining habitat-specific preference curves, reach-integrated template mapping and setting environmentally beneficial flow releases. Each of these techniques has some merit but needs significant refinement to achieve the level of (apparent) confidence that can be achieved in modelling flood levels or water transfers so that designs for sustainable habitat improvement can be presented clearly within the framework of the other river management objectives (Section 10.1).

Despite these constraints, the prospect for conservation-based river channel management based on restoration concepts appears good and may go some way to reversing the twentieth-century trend in environmental degradation that saw many rivers reduced to barren drains. The impetus for restoration stemmed from concerns with water quality and a desire to improve fishery habitat but has since broadened technically and in terms of public interest and involvement. Testimony to the public acceptance of restoration is illustrated by a growing anthropomorphism of the issue expressed as 'living rivers', 'healthy rivers' and 'space for the river'. Indeed, restoration has begun to involve community aspirations that extend far beyond the river channel and one prospect is that river restoration is the unifying practice that could allow river basin management to become truly (and finally) integrated (cf. Downs *et al.*, 1991). Although full restoration is not feasible because of changes in land use and climate and increases in catchment population since the 'pre-disturbance' period, restoration projects also fall short of the pre-disturbance target environment for practical and scientific reasons some of which have been explored in this section. Fundamentally, river restoration requires a design-with-nature approach based on a reformulated approach to management decision-making and techniques that 'work with the river' (see Chapter 7): Chapter 10 reviews progress towards this ideal.

10
Implementing 'Design-with-Nature'

Synopsis

Design-with-nature approaches to channel management emphasize strategic, catchment-scale planning and consultation, requiring a vision for action that reconciles management objectives with past, present and probable future physical conditions in the catchment, together with specification of attainable management targets (10.1). A variety of measures that 'work with the river' are available to pursue these aims (10.2) including catchment and corridor policies, methods for improving network connectivity, instream devices, channel reconstruction and methods for reinforcing the channel perimeter. There is a notable requirement for additional structured learning through project evaluation to determine how far these measures are achieving their specified targets in a sustainable way but it is apparent that, in moving towards design-with-nature, channel management in the last 30 years has undergone a revolutionary change in approach (10.3).

10.1 Planning management actions

Historically, river channel management activities arose in three ways. First, as a strategic response to the regional priorities for water resources exploitation and management. Traditionally, this involved 'heroic' hydraulic engineering of largely single-purpose construction schemes proposed and designed by technical experts to collect and/or distribute water while providing flood control (see Chapters 2 and 6). Second, management activity was prompted by a real or perceived risk of a significant channel 'hazard' (Chapter 5). This concern has generally been reactive (e.g. occurring immediately after a flood event) and enacted by a smaller management team consisting largely of civil engineers. Solutions were usually structural and addressed the immediate symptoms without explicit concern for long-term sustainability. The third prospect involved operational maintenance, a set of 'semi-strategic' corrective actions initiated and resolved internally by the management agency to ensure channel performance met their designated standards in terms of stability or flood control (see Chapter 2; Brookes, 1988). Maintenance activities occur both periodically and in response to 'problems' observed by management personnel during routine inspection, and are designed keep the river in its modified condition and curtail channel recovery processes that threaten to counteract a previous management operation. Maintenance activities are generally divorced from the process (and funding) that led to the original channel alterations (Williams, 2001).

In contrast, design-with-nature approaches strive to sustain the function and processes of the channel hydrosystem, integrate conservation actions with water resources and hazard avoidance objectives, preserve intact river systems and restore degraded systems (Chapter 9). As such, management of channel hazards and the need for operational maintenance should be resolved along with water resources priorities within a scheme of integrated river basin management (IRBM). This should reduce the need for reactive, risk-driven solutions because potential hazards will have been identified during the baseline channel surveys (Chapter 8), ensuring that strategic hazard management options are taken that avoid adverse reactions elsewhere in the hydrosystem. Operational maintenance should also be reduced because catchment-wide planning will broaden the available options and potentially allow source control of maintenance issues, while environmental assessment will have identified 'over-serviced' flood control channels where maintenance commitments can be relaxed (Darby and Thorne, 1995). In considering channel management to be a strategic operation, the planning framework is central in creating a practical 'vision' of management that reconciles divergent management goals and allows the specification of meaningful environmental targets. Once planned, suitable methods are required that work in harmony with fluvial processes to achieve the desired results (10.2). The progress made in achieving design-with-nature type solutions should be appraised using monitoring and project evaluation (10.3).

Creating the vision

Making effective and sustainable decisions for channel management requires a *vision* for management, reconciling community aspirations, the statutory concerns of the managing agencies and the environmental condition of the river system.

Best practice management is initiated by the formation of a multidisciplinary team charged initially with identifying the basic issues and goals of management (Figure 10.1; step 1). Potentially, the team may include a broad-based advisory group that identifies issues to management decision-makers who ultimately determine the project goals and priorities (Federal Interagency Stream Restoration Working Group (FISRWG), 1998). The team should consist of representatives of management agencies with statutory responsibilities towards the river environment and land use planning, stakeholders from the 'catchment community' (Woolhouse, 1994) including riparian landowners and indigenous peoples' representatives, catchment residents, funding experts, local industry and agriculture representatives, personnel from environmental groups and other concerned organizations (e.g. angling clubs) and technical experts whose role is to provide understanding on scientific and engineering matters. Technical expertise may be required from a variety of disciplines including surface and groundwater hydrology, fluvial geomorphology, soil science, terrestrial and aquatic ecology and biology, hydraulics, water quality, landscape architecture, economics, archaeology, recreation and land use planning including visual amenity, navigation and construction engineering (Gardiner, 1994; FISRWG, 1998). Experts may be drawn internally from the management agencies or local community but are likely also to include consultants and academics. Specialist facilitators may be required

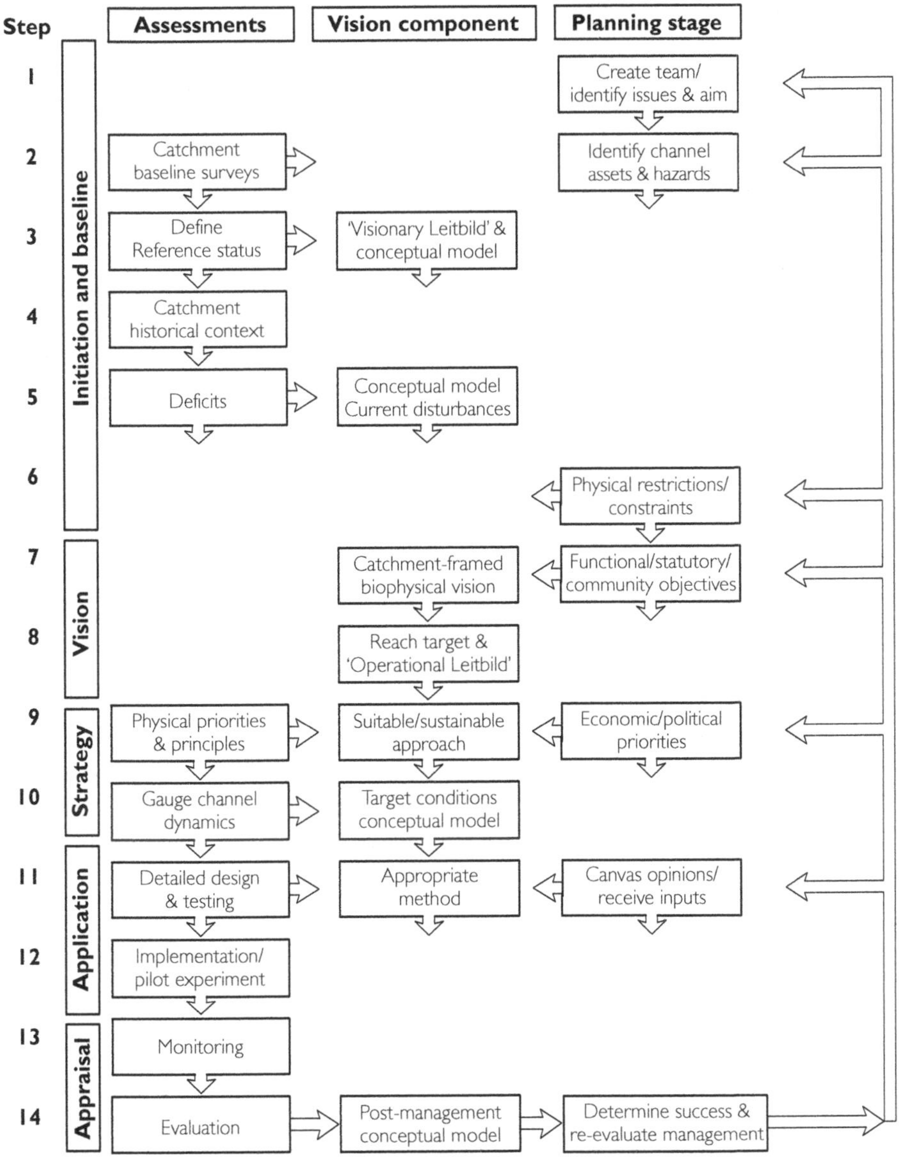

Figure 10.1 *A prospective, 14 step, best-practice framework for planning for river channel management (developed from Kondolf and Downs, 1996; Erskine and Webb, 1999; Ladson et al., 1999; Brierley and Fryirs, 2001)*

to convene public meetings, conduct public outreach and ensure information sharing (see FISRWG, 1998).

Identification by the core team of the basic management issues is complemented by catchment baseline surveys of channel condition (pp. 201) that determines reaches possessing similar channel assets, channel hazards and past management practices.

Undisturbed reaches with a high environmental value may be noted for preservation and identified as a natural, type-specific reference reach: a target vision that represents the full natural potential of the reach in the absence of economic or political constraints (the *Visionary Leitbild*; Jungwirth *et al.*, 2002)(step 3). Where they exist, reference reaches define an ideal target condition or a guiding template for management actions including as a reference standard to inform the goals of compensatory mitigation (Brinson and Rheinhardt, 1996). The reference reach should be represented via a conceptual model of process–form–habitat–biota linkages typical to the reach type.

The catchment baseline survey will also direct attention towards particular channel reaches that should be studied in their catchment-historical context to understand their evolution, condition and sensitivity, in order to judge the cause and nature of their 'problems' and prospects for natural recovery (p. 212) (step 4). This Level II environmental assessment sets the scene for determining the physical 'deficits' in the focal reaches, that is, to understand the extent of applied stresses and the corresponding channel response in cause-and-effect terms (step 5). Deficits should be depicted via conceptual models that illustrate the differences between the process–form–habitat–biota relationships and functions in the 'stressed' reach relative to an 'unstressed' (reference) reach in a similar geomorphological setting (example in Figure 10.2). Assessment should also be made of existing financial, political, institutional, legal and regulatory issues (FISRWG, 1998) that will constrain environmental prospects (i.e. achievement of the reference condition) (step 6). Constraints may include physical factors such as the continuation of flow regulation to provide water for irrigation or urban areas, the need to retain bank revetment in areas of impinging floodplain housing, and technical constraints related to available data and restoration capabilities (FISRWG, 1998). Constraints will also be related to the legacy of culturally valued components of the human landscape that restrict options for the present day; this may include the preservation of near-river buildings of architectural value or the maintenance of a water-based habitat that is dependent upon artificial conditions such as upstream of impounded water for a mill or lake. Constraints are likely to be numerous and highly influential on the channel management process.

These initial steps provide the basis for deriving a tangible, practical and ultimately achievable 'biophysical vision' of long-term objectives for the catchment (50–100 years) that reconcile public preferences, environmental aims and traditional river management concerns (step 7). Essentially, the vision statement should answer the question: 'what do we want the river to be like given the prevailing catchment boundary conditions?' (Brierley and Fryirs, 2001). It should include a catchment perspective on desirable conditions of flow and sediment transport, wildlife habitats and ecological functions and provide the basis for setting best practicable environmental target objectives for short-, medium- and long-term actions in individual reaches. Reconciliation between competing objectives may include preservation of cherished parts of the existing river landscape such as existing riparian buildings, or land uses, or even artificial reservoirs (Born *et al.*, 1998) that severely constrain management agency or conservation objectives. Likewise, perceived notions of desirable river aesthetics do not always correspond to high river conservation values (p. 244), there is often conflict between conservation concerns and recreational issues and the desire for managers to take action may conflict with the potential for natural recovery.

However, it is imperative that the channel management participants eventually develop a shared vision of a feasible future condition of the 'river channel' in its broadest sense, including the relationship of the reach with the river network, the river channel with its corridor and floodplain, and the floodplain with the catchment. Reach-level targets derived from the vision should be developed from a 'deficits analysis' that compares the conceptual model for current conditions in the reach against the probable condition and functions in the reach under the reference circumstances, factoring in pre-existing and expected future constraints and the shared management objectives to derive an integrative and time-dependent agenda for action for each reach (the *Operational Leitbild*; Jungwirth *et al.*, 2002) (step 8).

A strategy for prioritizing the project reach actions and assigning principles for management should be developed (step 9; see p. 245). Detailed, Level III, environmental assessments to gauge reach-specific channel dynamics and sensitivity (p. 217) will probably be required to determine whether the favoured management approach is warranted relative to the level of risk posed, whether the proposed scheme is appropriate to the conditions, dynamics and evolutionary characteristics of the reach, and as the basis for formulating detailed designs (step 10). For each proposed action, a conceptual model of post-project ecosystem functioning should be developed as the basis for comparison with the conceptual models of reference and contemporary conditions developed earlier (steps 3 and 5, respectively). This conceptual model, essentially a refined version of the reach-level targets, should stand both as a realistic reflection of the potential for the proposed management actions and as a statement of intended benefit from which measurable success criteria can be determined for use in project evaluation.

The application phase involves options evaluation: translating the most appropriate approach into one of several related methods that should be subjected to detailed design analysis, opinion canvassing, testing and re-design (Kondolf and Downs, 1996) until a series of measures are agreed upon (step 11). Project implementation may involve initiating new land use policies, adopting new operating rules for water collection and distribution, adding or removing sediment, adding or removing engineering structures or the engineering of a new channel course (step 12). Under an adaptive management perspective, implementation may involve several appropriate measures concurrently or begin with an experimental pilot project in order to maximize the learning experience and reduce the uncertainty inherent in implementing large-scale channel management operations (p. 193).

The final phase involves project appraisal based on periodic monitoring of the measurable success criteria (step 13) nominated following development of the post-project conceptual model. The monitoring should encompass a geomorphologically relevant time frame, which is probably in excess of ten years (Downs and Kondolf, 2002) and certainly sufficient to represent 'vital ecosystem attributes and … the structure, functioning and dynamics of complex systems' (Jungwirth *et al.*, 2002: 875). The monitoring period is a precursor to a contextualized evaluation of project success (step 14). Success can be gauged in terms of project performance in physical terms, based on a favourable comparison of a post-project conceptual model with the target conceptual model (step 10), in terms of favourable public opinion, and according to the learning experience achieved regarding the efficacy and efficiency of the management method (p. 229). Following project appraisal, the

A

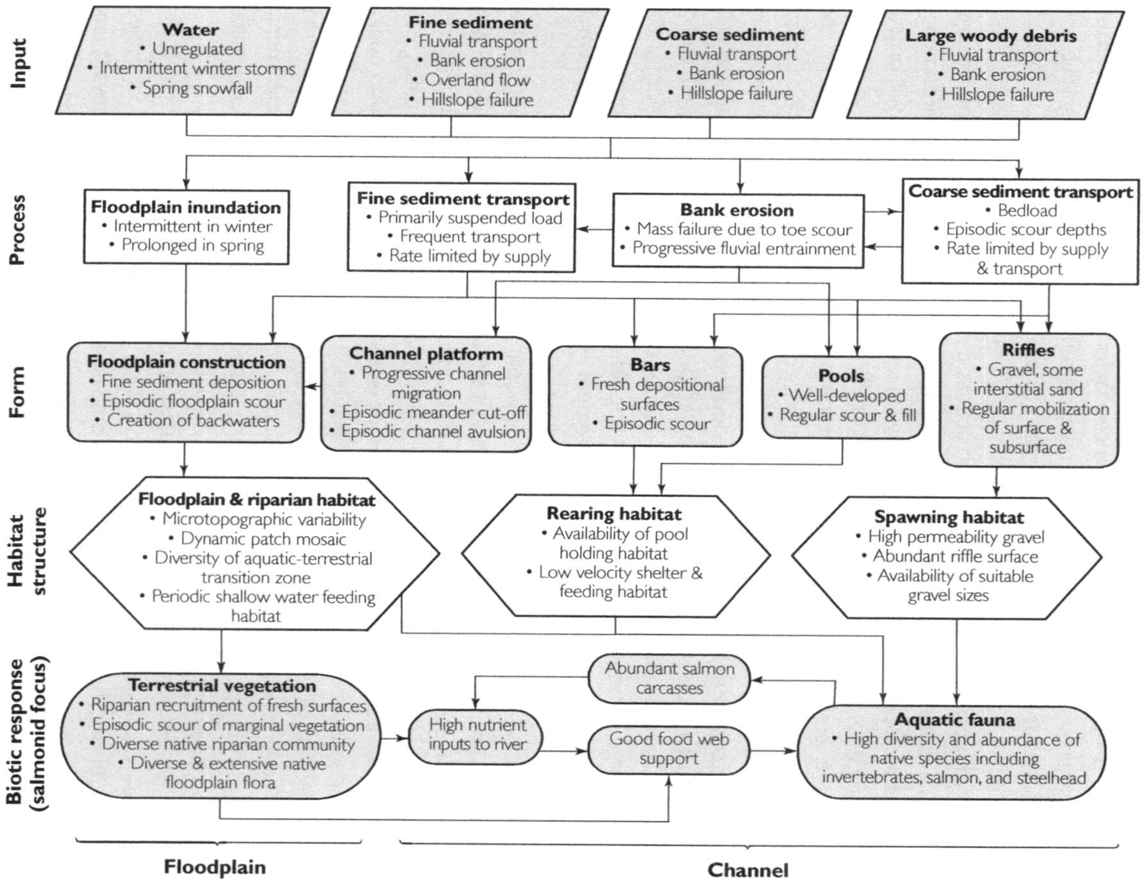
Input
Water
• Unregulated
• Intermittent winter storms
• Spring snowfall
Fine sediment
• Fluvial transport
• Bank erosion
• Overland flow
• Hillslope failure
Coarse sediment
• Fluvial transport
• Bank erosion
• Hillslope failure
Large woody debris
• Fluvial transport
• Bank erosion
• Hillslope failure
Process
Floodplain inundation
• Intermittent in winter
• Prolonged in spring
Fine sediment transport
• Primarily suspended load
• Frequent transport
• Rate limited by supply
Bank erosion
• Mass failure due to toe scour
• Progressive fluvial entrainment
Coarse sediment transport
• Bedload
• Episodic scour depths
• Rate limited by supply & transport
Form
Floodplain construction
• Fine sediment deposition
• Episodic floodplain scour
• Creation of backwaters
Channel platform
• Progressive channel migration
• Episodic meander cut-off
• Episodic channel avulsion
Bars
• Fresh depositional surfaces
• Episodic scour
Pools
• Well-developed
• Regular scour & fill
Riffles
• Gravel, some interstitial sand
• Regular mobilization of surface & subsurface
Habitat structure
Floodplain & riparian habitat
• Microtopographic variability
• Dynamic patch mosaic
• Diversity of aquatic-terrestrial transition zone
• Periodic shallow water feeding habitat
Rearing habitat
• Availability of pool holding habitat
• Low velocity shelter & feeding habitat
Spawning habitat
• High permeability gravel
• Abundant riffle surface
• Availability of suitable gravel sizes
Biotic response (salmonid focus)
Terrestrial vegetation
• Riparian recruitment of fresh surfaces
• Episodic scour of marginal vegetation
• Diverse native riparian community
• Diverse & extensive native floodplain flora
High nutrient inputs to river
Abundant salmon carcasses
Good food web support
Aquatic fauna
• High diversity and abundance of native species including invertebrates, salmon, and steelhead
Floodplain
Channel

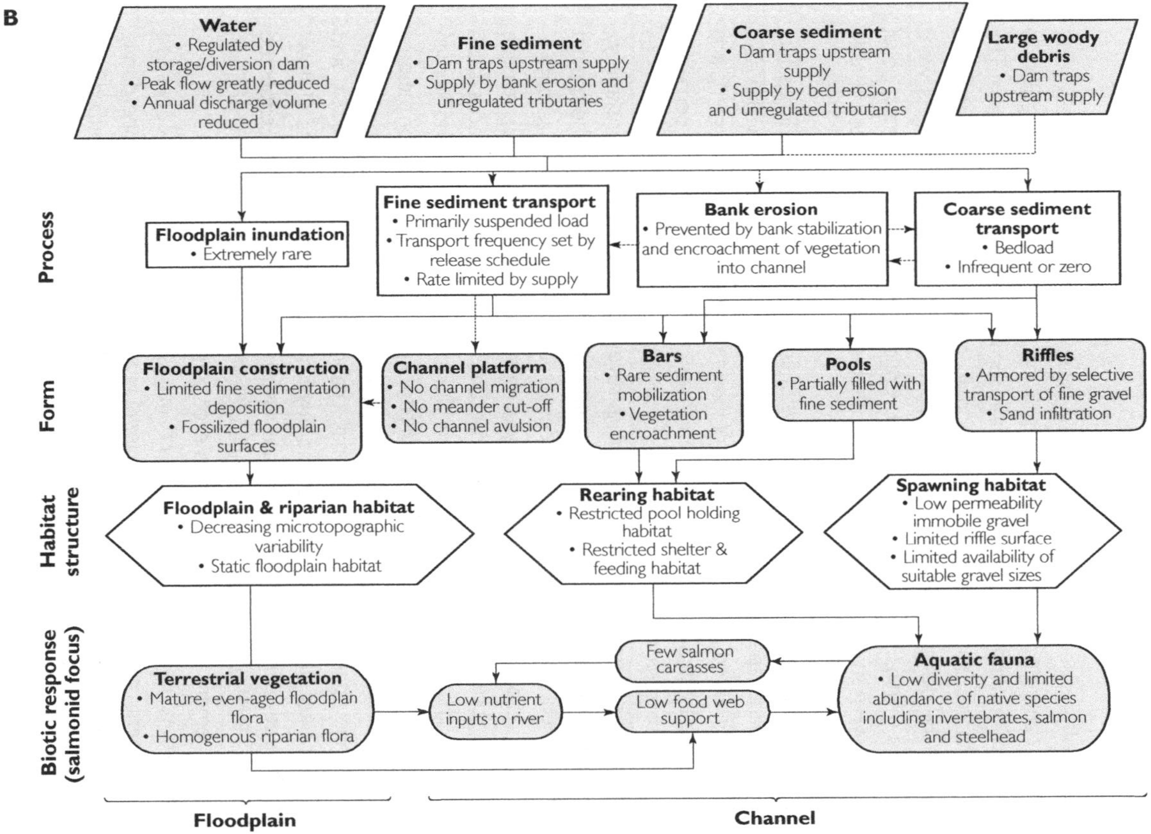

Figure 10.2 *Conceptual models illustrating process–from–habitat–biota relationships for a gravel-bedded river capable under natural conditions of supporting an anadromous salmonid fishery.* ***A*** *Processes and functions under undisturbed conditions;* ***B*** *impacts on processes and functions when flows are regulated by upstream impoundment (adapted from Stillwater Sciences, 2002)*

understanding achieved should be recycled in order to inform future management needs and ideals. The point of return into the framework will depend on the primary findings (see Figure 10.1) but will potentially affect future baseline conditions and any of the planning inputs. This emphasizes that the management vision is specific to a point in time so that, as physical condition and constraints change, the management response will change also.

Specifying environmental targets

The specification of precise, measurable targets for channel management ('success criteria') is integral to contemporary approaches (step 8 in Figure 10.1), both as a guide to suitable methods of implementation and as the basis for later post-project appraisal. The key performance target under traditional channel management was to ensure that the channel conveyed the design flood flow without appreciable erosion. Such targets were set on a 'predictive management' approach using hydraulic analysis that created a belief that the estimates were more precise than has subsequently been demonstrated (Baker, 2003; p. 110).

Potential impacts of predictive uncertainties were buffered using confidence limits, factors of safety, redundant capability and fail-safe designs: an approach that works well in rigid systems but is poorly suited to systems capable of long-term evolutionary change (Clark, 2002). Rarely were the estimates subjected to performance monitoring; instead, perceived inadequacies were corrected by maintenance operations or by requests for funding to permit supplemental engineering construction (e.g. revetments to constrain instability). Under contemporary approaches, management targets will vary according to the project objectives and may require indicators related to conservation achievements, hazard-minimization and water resources provision. The targets must not only be technically feasible and scientifically valid but also socially feasible, integrating human concerns into ecosystem functions to ensure that the link between conservation-based approaches and human quality of life is abundantly clear (Cairns, 2000). Indicators are required to monitor the projected social and managerial benefits of the project, including attributes of conveyance capacity and channel stability in 'multifunctional' channel environments (Downs and Thorne, 2000) but must also assess whether the physical and biological impacts of flow-driven, effects-centred management solutions are sustainable and deliver their expected environmental benefit. Evaluation of performance indicators is also the basis for periodically refining management actions as the system evolves (Clark, 2002; Figure 10.1).

The expected environmental benefit of project actions should have been documented as a post-management conceptual model of channel reach functioning developed during the design process (Figure 10.1). Ideally, conceptual models should include quantified links between their components but even simplified, qualitative conceptual models have great complexity related to the interaction between channel process, form, habitat and biota (e.g. Figure 10.2). Performance indicators derived from the environmental targets can be chosen from any level in the model but should ideally respond to change in a known way, respond rapidly to give early warning of adverse changes, and should simplify complex data to a level at which collection and communication are effective (Clark, 2002). As such, a spectrum of target setting prospects exists based on either 'top-down' or 'bottom-up' approaches to evaluation (Table 10.1). Top-down approaches focus on process and form as indicators of project success through catchment-level improvements. They can be characterized as holistic measures of ecosystem 'health' (e.g. Sparks, 1995) that do not require every internal structure and mutual inter-relationship within the ecosystem to be understood (Zonnerveld, 1990 in Lowell *et al.*, 2000). Because top-down approaches focus on aspects of hydrology or channel form that are easily measured and are reasonably robust, the indicators are generally empirical. For instance, in setting environmentally beneficial flow releases below dams, targets will centre on emulating aspects of the natural (pre-dam) hydrograph to maintain valued biological and geomorphological processes (p. 284). In improving channel form, targets are often related to reference channel conditions on assumption that 'the river is the best model of itself' (Shields, 1996), that is, that natural channel configuration is optimal for native flora and fauna. This may involve using 'carbon copy' indicators of the channel's historical condition, if the condition can be parameterized and is still appropriate to the contemporary catchment, using a natural analogue provided by a contemporary reference reach or using indicators that relate to boundary conditions provided by 'donor' data from other channels. Alternatively, 'physical biotopes' of flow and morphology types could be used to target a preferred mosaic of channel habitats (Newson *et al.*, 1998b; Newson and Newson, 2000).

In contrast, targets derived from bottom-up, reductionist performance indicators focus on the project biota and habitat as sentinels of collective ecosystem health. The approach demands a detailed understanding of the risk to each sentinel derived from accumulated stressor-effect data (Leuven and Poudevigne, 2002), and 'weight-of-evidence' approaches that combine field data with controlled field and laboratory experimental data have been promoted in this regard (Lowell *et al.*, 2000; Table 10.2). Indicators generally involve the occurrence frequency or distribution of valued habitats or charismatic mega-fauna or flora such as native fish and riparian tree species that are highly dependent on pre-disturbance conditions for their survival and abundance. Such an approach may provide the ultimate indication of project success (e.g. Botkin *et al.*, 2000) but practical constraints can limit its utility. For instance, other than the fundamental difficulty of targeting absolute numbers of (for example) native fish or trees that will result from a project, biological survival is often dependent on upstream and downstream factors that cannot be accommodated by the project. Related concerns include the annually variable impacts of disease, catastrophic fish mortality caused by accidental pollution events (e.g. chemical spills), complex foodweb interactions that vary the amount of available food and the effectiveness of prey on the target population, and the potential impact of flood and/or drought events on annual fish populations. Therefore, as biological indicators are more inherently variable than physical indicators (Clayton, 2002), project target setting often involves complementary measures of biological or habitat *potential* derived from simulation models (Table 10.1). For instance, habitat-based approaches based on a combination of target habitat parameters that define potentially 'good' habitat areas at different life-cycle stages for different species are often used. More recently, sophisticated models based on mechanistic information of individual fish preferences and on the year-to-year dynamics and survival of population of individual organism and population preferences have been developed (but have had little management application to date; Railsback and Harvey, 2002). Alternatively, a bottom-up

Table 10.1 A spectrum of methods for setting performance indicators for environmental conservation in river channel management

Approach	Subject of target	Method	Indications of Success
Top-down	Hydrology	High flow prescriptions in dammed rivers	Achievement of target flow magnitude, duration, timing and frequency. Replication of unregulated annual hydrograph (e.g. USFWS/Hoopa Valley Tribe, 1999).
	Channel form	'Carbon copy' of historical reference reach	Achievement of 'pre-disturbance' channel dimensions (e.g. Brookes and Sear, 1996).
		'Natural analogue' from contemporary reference reach	Achievement of 'comparison standards' from neighbouring reach, such as a template of velocity–depth pairs (e.g. Riverine Community Habitat Assessment Concept (RCHARC); Nestler *et al.*, 1993; Peters *et al.*, 1995).
		Relative parameterization when channel-specific target data are lacking	Achievement, or movement towards, reference parameters of 'channel naturalness' obtained from other rivers (e.g. Skinner *et al.*, 2004).
	Channel habitat	Mosaic of physical biotopes	Achievement of a reference assemblage of hydraulic habitats based on biotopes of flow and morphology combinations (e.g. Padmore, 1996; Newson and Newson, 2000).
		Mosaic of functional habitat	Achievement of a diversity of hydraulic and invertebrate habitats based on the assemblage of aquatic macrophytes and sediment types (e.g. Harper and Everard, 1998).
	Biota	Hydraulic habitat suitability	Achievement of a highly usable instream habitat based on the 'weighted usable area' of channel conditions using flow simulation models and empirically defined habitat preference curves (e.g. the Physical Habitat Simulation Model (PHABSIM) component of the Instream Flow Incremental Methodology; Bovee, 1982).
		Population-based biomechanical and habitat suitability	Achievement of optimal conditions of a combination of habitat conditions and individual preferences to maximize year-to-year population survival.

Approach	Subject of target	Method	Indications of Success
↑		Organism-based, biomechnical suitability	Achievement of fish preference conditions, or low mortality risk, based on mechanistic rules for fish movement, feeding and predator avoidance (Railsback *et al.*, 1999).
Bottom up		Flora/fauna surveys	Empirical evidence of improvement in biodiversity and/or biomass, e.g. in salmon-bearing rivers, a targeted improvement in the ratio of egg-to-smolt survival.

Table 10.2 Formalized set of criteria to generate weight-of-evidence risk assessment for large rivers

1	Spatial correlation of stressor and effect along gradient from more to less exposed areas
2	Temporal correlation of stressor and effect relative to time course of exposure
3	Plausible mechanism linking stressor and effect
4	Experimental verification of stressor effects under controlled conditions and concordance of experimental results with field data
5	Strength: steep exposure and response curve
6	Specificity: effect diagnostic of exposure to particular stressor
7	Evidence of exposure to contaminants or other stressors
8	Consistency of stressor-effect association among different studies within the region being studied
9	Coherence with existing knowledge from other regions where the same or analogous stressors and effects have been studied.

Source: Culp *et al.* (2000); Lowell *et al.* (2000).

system of template mapping can be used to determine habitat potential based on a mosaic of 'functional habitats' of aquatic macrophytes and sediment types, possibly then collected according to physical biotope (Harper *et al.*, 2000).

10.2 'Working with the river'

Having determined the objectives and specific environmental targets for channel management, the considerable challenge remains to achieve the targets using regulatory policies, science-based and engineering measures that work with the river, so that the management outcome is self-sustaining and approaches the design-with-nature ideal (see Chapter 7). Various measures commonly associated with each of the five generic management approaches (pp. 245–62) are outlined below in order of application preference. The measures are likely to require detailed (Level III) environmental assessments of channel dynamics as the basis for translating the preferred methods into detailed designs.

Catchment and corridor measures

Nonstructural measures are designed to ensure preservation of high-value habitats or to facilitate channel recovery processes in channels with sufficient energy and upstream sediment supply for recovery to occur (p. 251). However, where disturbances (e.g. straightening) have initiated a complex set of geomorphic processes that will eventually propagate regionally causing new compound channel cross-sections (e.g., Graf, 1977), natural recovery processes may threaten nearby rare habitats and cause significant risk to bridges, houses or agricultural land through incision and channel widening. In these cases, nonstructural methods may need supplementing by structural measures. In deriving the right balance of measures, consideration should be given to the level of risk posed by the natural trajectory of channel recovery against the potential risk posed by holding the channel in an out-of-equilibrium condition by structural means.

Catchment-based policies

Catchment-based land management policies focus on benign neglect and best management practices to restore or prevent the disturbance of natural water and sediment regimes. Under benign neglect, the channel is expected to recovery naturally by allowing the land to fallow. This facilitates the regeneration of native vegetation, allowing the return of pre-disturbance rates of runoff and sediment production and the channel to recover physically. For instance, rivers of the Pere Marquette catchment in Michigan, USA began to recover following collapse of the timber industry in the early 1900s but designation of large parts of the catchment as the Manistee National Forest in 1938 aided recovery further and the river is now designated as both a natural and scenic river with federal policies to manage the sports fishing and canoe industries and restrict commercial and residential development (National Research Council, 1992: 211–12). Utilizing benign neglect requires good communication on the part of channel managers to the public and funding agencies to ensure that the measure is not perceived as 'doing nothing'. It also requires detailed environmental assessment in order to gauge the potential for channel recovery and to predict the final form of the channel: in instances where disturbance has caused channel incision, benign neglect may not result in the recovery of the pre-disturbance channel morphology within management timeframes (cf. Fryirs and Brierley, 2000, Figure 8.7). It may be necessary to remove inchannel structural controls if they exist. Benign neglect policies are only practicable in catchments of low population density whereas best land management policies can be applied in heavily populated areas to reduce the requirement for direct channel engineering works to achieve river channel management objectives. The policies generally involve source control measures designed to reduce the volume and peakedness of storm runoff or 'downslope' measures to slow storm runoff and intercept eroded sediment (Table 10.3).

Catchment-based water management policies that include protection of the channel hydrosystem are particularly important (and, thus, politically contentious) in water-scarce

Table 10.3 Best practice land management policies for reducing accelerated runoff and sediment delivery to river channels

Land use	Measure	Purpose
Urban	Temporary fencing	Prevent fine sediment delivery to channel during construction
	'Soakaways': grassed flushes, gravely ditches	Slow urban runoff, allow greater infiltration, reduce fine sediment delivery
	Retention/detention basins, including constructed wetlands	To regulate and delay urban runoff and sediment delivery. In urbanizing areas, basins are often designed to retain flow volumes at pre-urban levels
Agriculture	Contour ploughing, terracing	Reduce soil erosion caused by sheetwash, rilling and gullying
	Ground cover crops	Reduce soil erosion
	Settling basins with drop pipes	Encourage sediment deposition prior to entering channel network, drop pipes prevent gullying at point of entry to channel
	Hedgerows or trees between fields	Increase aerodynamic surface roughness and enhance deposition of wind-eroded soils
	Constructed wetlands/ embayments	Filter fine sediment and chemicals from agricultural drains prior to discharge to channel
	Disconnected ditches	Reduce increased runoff rates and erosion potential from drained fields
Commercial forestry	Brush cover over felled areas	Reduces rainsplash erosion and reduces downslope sediment delivery rates
	Disconnected ditches	Reduces increased runoff rates and erosion potential from drained hillslopes
	Road management	Best practice road planning, design, location, construction, drainage and maintenance, including at stream crossings, to reduce sediment derived from road fill failures, surface erosion of gravelled roads, sediment and LWD plugging of culverts and consequent failures
	Road de-commissioning	Allows ploughing and re-contouring of hillslopes along obsolete roads

Source: FISRWG (1998).

environments and may result in re-regulation of reservoir flows or dam removal. In many countries river management authorities now have statutory obligations to fulfil conservation objectives as part of their duties (e.g. p. 181) and in some countries, such as the USA, flow allocation to meet instream ecosystem objectives is increasingly the result of legal resolution. The benefits of creating a legal basis for instream flows include a clearly legitimated and articulated flow right and the ability to challenge later proposals to alter stream conditions from those existing when the right was first established (Gould, 1977; Lamb and Doerksen, 1990 in National Research Council, 1992). In addition, Federal Reserved Water Rights for instream flows can be established to protect present *and future* flows in national forest, parks, wildlife refuges and wild and scenic rivers. In Europe, the incoming Water Framework Directive (European Community, 2000) is centred on the general protection of aquatic ecology, and specific protection of unique and valuable habitats, drinking and bathing water sources. It therefore encompasses notions of 'strong' sustainability (Newson, 2002) in that the policies are driven by ecosystem demands rather than by human demands (cf. Pearce, 1993). All surface waters are required to reach 'good ecological status' by 2015 in terms of their biological community, hydrological and chemical characteristics, requiring emphasis on the assessment and monitoring of river channel features relevant to ecosystem protection (see Chapter 8). The directive (2000/60/EC) is likely to have the greatest impact on river channel management in European Union nations since legislation requiring environmental assessment in 1985 (directive 85/337/EEC).

Corridor-based strategies

In larger catchments, or those with multiple and diverse land uses, where catchment-based policies are not feasible, floodplain or corridor management strategies may be used. Stemming from attempts to curb the economic losses associated with floods, floodplain management policies have generally involved developing land upstream of flood-prone areas for flood storage, the use of regulatory land use 'zoning' or structure relocations to reduce the inherent risk to human infrastructure ('vulnerability modification'; Smith, 1992), the purchase of corridor lands or water rights explicitly for conservation purposes, and leasing of 'flood easements', corridor areas that are permitted to flood regularly. While the objectives of floodplain management plans are generally related to maximizing human safety and minimizing economic losses, the corollary of such objectives are often of benefit to retaining or recovering the physical integrity of river channels by promoting lower intensity land uses near to the channel. For instance, in New South Wales, Australia, floodplain policies included planning and managing flood-liable land 'in a manner compatible with the assessed frequency and severity of flooding' while an integral part of the management process was to identify major ecological assets in the floodplain and their threats, and to propose measures to protect these assets (New South Wales Government, 1986).

Where the river management agency has virtually no jurisdiction beyond a narrow strip of land bordering the channel, there are physical measures that can be used to protect the river margin environment. Measures vary in lateral spatial extent from the channel corridor to the toe of the channel bank. Buffer strips were promoted initially for their ability to filter fine sediments suspected of carrying pollutants, and to allow dissolved chemicals to be taken up by plants, but have since been promoted for conservation management purposes including the restoration of native shrub or woody vegetation beside the channel and the initiation of wetland habitats (Figure 10.3). Defining the width of a river corridor based on functional water quality, ecological or geomorphological criteria is problematic and suggested values vary from 1 m to over 200 m (Large and Petts, 1994) partly because they are scale-dependent on channel width. From a hydrological perspective, the corridor is often related to the 'flood envelope' of inundation caused by a flood of a known return period (e.g. the 100-year flood) and is thus derived from flood frequency statistics in combination with application of a hydraulic model. Alternatively, where livestock pressure is the primary cause of channel degradation, restricting livestock access to the channel through fencing is beneficial and in Otter Creek, Nebraska, USA, within three years of Nebraska Game and Parks Commission leasing the headwaters and excluding livestock

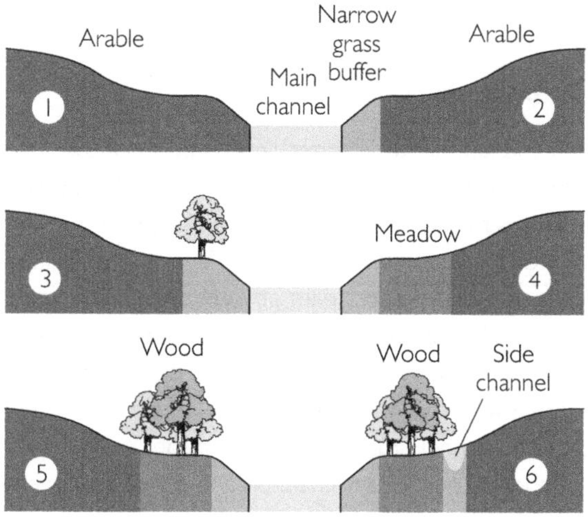

Function	*Maximum score*	*Scenario: increasing buffer complexity (score)*					
		1	*2*	*3*	*4*	*5*	*6*
Biotic diversity/ environmental enhancement	*3*	*0*	*0/1*	*1/2*	*1/2*	*2*	*3*
Overland flow control	*3*	*0*	*1*	*1*	*2*	*3*	*3*
Subsurface flow control	*3*	*0*	*0*	*0/1*	*0*	*1–3*	*3*
Bank stability	*3*	*0*	*1*	*2*	*1*	*3*	*3*
Water storage capability	*3*	*0*	*0*	*1*	*1/2*	*3*	*3*
Recreation provision	*3*	*0*	*1*	*2*	*2*	*3*	*3*
Aesthetic value	*3*	*0*	*1*	*2*	*2*	*3*	*3*
Total		*0*	*4–5*	*9–11*	*9–11*	*18–20*	*21*

Figure 10.3 *Theoretical benefits of vegetated buffer strips in the management of river corridors, scored according to conditions ranging from no buffer present to the existence of an ecologically complex buffer. Scores are attributed to the performance of the buffer across seven different functions (Large and Petts, 1994)*

access, the stream had narrowed, stream banks stabilized, pools formed, sand in the gravel spawning beds was reduced and the water temperature was lower (Van Velson, 1979, in National Research Council, 1992). Overall, the conditions for native fish populations were far more favourable. However, fencing can be expensive for private landowners and the fencing type needs to be chosen with care to ensure that wildlife access is not restricted along with livestock. By the water's edge, bankside tree planting can be a highly productive means of habitat improvement (Ward *et al.*, 1994). In flood control channels in the northern hemisphere, trees should be located on the southern bank to maximize shade and downstream of a meander inflexion point where the maximum flow velocities hug the outer northern side of the channel (Downs and Thorne, 1998a). Single stem rather than multistem species may be preferred, planted parallel to the flow so as to present the minimum obstruction to the passage of flood flows. Careful management may be necessary where overhanging and decayed branches could impede flood flows, whereas in channels without stringent flood control concerns, the fall of branches may be one of the major benefits of bankside trees.

Improving network connectivity

Improving longitudinal and lateral flow connectivity throughout the channel network is an environmentally preferred option in regulated river systems (National Research Council, 1992) because it addresses management concerns from a process-led and potentially network-wide perspective (Section 9.3.3). Partial flow de-regulation by prescribing environmental high flow releases downstream of dams and abstractions can assist in aligning the frequency of flows of a particular magnitude, duration and timing more closely with environmental targets (Section 10.1.2), but is subject to operational conflicts (Kondolf and Wilcock, 1996). Alternatively, removing superfluous weirs and other run-of-the-river obstructions, setting back channel embankments from the river's edge, removing or lowering of barriers to reconnect secondary channels and removing large dams may result in full deregulation of longitudinal and lateral flow processes and the reinstatement of sediment transport continuity, but may conflict with water resources and hazard management objectives.

Environmental high flow releases

Prescribing environmental high flows is an extension of the concept of setting minimum acceptable 'compensation' flows in regulated rivers. These were commonly derived as a statistical threshold from the pre-dam flow duration curve such as the 95th percentile, a percentage of the Average Daily Flow or as the mean annual minimum 7-day flow frequency statistic (Petts and Maddock, 1994). Scientific understanding of the critical role of flood events, conceptualized as the 'flood pulse advantage'(Bayley, 1991; see Chapter 3) highlighted the environmental deficiencies of minimum compensation flows (Hill *et al.*, 1991) and shifted attention towards prescribing a varied flow regime, manipulating components of the flood hydrograph, such as the baseflow, rising limb, recession limb and peak discharge, to suit conservation needs. Prescribed high flows can be defined according to their hydrologic return period, their impact on sediment transport processes or on the channel morphology (Reiser *et al.*, 1989) and can be classified according to their approximate magnitude and purpose (Table 10.4). In general, there is the prospect of targeting a wider range of channel process and habitat improvements using the higher flow types, but they may generate greater conflict with other management objectives, especially related to flood control and bank erosion on developed floodplains, and irrigation and hydroelectric purposes in water-scarce environments. High flows causing sediment transport may also have the long-term effect of reducing the volume of available spawning gravels in rivers that are divorced from their upstream sediment supply, requiring careful analysis to minimize gravel loss and, potentially, supplementary measures such as periodic gravel augmentation (Kondolf and Wilcock, 1996). Environmental high flow parameters of magnitude, duration, frequency and timing can be defined either by the hydraulic habitat requirements of the targeted flora and fauna or by the geomorphological attributes

Table 10.4 Classification of prescribed high flows according to their magnitude and purpose

Flow class	Approximate magnitude	Purpose
Depth-maintaining	Various – from elevated baseflows to overbank flows	• benefit fish passage, holding and favourable rearing • assist the recruitment of riparian vegetation • provide the 'flow pulse' to secondary channels (Tockner, *et al.*, 2000)
Bed-maintaining	Moderately high inchannel flows	• scour pools of fine sediment • prevent interstitial filling and remove surficial fine sediment from gravel beds • maintain the 'looseness' of coarse sediment on riffles to benefit fish holding and spawning
Morphology-maintaining	Bankfull or near-bankfull flows	• scour fine sediment from the thalweg and promote marginal accretion in sand-bed rivers • scour fine sediment ingress and preserve channel width in gravel-bed rivers • promote channel migration to maintain aquatic and riparian habitat diversity
Floodplain-maintaining	Overbank flows	• promote sediment deposition and erosion on floodplains to encourage multi-age riparian habitats through bank erosion, avulsions and meander cut-offs

Source: Downs *et al.* (2002b).

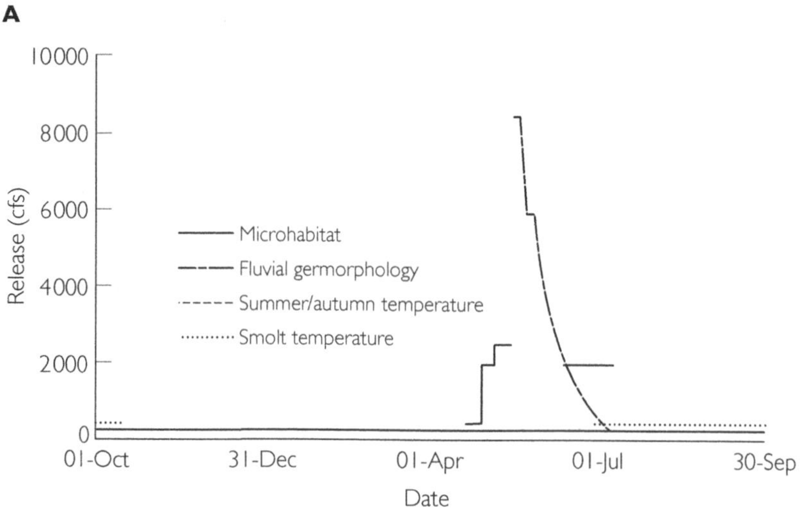
A
10000
8000
6000
4000
2000
0
Release (cfs)
Microhabitat
Fluvial germorphology
Summer/autumn temperature
Smolt temperature
01-Oct
31-Dec
01-Apr
01-Jul
30-Sep
Date

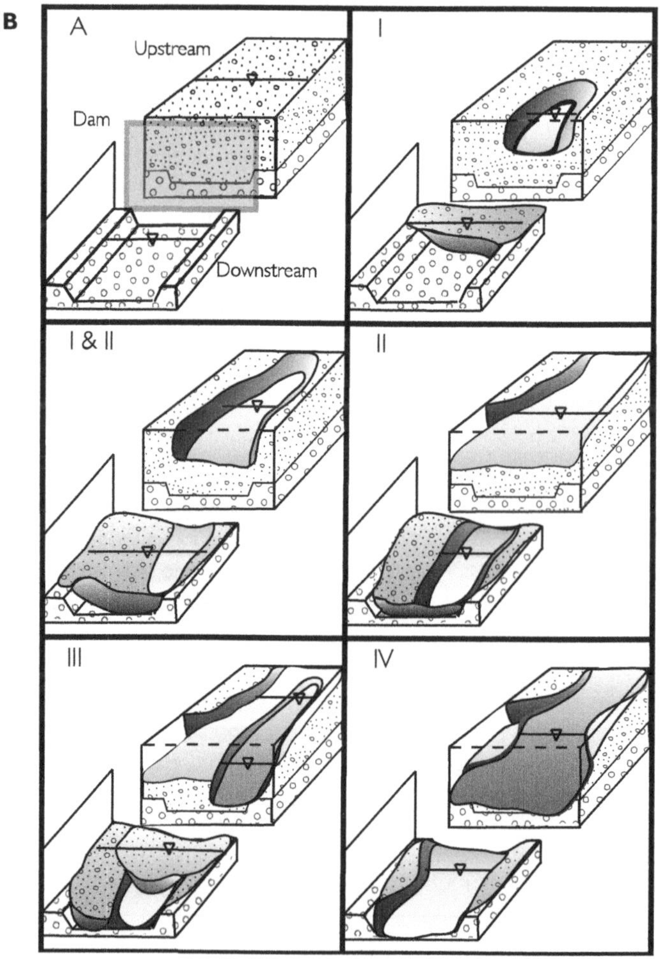
B
A
Upstream
Dam
Downstream
I
I & II
II
III
IV

required of the channel. The result is a composite hydrograph of seasonal flow requirements that emulates components of the pre-dam hydrograph. On the Trinity River, in northern California, a composite hydrograph was constructed from environmental target criteria (Figure 10.4A) and the necessary frequency of achieving each parameter defined so that higher magnitude flow release components were targeted only for wetter years (United States Fish & Wildlife Service (USFWS)/Hoopa Valley Tribe, 1999).

Overall, there are insufficient examples to date of carefully planned and targeted high flow prescriptions to categorically prove their claimed utility. Problems are exacerbated when trying to superimpose high flow prescriptions over legislation that pre-dated such concerns, such as for the Snowy River, Australia (Erskine *et al.*, 1999). Practical application seems likely to require careful compromise between management objectives and supplementary techniques such as gravel augmentation and morphological reconstruction of the channel in order to target the desired environmental objectives (USFWS/Hoopa Valley Tribe, 1999; and see Chapter 9). Overall, high flow prescription is an exercise in creating an inventive flow regime that interacts with a highly modified sediment budget to produce a sustainable and ecologically acceptable river channel perimeter, but one that has few naturally formed counterparts.

Reconnecting backwater channels

Many lowland rivers have been regulated progressively by channelization measures involving confinement of flow to a single channel of a formerly multichannel system. Partial deregulation of flows in this case involves restoring lateral flow connectivity by setting back flood control embankments to allow more frequent floodplain inundation, and/or by modifying or removing barriers to relic channels to allow the re-establishment of multichannel flows, especially in the large rivers. These strategies are part of a growing movement towards long-term 'floodplain restoration' (Brookes, 1996a; De Waal *et al.*, 1998; Adams and Perrow, 1999; Schiemer *et al.*, 1999) and for which few large-scale examples exist (Buijse *et al.*, 2002).

Setting back of flood control embankments generally involves either purchase or leasing ('flood easements' in the USA) of riparian land. This action provides multiple channel management opportunities including greater freedom for the river to meander, the establishment of high quality riparian habitats within the floodway, the potential for additional flood capacity to benefit flood control for adjacent and downstream settlements and the prospect of constructing lower embankments that reduce the risk to adjacent set-

Figure 10.4 A *Illustration of a composite hydrograph designed to meet multiple habitat objectives using strategic flow releases at different periods of the year. This example designed to meet microhabitat, fluvial geomorphic, summer/autumn temperature and smolt temperature management objectives during a wet water year for the Trinity River, northern California (USFWS and Hoopa Valley Tribe, 1999).* **B** *Prospective channel evolution upstream and downstream following the removal of a low-head dam in which sand and gravel sediments are stored (Wooster, 2002). Evolutionary sequence is based on flume study. Typically, two phases of vertical incision are initiated (phases II and V) resulting in two downstream sediment pulses. Reproduced with permission of J. Wooster*

tlements should the embankment fail or be overtopped. Riparian habitat creation will require the embankments to be set-back sufficiently so that slackwater habitats in the floodway still exist during flood flows, and in areas of previously modified drainage and topography (e.g. laser-levelled fields), consideration is required of the drainage pathways of receding flood waters to ensure that fish stranding is minimized (Somner *et al.*, 2001).

The reconnection of relic channels is an increasingly popular option in formerly meandering or multichannelled large lowland rivers that have been straightened or 'trained' to ensure a narrow navigation channel. In the Kissimmee River, Florida, the original tributary sloughs were not infilled following channelization in the 1960s, allowing them to be reconnected and new weirs constructed across the mainstem channel to encourage flow into the former courses (Toth, 1996; Toth *et al.*, 1998). In the rivers Rhine and Danube (and Rhône, Henry *et al.*, 1995), reconnection of secondary channels is one of a range of adopted approaches that include connecting isolated water bodies and creating new secondary channels in locations where floodplain modification and safety risks do not allow relic courses to create a functional lateral flow gradient (Buijse *et al.*, 2002; Table 10.5). Management considerations may require that the reconnected secondary channels do not receive continuous flowing water and in these cases control structures such as notched weirs, submersible dikes, sluices and culverts may be necessary to reconcile the objectives for the mainstem and secondary channels. However, the functional challenge in these cases is to ensure that sufficient flow occurs to prevent stagnant conditions (Schiemer *et al.*, 1999) but not so much that deposition occurs in the main channel (Schropp and Bakker, 1998). In other cases, a downstream barrier may be used to create breeding and rearing areas for lentic species (Cowx and Welcomme, 1998) although the prospect of losing the channel to fine sediment infilling should be evaluated. Fine sediment deposition in disconnected secondary channels may require dredging if the aim is to maintain 'oxbow' type habitats within a largely channelized river setting.

Weir removal

Improved longitudinal flow connectivity can be achieved by removing obstacles such as low weirs, or redundant drainage or irrigation structures, providing improvement to natural landscape aesthetics, allowing fish access to disconnected spawning or rearing grounds. In Denmark, more than 500 obstacles to fish and invertebrate passage, including many low weirs, had been removed by 1995 (Nielsen, 1996). Weir removal should follow a fisheries appraisal to see whether the weir provides benefits to native species. For instance, native species possess an ability to climb, on adapted lower fins, over all 44 weirs in the lower reaches of the Waters of Leith, Dunedin, New Zealand, whereas the passage of introduced salmonids is restricted by their ability to jump five of the weirs (see Downs and Caruso, 2000). There may also be heritage value attached to artificially maintained weir-based habitats such as on River Cole, Wiltshire, England, where the Coleshill Mill sluices and dam boards were restored to allow greater water level control in order to protect ancient willows and water table dependent wetland habitats to the right bank of the main river (Holmes, 1998). Alternatives to weir removal include the introduction of baffles on the bed of concrete channels to reduce velocities, narrow slots cut into weirs to attract fish and reduce barrier heights, and the construction of 'leaky' weirs of boulders intended

Table 10.5 'Characterisation of secondary channel restoration projects along the Rivers Rhine and Danube. Channels have been rehabilitated by re-opening the upstream end, connecting isolated water bodies to the main channel, or by creating new secondary channel. Discharge values refer to flow through secondary channel at mean water level' (Buijse *et al.*, 2002: 899)

Location	**Year**	**River**	**Country**	**Restoration measure**	**Upstream connectivity (days yr^{-1})**	**Discharge (% of main channel flow)**	**Length (km)**	**Width (m)**	**Sediment trap**	**Reference**
Vén Duna	1998	Danube	Hungary	Re-opening	345	3	4.3	50–100	No	(Marchand, 1993 feasibility study)
Regelsbrunn	1998	Danube	Austria	Re-opening	220	2	10	Ca. 100	No	(Schiemer *et al.*, 1999)
Opijnen	1994	Rhine	Netherlands	Re-opening	Permanent	0.5	1	10–150	No	(Grift *et al.*, 2001; Simons *et al.*, 2001)
Beneden-Leeuwen	1995, 1997	Rhine	Netherlands	Connection	Permanent	0.3	2.5	10–30	Yes, inlet	(Grift *et al.*, 2001; Simons *et al.*, 2001)
Gameren 1	1996	Rhine	Netherlands	Creation	325	0.8	1	40–80	No	Jans (unpublished data)
Gameren 2	1996	Rhine	Netherlands	Creation	100	<0.3	0.5	20–60	No	Jans (unpublished data)
Gameren 3	1999	Rhine	Netherlands	Connection	Permanent	1.5	2	40–150	Yes, at ¾ length	Jans (unpublished data)

to raise tailwater levels and create sufficient acceleration space for jumping. Where removal is preferred, new barriers may be created electronically via lights, pulses or sound to achieve specific fisheries management targets, especially in deeper rivers (FISRWG, 1998).

Large dam removal

Dam decommissioning and removal creates the prospect of restoring both the network flow and sediment transport regimes (in the absence of other regulating structures) thus allowing natural recovery processes to rehabilitate the river channels upstream and downstream of the former dam. Dam removal is now being pursued for river restoration purposes (American Rivers *et al.*, 1999), although the impetus for dam removal has historically involved the cost of repairing privately operated dams that were deemed unsafe (Shuman, 1995). In the USA, since the 1940s over 400 dams have been removed (Shuman, 1995), of which almost 200 dams were removed during the 1990s (American Rivers *et al.*, 1999). Over 50% of the removed dams are less than 5 m high. Dam removal requires a long-term planning and management perspective (National Research Council, 1992; Wunderlich *et al.*, 1994) and is likely to be highly contentious (Shuman, 1995) as local property values and recreational industries may centre on the impounded lake and, downstream, encroachment of human activity onto the regulated floodplain may now create a far greater flood risk than existed pre-dam. Questionnaire surveys in communities around 14 dams removed in Wisconsin, USA revealed that many stakeholders perceived a sense of loss related to recreational opportunities, fish and wildlife habitat and to nostalgic aesthetic values (Born *et al.*, 1998). The same perceptions were also cited by others as 'gains' from the dam removal, along with improvements to safety and the reduction in ongoing financial burdens related to dam maintenance.

The primary environmental concern with dam removal is often for the fate of the accumulated sediment stored behind the dam that, when released, may cause downstream flood risk to be increased and fish spawning habitats to be smothered. Often greater than one-third of the total stored sediment is released within one year of dam removal (Doyle *et al.*, 2000b). However, for five streams in southern Wisconsin (Lenhart, 2000, in Doyle *et al.*, 2000b), discharges calculated at different periods after dam removal suggested that the statistical recurrence interval of the bankfull discharge six years after dam removal was 2–31 years, and 40 years after dam removal was 6–15 years, suggesting that volume of deposited sediment following upstream release has not been sufficient to offset the amount of incision caused following dam closure (e.g. Williams and Wolman, 1984). Research indicates that, particularly for coarse-grained material, the stored sediment disperses downstream, rather than progressively translating as a 'sediment wave' thus increasing flood risk only at locations adjacent to the dam site (Lisle *et al.*, 2001), but potentially causing long-lived aggradation. Upstream, removal of the dam and erosion of the sediment wedge is liable to cause rejuvenation of the system leading to phased channel incision and widening not unlike that depicted in the Channel Evolution Model (Doyle *et al.*, 2002; Wooster, 2002; Figure 10.4B) in which case downstream sediment supply will vary in time according to the phase of evolution (Pizzuto, 2002; Doyle *et al.*, 2003b). A spectrum of environmental issues associated with dam removal is summarized in Table 10.6.

Table 10.6 Summary of environmental issues related to dam removal

Category	**Issue**
Sediment transport	Dynamics of sediment transport to downstream reaches during and following dam removal. Changes to the river channel morphology above and/or below the dam during and after removal. Potential for downstream beach-building.
Hydrology and water quality	Potential for the pulse release of toxic sediments stored behind the dam to outstrip the assimilative capacity of the river during and following dam removal. Degradation of water quality during the drawdown for dam removal, including drinking water supplies. Impact on the elevation of the regional groundwater table Potential for removal to lead to a net gain or loss in wetland area. Requirement for floodplain regrading to ensure floodplain function following removal.
Ecology	Potential for removal to enhance the recovery of threatened or endangered species. Potential for removal to result in changes to populations of unwanted invasive species or to restore native species Timeframes for recovery of macroinvertebrates, fish and vegetation communities according to the mode of dam removal.
Catchment context	Environmental considerations of the dam and its removal relative to environmental conditions elsewhere in the catchment. Prospect that catchment changes since dam construction will prevent ecosystem restoration goals being achieved.

Source: Developed from Shuman (1995); Stanley and Doyle (2002); The Aspen Institute (2002); The Heinz Center (2002); Doyle *et al.* (2003a).

Responding to the lack of long-term evidence for dam removal impacts, various removal scenarios have been proposed including complete removal in one action, staged breaching and partial removal. Several early dam removals used complete removal using explosives (including the 17-m-high Sweasey Dam, California in 1970; Winter, 1990) but in future staged breaching is likely to be preferred, in an attempt to ensure that large volume sediment releases do not coincide with the spawning season of anadromous fish (Doyle *et al.*, 2000). In addition, it is often proposed that some or all of the accumulated sediment is removed by excavator prior to the dam's removal, rather than allowing the river to incise its way through the sediment.

In addition to environmental and engineering issues, there are a suite of other issues to consider in advance of dam removal. These include public safety and security, past and potential future dam uses, the impact of dam removal on other regional and catchment management plans, legal requirements, social values, stakeholder education and community interests, economic responsibilities and funding sources, the decision-making process, funding and the political context (The Aspen Institute, 2002; The Heinz Center, 2002).

Instream measures

Instream measures (small structures, bedform recreation, cover devices, sediment traps) have been used to mitigate and enhance river channel management operations since the early 1980s (e.g. Shields, 1982) while conscious use of measures such as boulder placement to improve fish habitats can be traced back much further (e.g. Towner-Coston, 1936). Measures have been applied mostly to reduce bank erosion problems by diverting flow and to improve degraded fish habitats by increasing channel morphological diversity (p. 255). Structures can be designed to eventually be destroyed or buried by the processes of channel improvement that they instigate, leaving behind a channel with few evident structures as part of a strategy of 'prompted recovery' (Downs and Thorne, 1998a) or 'assisted natural recovery' (Newson *et al.*, 2002). Instream measures are appropriate where the management authority requires a reasonably rapid, short-term improvement to an environmentally impoverished channel but has only limited powers over land use practices or wishes to initiate a process of channel recovery, but they are not a substitute for a fully functional, vegetated stream corridor within a well-managed watershed (FISRWG, 1998; 8–70).

The challenge in prompted recovery strategies is to cause intentional erosion or deposition by altering local energy conditions (Hey, 1994a), using measures that modify reach-scale flow hydraulics and sediment transport in the context of prevailing catchment flow and sediment transport regimes to ensure long-term benefits (e.g. Downs and Thorne, 1998). If the objective is channel stability, Level III channel assessments (p. 217) will be required to determine how to tackle the issue most effectively. For instream habitat improvements that target native fish, site selection requires knowledge of fish populations and prevailing physical habitat conditions set against an understanding of the causes of habitat degradation, the physical requirements for an improved fish habitat and corridor constraints on restoration activities (Newbury and Gaboury, 1993; FISRWG, 1998). Where structures are used, they should be constructed of native materials and meet the targeted shortfalls between the physical habitat conditions and the ecohydraulic requirement of the native fish populations, and the chosen designs tested for their impact on the hydraulics of flow and sediment transport, with their size, shape and position adjusted to provide the compromise between competing management objectives. For instance, in flood defence channels, structures should be low in height so that their impact on the depth of flood flows is minimised (Downs and Thorne, 2000). Instream structures can easily work *against* the river if applied incorrectly. Structures such as deflectors and log-weirs require careful design and 'keying in' to the bank or bed so that they do not suffer erosion that causes the structure to fail. In addition, boulder-built weirs should possess an arched, double layer construction (with the lower rocks projected less than 15% of their diameter into the flow) to provide sufficient self-reinforcement so that boulders do not roll into the plunge pool created downstream of the structure, and sufficient interstitial space between the boulders to prevent sediment deposition (Rosgen, 1996). It is often possible to install a selection of measures as the basis for providing information for use in later projects. Following implementation, channel improvements will be progressive as the measures depend upon high flow events to drive the intended processes of erosion and/or deposition, and so periodic monitoring is required to gauge success and assess the necessity for structure maintenance.

Instream structures

Deflectors, weirs, sills and vanes are commonly used either to promote morphological diversity in channels previously channelized for flood defence, to deflect flow away from eroding banks or to create scour in zones subject to sedimentation. Potentially, all three objectives may be tackled simultaneously. The challenge for causing a sustained improvement is to design and position the structures to accentuate natural flow processes. Figure 10.5 schematically indicates the performance of a selection of instream devices. Flow deflectors reduce channel width, causing local flow acceleration and accentuated skew in secondary flows that creates both a pool off the tip of the structure and slack water deposition zones in the lee of the device downstream (or upstream and downstream where deflectors are angled upstream). The detail of the effect is determined by the number of structures, the structure's design, size, shape, position in the reach and location in the river network (Hey, 1994a; Kuhnle *et al.*, 1999, 2002; Skinner, 1999). In higher-energy rivers, careful design of the device is required to prevent erosion where flow is directed at the bank. At low flows, this may occur opposite the device, while at high discharge, flows overtopping the device may be refracted into the host bank. To minimize the latter effect, most flow deflectors should be constructed to the approximate low flow height, so that their impact is progressively minimized as flows increase (Figure 10.6A). Deflectors to increase habitat diversity while reducing bank erosion usually project from the eroding bank tangentially into the river flow to disrupt secondary currents, reducing the erosion potential and causing deposition between the deflectors (Shields *et al.*, 1995a, 1998b). Low weirs should be spaced relative to channel width and extend either across the full channel width or a large proportion of it (Figure 10.5). Where sufficient gradient exists, weirs improve instream habitat by facilitating the development of plunge pools or scour pools (partial width weirs). Sediments will be deposited in the backwater zone upstream of the structure and scoured sediment downstream will be redeposited locally (Hey, 1994a). Submerged vanes are straight structures that may or may not be attached to the channel bank and that cause overtopping flows to plunge towards the channel bed causing erosion and the creation of a plunge pool, depending on their arrangement (Hey, 1994; see Figure 10.5). They are unobtrusive but less frequently used than deflectors and weirs, partly because of concerns for boat safety in navigable rivers.

Recreating local bedforms

Environmental loss in rivers often includes both a loss of pool and riffle bedforms and the natural surface of coarse sediment. In channelized lowland rivers, the homogeneity of the

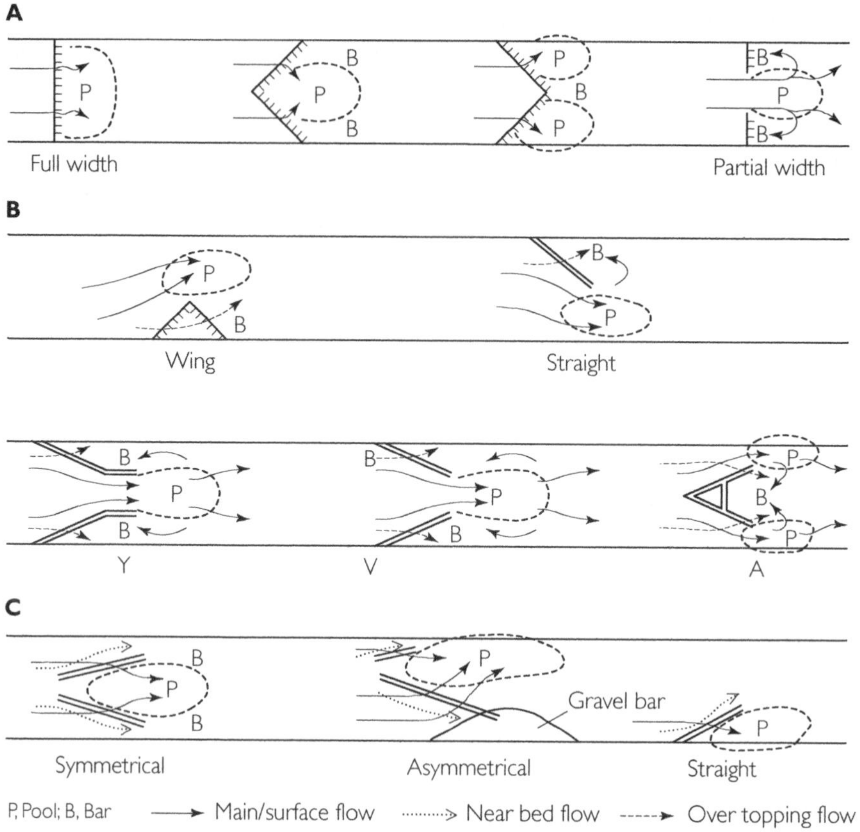

Figure 10.5 *Schematic representation of the effects on flow of various instream structures used for improving stream habitat:* ***A*** *weirs and dams;* ***B*** *deflectors;* ***C*** *vanes (Hey, 1994a)*

channel following engineering is compounded by low stream power that prevents recovery except for the blanket deposition of fine sediment across the channel bed that alters the invertebrate community and consequently affects predators such as fish. In regulated rivers that have been channelized or subject to instream gravel mining, the channel form is essentially paralysed by reduced flows and sediment supply, irrespective of the potential for recovery, and habitat for native fauna is reduced. In such cases, instream measures often include the reinstatement of bedforms and supplementation of coarse sediment supply. Reinstated pools and riffles should mimic the form and spacing of reference features in neighbouring reaches. Where no reference situation exists, and in straightened channels, the riffle spacing is usually determined as approximately 5–7 bankfull channel widths to correspond to the hydraulic geometry of natural channels (Leopold and Wolman, 1957). This relation should be verified locally, for instance, in Britain, the spacing ranges between 3 and 10 channel widths (Brookes, 1988). In resectioned flood control channels, channel top-width will greatly exceed the natural bankfull flow width and this should be accommodated in design. Various recommendations for riffle design have been developed and depend on the channel materials and gradient (e.g. Newbury and Gaboury, 1993; Hey, 1994a). For example, in recreating riffles on the River Wensum, UK, riffles were designed at a slight angle across the river to direct flow towards the outer bank downstream to assist in generating secondary flows, with a crest elevation that causes some ponding of low flows and with a shallow downstream face to provide a significant riffle surface area (Hey, 1994a; Figure 10.6B). Where there is no coarse sediment supply from upstream, it will be necessary to reinstate coarse material sized to remain static under all flow conditions, unless regular coarse material augmentation is feasible. If there is still an upstream supply of fine sediment, the substrate should be sized to ensure that sufficient interstitial flow velocity exists to prevent fine sediment deposition from filling the interstitial spaces. Where such 'mechanically suitable' sediment is required, it should be analysed for its suitability to native ecology.

Cover devices

Fish refuge in a majority of natural river corridors is ensured by the morphological diversity of the channel in conjunction with shelter for protection from high flow velocities, predators and high water temperatures. Natural channel cover is provided by shading bankside trees and their roots, the input of fallen trees as large woody debris (LWD), boulder clusters and marginal reed stands, depending on the channel type. Many channelization projects removed such cover habitat in order to maintain flood flow capacity and the resultant channels lack fish refuges. Plans for reinstating cover may need analysing in the context of other management objectives. Bankside tree planting is an effective long-term measure (and will eventually provide LWD) and the trees should be positioned to maximize shade and minimize impacts on flood conveyance (Downs and Thorne, 2000). Reed planting is best suited to the shallow marginal zones of low gradient rivers in which the water is less than 0.5 m deep and velocities are less than 0.2 m s^{-1} so as to avoid erosion (Coppin and Richards, 1990), and has the additional benefit of trapping fine sediments and narrowing the low-flow channel. It may be necessary to reconfigure the bank toe to provide these conditions, and to check that reed colonization across the entire channel bed can be avoided. In coarse-bedded streams, 'random' boulder placement is used to improve fish habitat by creating slow flow shelter zones in their lee. More recently, the reinstatement of large woody debris into streams, including their root bole, has been used to create hydraulic complexity that benefits cover, but that also improves habitat diversity and supplements organic material input to the channel. Large logs and half-logs have been arranged into a range of fully and semi-anchored and floating fish cover devices (e.g. Rosgen, 1996) and box-type structures ('lunkers') created. Lunkers are likely to be most successful in gravel-cobble channels where there is little fine sediment to fill the structure and where they are placed to create a deep recess on the outside of meander bends (FISRWG, 1998). Construction materials include wood and stone and even concrete mouldings or large diameter pipes. Most are finished with a soil and vegetation cover. In terms of the natural physical environment there is little to commend these latter devices, although one plausible justification is that the devices can provide cover until longer-term cover strategies, such as bankside tree planting, have become established.

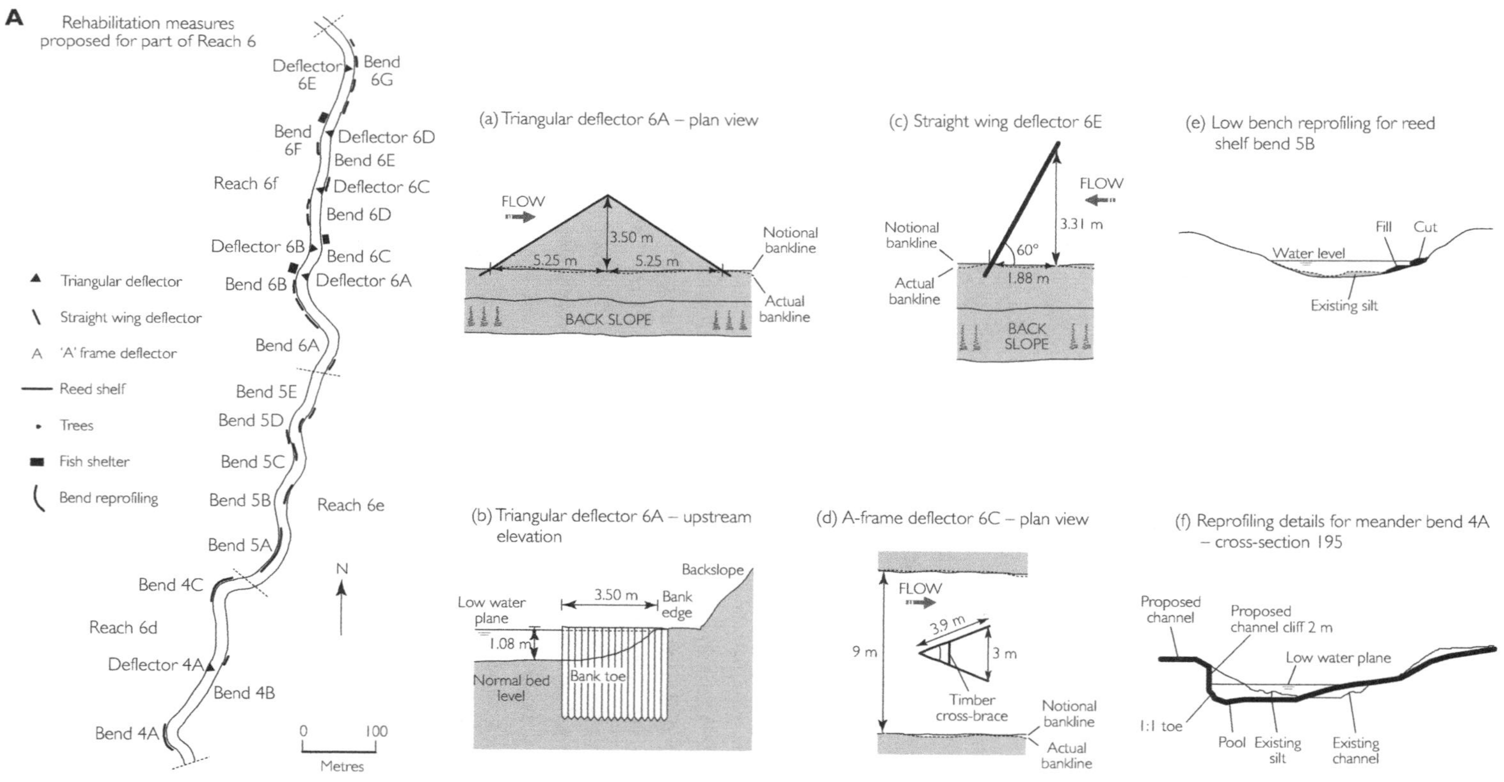
A
Rehabilitation measures proposed for part of Reach 6
Triangular deflector
Straight wing deflector
'A' frame deflector
Reed shelf
Trees
Fish shelter
Bend reprofiling
Deflector 6E
Bend 6G
Bend 6F
Deflector 6D
Bend 6E
Reach 6f
Deflector 6C
Bend 6D
Deflector 6B
Bend 6C
Bend 6B
Deflector 6A
Bend 6A
Bend 5E
Bend 5D
Bend 5C
Bend 5B
Reach 6e
Bend 5A
Bend 4C
N
Reach 6d
Deflector 4A
Bend 4B
Bend 4A
0
100
Metres
(a) Triangular deflector 6A – plan view
FLOW
3.50 m
5.25 m
5.25 m
Notional bankline
Actual bankline
BACK SLOPE
(b) Triangular deflector 6A – upstream elevation
Backslope
Low water plane
3.50 m
Bank edge
1.08 m
Normal bed level
Bank toe
(c) Straight wing deflector 6E
FLOW
3.31 m
Notional bankline
60°
1.88 m
Actual bankline
BACK SLOPE
(d) A-frame deflector 6C – plan view
FLOW
3.9 m
9 m
3 m
Timber cross-brace
Notional bankline
Actual bankline
(e) Low bench reprofiling for reed shelf bend 5B
Fill
Cut
Water level
Existing silt
(f) Reprofiling details for meander bend 4A – cross-section 195
Proposed channel
Proposed channel cliff 2 m
Low water plane
1:1 toe
Pool
Existing silt
Existing channel

B

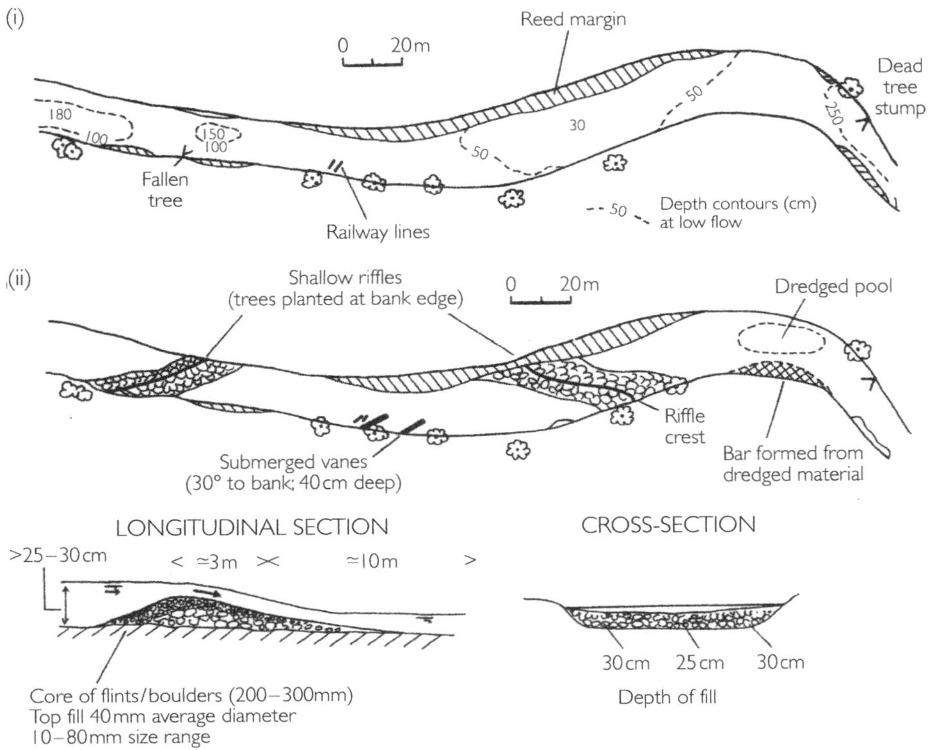

Figure 10.6 A *Planform arrangement and design details for the deployment of instream deflectors to improve channel habitat diversity: River Idle, UK (Downs and Thorne, 1998a);* ***B*** *conceptual design details for the re-creation of pools and riffles at Lyng, River Wensum, UK: (i) geomorphological map of original channel; (ii) proposed pool–riffle sequence (Hey, 1994)*

Sediment traps

Instream sediment traps are designed to intercept 'excess' sediment transport that would otherwise be deposited downstream to the detriment of management objectives. One option is to trap fine sediment by reed planting (see above); most other traps consist of an excavated section of channel bed that collects sediment prior to periodic removal by the management authority. In boulder-bedded rivers, another option is to construct a sill projecting from the channel bed to prevent transport of bed material. Sediment traps are usually a local, symptomatic solution to sedimentation but a strategic use intended for environmental benefit occurred on the Upper River Wharfe in Yorkshire where two sediment traps were installed to reduce bed aggradation in a zone of natural aggradation upstream of the Wharfe's confluence with the Cray Beck (Hey and Winterbottom, 1990). Positioning the traps upstream of the confluence ensured a continued (but lower) rate of sediment transport to prevent bed armouring and fisheries degradation while the traps succeeded in reducing the frequency of overbank flooding, although there is an ongoing commitment to empty the traps and the potential for channel adjustment to occur owing to the reduced sediment supply downstream.

Morphological reconstruction

Morphological reconstruction can be a less desirable channel management solution than the previously outlined measures for reasons of being form-led rather than process-led, for being highly disturbing to channel habitats during and for some period following implementation, and for the tendency towards homogeneous designs using revetments that perpetuate the view of channel stability as a static property (p. 249). However, in lowland channels with little prospect of natural recovery, the approach may be the only feasible option within management time frames. In this case, options include restoring sinuosity in formerly straightened rivers, creating a two-stage channel either to provide sustainable flood control or to arrest channel incision, and resizing flood control channels that are oversized and subject to sedimentation.

Restoring sinuosity in small lowland streams in Europe has been practiced since the early 1980s, particularly in Germany (Glitz, 1983; Binder *et al.*, 1983; Kern, 1992) and in Denmark (Brookes, 1987b; Friberg *et al.*, 1994). Meander restoration also formed the basis for the three EU-funded demonstration restoration projects during the 1990s in England and Denmark (Nielsen, 1996; Holmes and Nielsen, 1998; Holmes, 1998; see Chapter 7; Figure 10.7). In contrast, re-meandering efforts in the USA have sometimes involved more highly dynamic channels with significant sediment transport loads that have resulted in unexpected channel adjustment after implementation: examples include Deep Run, Maryland (Smith, 1997), Uvas Creek, California (Kondolf *et al.*, 2001) and Whitemarsh Run (Soar and Thorne, 2001). Two-stage channels are generally created by lowering the floodplain bordering an incised channel to provide additional flow capacity while creating a low-flow channel to 'natural' bankfull channel dimensions to prevent sediment deposition and the excessive growth of aquatic macrophytes. On the River Roding, near London, England, floodplain lowering was intended to provide additional flood protection to a low-lying settlement (Sellin *et al.*, 1990) although much of the additional floodplain flow capacity was lost through vegetation growth. Conversely, Miller Creek, California, was a

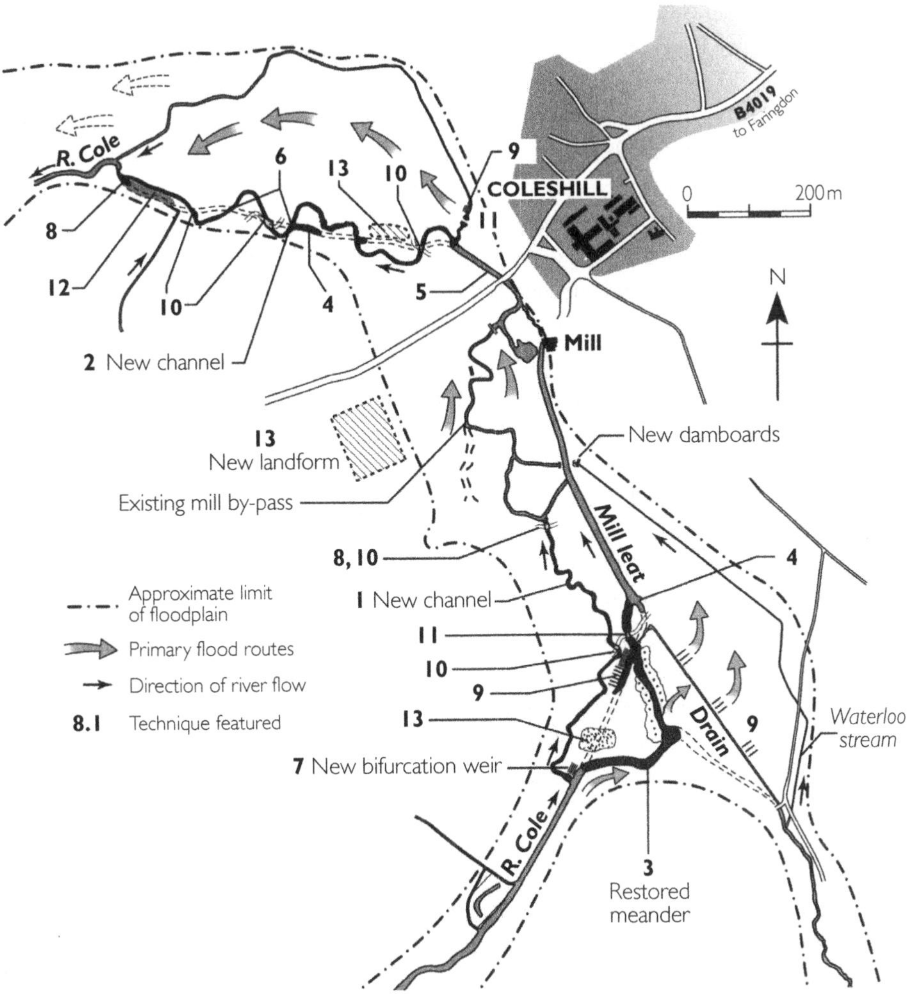

Figure 10.7 *Planform arrangement of features in the morphological reconstruction of the River Cole, central-southern England (compare with photographs in Figures 7.5 and 9.6A). Techniques included:* ***1*** *new meandering channel upstream of mill;* ***2*** *new meandering channel downstream of mill;* ***3*** *single meander in Mill Leat;* ***4*** *new backwaters in redundant river channels;* ***5*** *new aquatic ledges;* ***6*** *short-term bank revetments;* ***7*** *bifurcation weir and spillway;* ***8*** *drop weirs in bed;* ***9*** *floodplain spillways;* ***10*** *fords and livestock access;* ***11*** *new crossings;* ***12*** *new reedbed;* ***13*** *new landforms*

deeply incised channel and the decision was taken to significantly lower the floodway corridor to provide flood protection while reducing the erosion forces present during flood flows in the incised channel (Haltiner *et al.*, 1996). Resizing of flood defence channels may involve either installation of instream structures such as deflectors to reduce the effective channel width (see above), the construction of a structural bench along one bank of the channel to create a two-stage cross-section or full reconstruction similar to meander restoration.

The design challenge for morphological reconstruction is to incorporate aspects of the *dynamic* nature of fluvial geomorphological environments as 'geomorphic engineering' (Coates, 1976). To incorporate channel dynamics into channel design requires at least three criteria to be met, including (i) a catchment-wide context to design (Hasfurther, 1985; Kondolf and Downs, 1996), (ii) an allowance for design variability between cross-sections, within natural boundary conditions (Soar *et al.*, 1998) and (iii) an allowance for recent and future environmental change over longer time spans. The third of these concerns is complex and named previously as one of the challenges for channel management (p. 145). It may involve 'second guessing' factors such as future climate parameters and the requirements imposed by other management objectives, such as flood control (Downs and Thorne, 2000). Sustainable channel design also requires knowledge of valley soils, appropriate plant communities and the opportunities for complementary nonstructural policies (FISRWG, 1998).

Recent initiatives to provide full channel design parameterization include a procedure for managing flood control in steep-gravel bed streams in the UK (Hey and Heritage, 1993) and approaches for providing stable meandering channels for river restoration (Hey, 1994b; Shields, 1996; FISRWG, 1998; Soar and Thorne, 2001). Single-thread meandering alluvial channel design can use either an 'alignment-first' or 'slope-first' process (Shields, 1996; Table 10.7). Alignment-first approaches are generally based on either 'carbon copy' of pre-disturbance meander patterns, natural analogues obtained from contemporary reference reaches, empirical relationships of meander dimensions based on channel width and discharge, or basin-wide analysis of meander dimensions in unconfined settings (Shields, 1996). Average channel slope is derived as a by-product of the estimated parameters. Slope-first approaches are based on regional hydraulic geometry parameters (e.g. Hey and Thorne, 1986) or by computing values required to convey the design water and sediment discharges (e.g. Copeland, 1994). Values for the mean width and flow depth of the design discharge are also required and can be achieved using empirical methods based on regime theory and hydraulic geometry (see Chapter 2). Alternatively, 'analytical' methods for stable channel design can be used to solve equations that describe the entire channel perimeter using governing equations of flow continuity, flow resistance and sediment transport in combination with simplifications, assumptions and empirical factors to solve for unknown parameters (Shields, 1996). It is also possible to use statistical error bands in hydraulic geometry relationships to allow channel variability to be incorporated integrally into restoration designs (Soar and Thorne, 2001). Once validated through computer modelling, details of the channel shape and longitudinal pool–riffle spacing can be estimated from values reported in the research literature.

The stability of the resultant design is usually checked by simulating estimates for some combination of flow velocity, average boundary shear stress, bank stability, stream power or sediment transport (Shields, 1996). However, this may encourage a 'static stability' view of the channel rather than a view of 'dynamic stability' and lead to the unnecessary use of channel bed and bank revetments that are of long-term detriment to the potential environmental gains of the restored channel. Ultimately, structural reinforcement can be minimized by ensuring a wide river corridor in step 2 of the design (Table 10.7). One means of ensuring that channel design is based on geomorphic engineering principles is to

Table 10.7 Steps in designing a stable meandering channel for use in morphological reconstruction

Step	Task	Tools
1	Describe the physical aspects of the catchment and characterise its hydrological response	Use baseline environmental and historical assessment of the river channel (pp. 205–11)
2	Select a preliminary right-of-way for the restored stream channel corridor and compute valley length and valley slope	Compare with constraints
3	Determine the approximate bed material size distribution for the new channel	Use samples from the existing channel, from a reference reach or from excavation of a pre-disturbance channel, as appropriate
4	Select a design discharge of range of discharges	Use hydrologic and hydraulic analysis based primarily on flows estimated to control the natural channel geometry (i.e. 'bankfull' flow)
5	Predict stable planform type (e.g. straight, meandering or braided)	Use historical analysis (p. 212)

Choose preferred approach

Step	Approach A (Shields, 1996)		Approach B (Hey, 1994b)		Approach C (Fogg, personal communication 1995)	
	Task	**Tools**	**Task**	**Tools**	**Task**	**Tools**
6	Determine meander geometry and channel alignment	Empirical formulae for meander wavelength, adaptation of measurements from pre-disturbed conditions or nearly undisturbed reaches	Determine bed material discharge for design channel at design discharge, compute bed material sediment concentration	Analyse measured data or use appropriate sediment transport function and hydraulic properties or reach upstream from design reach	Compute mean flow, width, depth and slope at design discharge	Regime or hydraulic geometry formulaes with regional coefficients
7	Compute sinuosity, channel length and slope	Channel length – sinuosity × alley length Channel slope = valley slope/sinuosity	Compute mean flow, width, depth and slope at design discharge	Regime or hydraulic geometry formulae with regional coefficients, or analytical methods (e.g. White *et al.*, 1982, or Copeland, 1994)	Compute or estimate flow resistance coefficient at design discharge	Appropriate relationship between depth, bed sediment size and resistance coefficient, modified based on expected sinuosity and bank/berm vegetation

Table 10.7 *continued*

Step	Task	Tools	Task	Tools	Task	Tools
8	Compute mean flow width and depth at design discharge		Compute sinuosity and channel length	Sinuosity – valley slope/ channel slope. Channel length – sinuosity × valley length	Compute mean channel slope and depth to pass required discharge	Uniform flow equation (e.g., Manning/Chezy), continuity equation, and design channel cross-sectional shape; numerical water surface profile models may be used instead of uniform flow equation
9	Compute riffle spacing (of gravel bed), and add design details	Empirical formulae, observation of similar streams, habitat criteria	Determine meander geometry and channel alignment	Lay string scaled to channel length on a map (or equivalent procedure) such that meander arc lengths vary from 4 to 9 channel widths	Compute velocity or boundary shear stress at design discharge	Allowable velocity or shear stress criteria based on channel boundary materials
10	Check channel stability and reiterate as needed	Check stability	Compute riffle spacing (if gravel bed), and add design details	Empirical formulae, observation of similar streams, habitat criteria	Compute sinuosity and channel length	Sinuosity – valley slope/channel slope. Channel length – sinuosity × valley length
11			Check channel stability and reiterate as needed	Check stability	Determine meander geometry and channel alignment	Lay out string scaled to channel length on a map (or equivalent procedure) such that meander arc lengths vary from 4 to 9 channel widths
12					Check channel stability and reiterate as needed	Check stability

Source: Modified from Shields (1996: 26–29), FISRWG (1998: 8–31).

assess design stability using reach-scale sediment budgets (Soar and Thorne, 2001). Under this system, the sediment transport rate from the upstream channel reach is estimated via a magnitude–frequency analysis of sediment transport potential (Wolman and Miller, 1960; Andrews, 1980) and compared with the project and downstream reaches in order to estimate the tendency for net aggradation or incision. Channel design is based on achieving a 'zero sediment flux' through the region (or accommodating the natural flux differences). The advantages of the approach include the fact that the reconstructed channel should be essentially stable in the short to medium term (whereas erosion and deposition in numerous reconstructed channels point towards a significant net sediment flux) and that the channel is designed according to the prevailing (and, potentially, future) catchment context of the flow and sediment transport rather than on possibly unsustainable pre-disturbance conditions. Under this scenario, the channel stability assessment becomes the driving force for channel design (Figure 10.8).

Reinforcing the channel perimeter with minimum environmental impact

Ongoing commitments to the protection of land and infrastructure close to river channels often drives a requirement for ensuring a fixed, immobile channel using structural reinforcement of the channel's bed and banks despite the generally adverse environmental impacts of this approach (p. 259). Structural reinforcement remains likely to be used as a supplementary measure when management objectives cannot be fully achieved by the more environmentally sustainable measures outlined in the sections above.

Bed protection

The control of channel incision is increasingly recognized as a fundamental issue for channel management (e.g. Darby and Simon, 1999), partly because the process tends to result in a highly dynamic channel morphology. Long-term options for management include benign neglect whereby the channel is left to recover to a newly stable geometry (p. 280), possibly assisted by natural coarsening of channel bed materials or by bedrock exposure to limit bed erosion, and catchment measures to reduce runoff (Bravard *et al.*, 1999a). Often, however, a faster solution is required because incision threatens significant erosion of riparian land and the integrity of bridges and other structures and so is economically and politically unacceptable. In such cases, options for preventing incision need to be set against the potential threat of future channel instability caused by holding the channel in a nonequilibrium condition. For instance, maintaining a threatened bridge cross-section can be counterproductive if the flow through the cross-section causes downstream scouring that induces a new wave of upstream degradation (Simon and Darby, 1999). Effective management of channel incision demands careful siting of grade control measures. This is facilitated by the fact that channel incision usually passes through a consistent, predictable sequence of channel forms in time allowing measures to be targeted according to the 'stage' in the incision evolution process (Schumm *et al.*, 1984; Simon, 1989). As such, grade control may be used both to prevent bed erosion and to prevent further channel widening through bank erosion. It may also be used conjunctively to improve instream habitat variability in channelized rivers (Section 10.2) and, whereas bed structures were often an inadvertent barrier to fish migration in the past, they are now being used consciously to

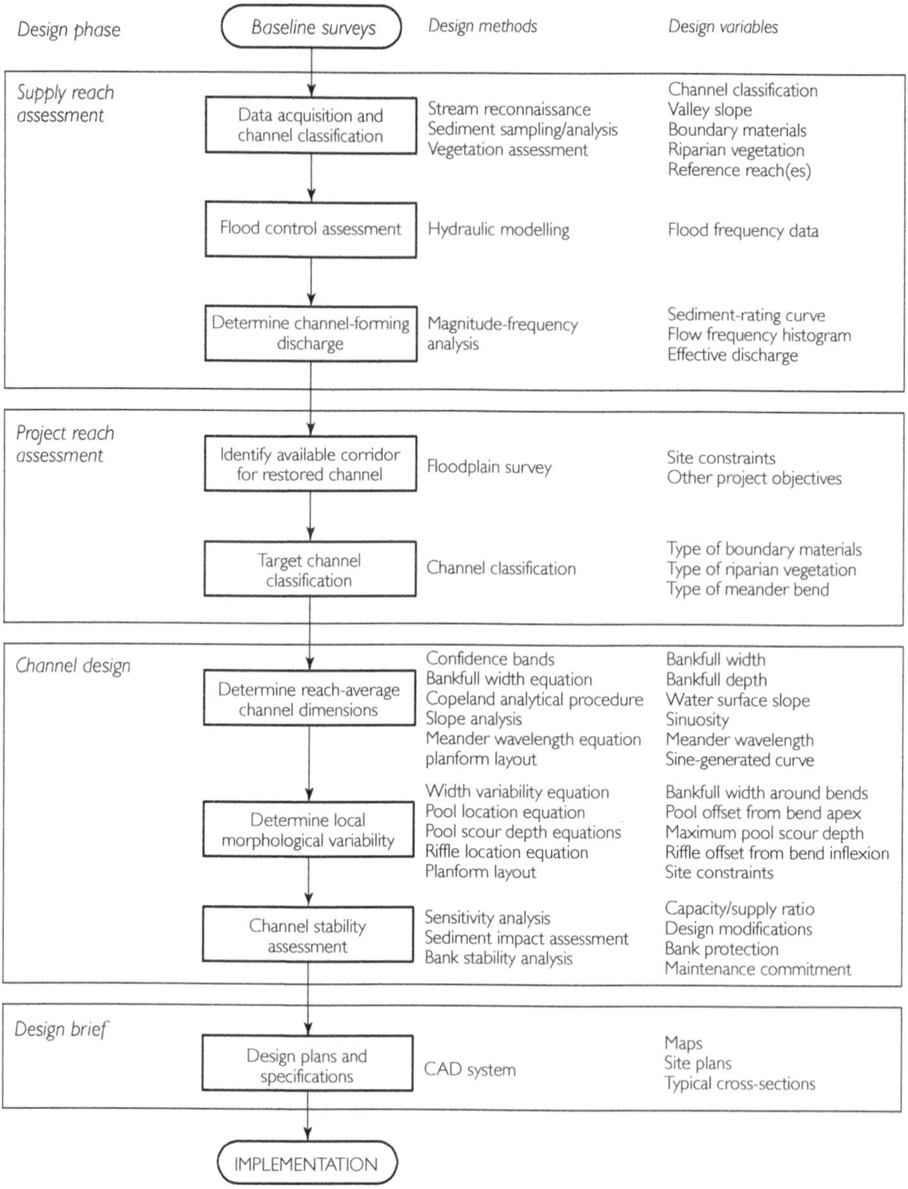

Figure 10.8 *Best practice design procedure for single-thread sinuous rivers based on equilibrating sediment transport upstream, through and downstream of the project reach as a core part of the project design (Soar and Thorne, 2001)*

prevent non-native fish migration (FISRWG, 1998).

Grade control is generally achieved using either grade control structures, armouring of the channel bed, increasing bedload supply or morphological reconstruction of the channel (see Table 10.8). Knowledge of channel slope and bed materials, and a regional assessment of the stage in channel evolution is critical. Grade control structures are best

Table 10.8 Example options for grade control in incising streams

Grade control options	Examples	Purpose
Structures	Check dams/weirs	Localizes energy expenditure at the toe of the dam and allows lower slope in the reaches between the dams
	Low head sills	Replaces knickpoint with a series of smaller resistant drops to reduce erosion potential
	Block ramps	Knickzone in steep channels is covered in a surface of large blocks to increase bed roughness and dissipate energy
Bed armouring	Artificial armour	Bed material is replaced by coarser grains that are more resistant to erosion
	Large blocks	Placement of intermittent large blocks provides additional channel roughness and energy dissipation around the block
	Channel lining	Use of artificial substrate, e.g. concrete to increase bed resistance to erosion
Increased bedload supply	Destabilize hillslope	Reactive hillslope sediment supply processes to supplement channel supply
	Gravel augmentation	Regular injection of sediment supplements bed sediment supply
Channel widening	Compound channel creation	Reduces unit stream power and shear stress

Source: Based on Bravard *et al.* (1999a); Jaeggi and Zarn (1999); Watson and Biedenharn (1999).

placed upstream of a migrating knickpoint with suitable foundations to prevent their demise as the knickpoint reaches the structure. They are less suitable in an active knickpoint zone where rapid channel widening may outflank the structure and are generally not appropriate downstream of the knickpoint because bed aggradation is a more likely adjustment.

The type of grade control depends in part on the sediment calibre of the channel and the intensity of the incision. Aesthetically, native materials are preferred. It is usually preferable to manage a large drop using several low structures rather than one large structure, both to facilitate fish passage and to reduce the chance of catastrophic structural failure

(Haltiner *et al.*, 1996). Where multiple structures are used, they should be spaced so that the deposition from one structure does not affect the structure upstream and designed so as to prevent significant sediment trapping that might promote downstream incision (Watson and Biedenharn, 1999; Figure 10.9A). It may also be possible to space the structures to simulate a step-pool sequence in steeper channels (step-pools commonly occur in channels with gradients of 0.0300–0.1000; Montgomery and Buffington, 1997). In gravel and cobble-bedded rivers subject to light incision, grade control may be achieved by adding coarse substrate to riffles or by constructing low weirs from logs, boulders or blockstone (e.g. Rosgen, 1996; FISRWG, 1998). Structures set diagonally across the channel or straight structures with alternating notches can be used to prompt a sinuous low flow 'thalweg' and for fish passage. Check dams may be necessary in areas of severe incision. For instance, in the highly erodible loess regions of mid-western USA, channel straightening for agricultural drainage has resulted in multiple knickzones, some exceeding 5 m in total height and with annual migration rates of up to 12 m (based on short-term monitoring; Simon and Thomas, 2002), and capable of causing a four-fold increase in channel width (Downs and Simon, 2001). Erosion control begins with the use of corrugated 'drop pipes' to channel overland flow from fields into the incised channel without extending incipient rills and gullies. In-channel measures consist of hard engineering of 'low drop' structures constructed as concrete capped sheet-pile weirs with stone lined stilling basins. Where the drop height exceeds 1.8 m, 'high drop' structures of reinforced concrete weirs with stilling basins and baffle piers are used (Shields *et al.*, 1995; Figure 9.7D). The sheet piling provides stability and prevents seepage flows under the structure (an alternative is to use impervious fill material) and the stilling basin and baffle plate act to dissipate downstream energy (Watson and Biedenharn, 1999; Figure 10.9B). High drop structures in particular are likely to be environmentally detrimental and should be used only where absolutely necessary.

Bank protection

In terms of retaining environmental value, it is almost universally accepted that the extent of bank revetment must be kept as low as possible and considered within a range of possible alternatives. A recent design perspective is afforded by 'appropriate bank management' (Thorne *et al.*, 1996b), a procedure that begins with a field survey to assess the cause, severity and extent of the bank erosion problem, the mechanics of bank retreat (Lawler *et al.*, 1997) and the implicit acknowledgment that bank erosion should be allowed to continue unless it poses an intolerable risk. A complementary approach is to use historical analysis, overlaying aerial photographs to superimpose channel courses through time to depict the areas of greatest hazard (Graf, 1984, 2000; Piégay *et al.*, 1997). Where intervention is necessary, an 'active' solution should be sought initially and structural protection used only when this approach is impracticable. Active bank management measures range from keeping a watching brief, especially if intervention may trigger less desirable consequences elsewhere in the river (although this perspective may be alien to many river managers and riparian landowners; Thorne *et al.*, 1996b), practising managed retreat where it is feasible to move the threatened activity to a safer location, and managing the problem at cause through solutions upstream or downstream of the erosion location. If

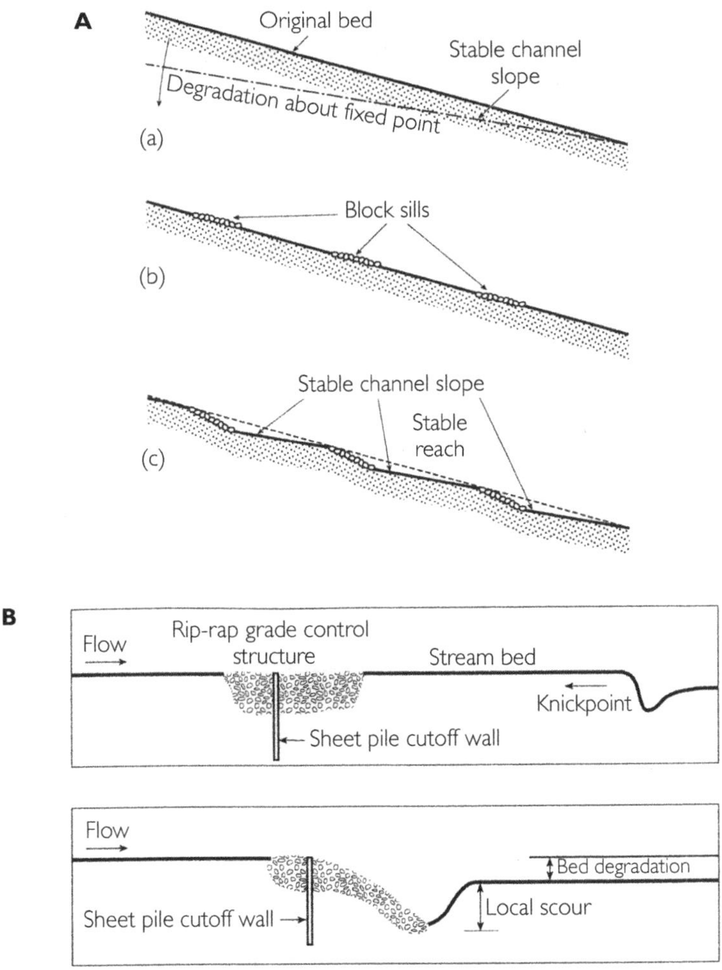

Figure 10.9 A *Conceptual design for block sills to resist channel bed erosion by providing hard points on the bed of the channel (Watson and Biedenharn, 1999, following Whittaker and Jaeggi, 1986). Deployment is usually most effective as a series of low drops.* ***B*** *Conceptual design for a rip-rap grade control structure with a sheet pile cutoff wall (Watson and Biedenharn, 1999). Rip-rap is 'launched' into the scour hole as bed degradation nears the wall and protects the wall from undermining to prevent continued upstream migration of the knickpoint.*

this approach is insufficient and structural intervention is justified, the scope, strength and length of bank reinforcement should be matched with the cause, severity and extent of the problem bearing in mind the responsibility to balance the conflicting management goals of efficacy, economy, engineering and environment (Thorne *et al.*, 1996b; Table 10.9).

Where justified, there are a wide variety of structural bank protection measures ranging from 'soft' to 'hard' and with relative environmental and engineering advantages and disadvantages (Table 10.10). Most protection measures are revetments, although indirect

Table 10.9 Management options for selected bank retreat processes

Bank retreat process	Management consideration	Best management options	
		Nonstructural	**Structural**
Bank erosion			
Parallel flow (fluvial entrainment)	Usually natural but rate may be accelerated by human actions	Allow adjustment or control of identifiable cause	Revetments for surface armouring or deflectors to reduce intensity of attack
Impinging flow (fluvial entrainment)	Usually indicates a poorly aligned bank, unstable channel or flow deflection by an obstruction	Allow adjustment or obstruction modification or removal	Revetment or realign channel
Rills and gullies (surface erosion)	Usually caused by human activities that increase surface runoff and/or reduce erosion resistance	Control of identifiable cause such as limiting access, improving drainage, restoring vegetation, preventing mechanical damage	Soakaways, pipes, lined channels or drop structures to control surface drainage
Piping (seepage erosion)	Natural due to seepage and bank stratigraphy, or driven by human activity or engineering works	Allow adjustment or control of identifiable cause	Granular, geotextile or vegetative filter to allow free subsurface drainage without soil loss
Bank failures			
Shallow slides	Damages bank surface and vegetation. Occurs at various scales and for various reasons	Allow adjustment; may involve large areas where slide angle is low. Manage human causes where possible	Regrade bank to lower angle and protect the toe
Rotational slope	Results from deep instability inside the bank: disrupts bank profile and may destabilize the channel	Often linked to incision: requires wide set-aside corridor or engineering structures	Regrade bank to lower angle, protect the toe and improve drainage
Slab-type failure	On steep undercut banks due to tensile forces inside the soil: disrupts bank profile and may destabilize the channel	Allow adjustment or manage cause of bank undercutting	Regrade bank to lower angle and protect the toe

Bank retreat process	Management consideration	Best management options	
		Nonstructural	Structural
Cantilever failure	On bank undermined by erosion of a weak layer lower in the bank	Allowed adjustment or manage cause of bank undercutting such as boatwash	Armour bank to protect weak layer, install filter to prevent piping and re-vegetate to increase soil tensile strength
Bank weakening factors			
Leaching	Driven by strong seepage in clay soils	If natural, allow retreat. If assisted by pollution, manage the cause	Strengthen soil using grout, resin or soil cement if cost-effective
Trampling	By uncontrolled farm animals or heavy pedestrian activity	Control access and provide education	Geogrids or cellular blocks for surface protection while allowing vegetation growth
Riparian vegetation removal	Reduces the mechanical and hydrological strength provided by vegetation	Allow natural renewal or re-plant and limit access	Fencing to limit access and encourage vegetation re-growth
Mechanical damage	By uncontrolled or unadvised access close to the bank edge	Control access and provide education	Soft, hard or hybrid revetment matched to the intensity of mechanical damage
High pore-water pressure	Natural following drawdown in poorly drained banks, or due to impacts of subsurface drainage	If natural, allow adjustment. Otherwise eliminate cause of poor drainage	Granular, geotextile or vegetative filter or pipes to allow free subsurface drainage without soil loss

Source: Modified from Thorne *et al.* (1996b).

protection using submerged vanes to disrupt secondary flows that drive bank erosion has been trialled in rivers with stable bed levels (Paice and Hey, 1989; Hey, 1994a; Newson *et al.*, 1997). Soft revetment solutions may involve bioengineering solutions (see Gray and Sotir, 1996) using native live materials such as reeds at the bank toe, grasses, reeds, staked trees and fascines on the eroding bank surface or tree planting at the bank top, and/or the use of biodegradable materials such as brush mattresses, coconut fibre roll and buried rootwads and logs (including 'engineered log jams'; Abbe *et al.*, 1997) to provide bank resistance (Rosgen, 1996; FISRWG, 1998; Figure 10.10). Supplementary resistance may be

Table 10.10 Engineering and environmental performance of bank protection works

Engineering and environmental performance			
Options	**Advantages**	**Disadvantages**	**Design limitations**
Revetment			
Grass			Lowest erosional
Grass and jute		resistance	
Grass and geotextile	Natural and		
Reeds	unobtrusive		
Willows/osier			
Ash			
Gravel and geotextile			
Matic and geotextile	Natural but more		
Woven wooden fence	obtrusive. Vegetation		Chosen option
Rip-rap	can become		should be softest to
Rip-rap over geotextile	established		prevent failure
Gabion mattresses			
Gabion baskets		Artificial but some	
Grouted rip-rap		vegetation can	
Grouted gabion baskets		become established	
Jointed cellular blocks			
Cable-tied blocks			
Wooden boards			
Sheet piling		Artificial and	Highest erosional
Masonary wall		cannot become	resistance
Concrete wall		vegetated	
Hydraulic controls			
Submerged vanes	Natural bank retained, vanes are submerged and unobtrusive. Instream habitats enhanced	Not appropriate where systematic bed degradation or aggradation	Suitable for bank erosion control in meander bends on rivers which are in regime

Source: Hey (1996: 97).

provided using a biodegradable geotextile fabric to cover the bank (e.g. coir matting) or a synthetic material to provide integral bank support (e.g. geo-grids) while the vegetation develops. Cattle should be excluded from the bank as vegetation establishes and measures must be keyed sufficiently into the bank toe to protect against mass bank failures. Hard engineering measures are warranted only where the potential consequences of erosion are severe and the at-risk infrastructure cannot be moved. Measures include sloping revetments such as rip-rap, sand bags, paving slabs or tiered gabion baskets or, in highly constrained situations, the use of vertical walls of cement, steel, mortared stone or wood. Rip-rap is often considered a more environmentally friendly hard revetment because of the prospect of combining it with live vegetation ('hybrid bioengineering') but requires careful assessment of the void space between the rocks to balance the competing needs for space to allow vegetation growth but closeness to prevent bank erosion behind the

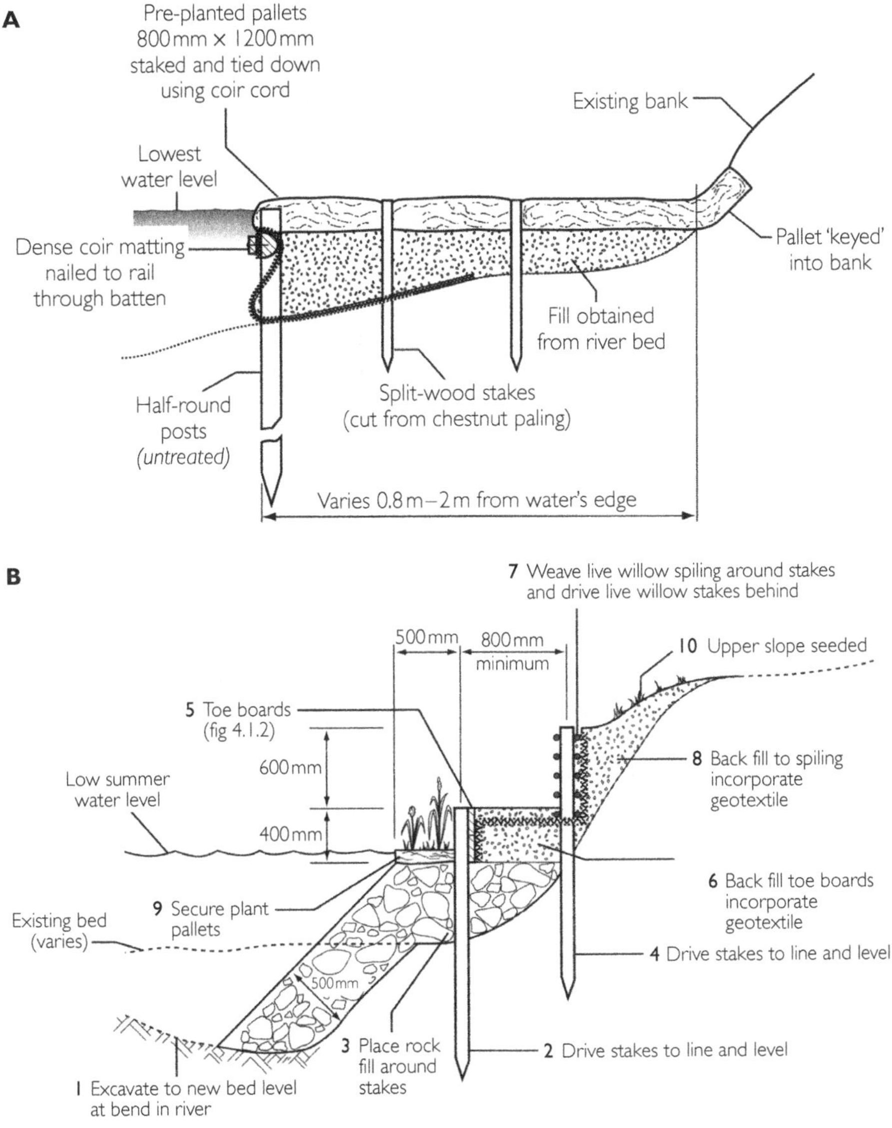

Figure 10.10 *Conceptual design details for 'soft' engineering measures to protect river banks in lowland environments (River Restoration Centre, 1999). All allow vegetation to cover the revetment in time.* ***A*** *Creation of ledges using stakes and coir matting to control bank toe erosion and for marginal planting banks.* ***B*** *Creation of willow piling revetment supported by wooden stakes and rock at the bank toe to support steep channel banks.* (continued)

C

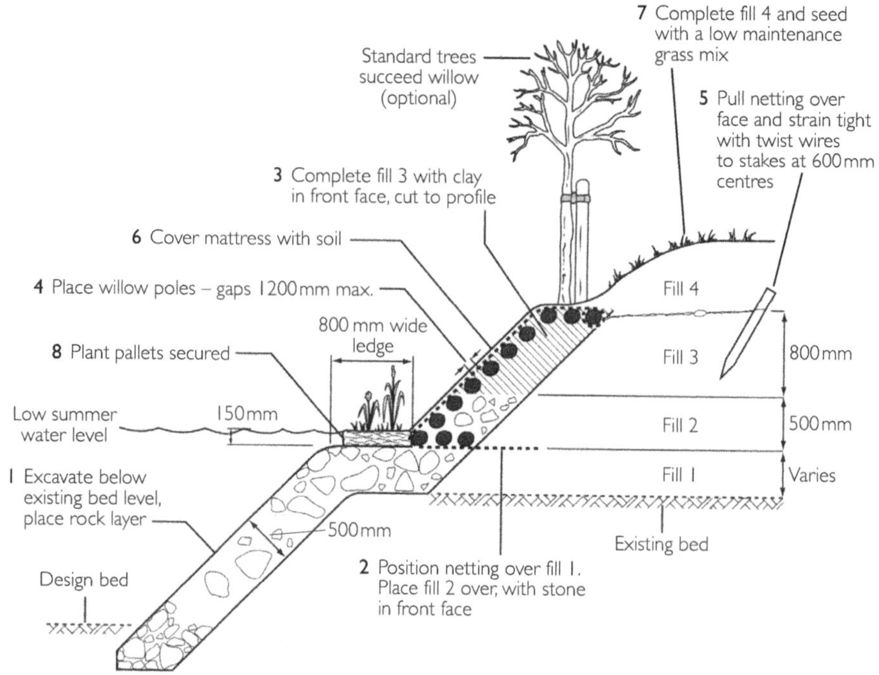

D

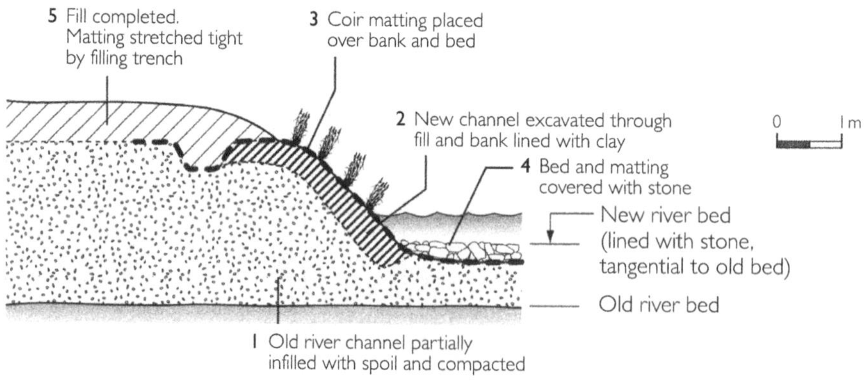

Figure 10.10 (continued) *Conceptual design details for 'soft' engineering measures to protect river banks in lowland environments (River Restoration Centre, 1999). All allow vegetation to cover the revetment in time.* ***C*** *Creation of anchored willow mattress revetment supported by rock at the bank toe and suitable for graded river banks.* ***D*** *Use of coir revetment to support newly created channel banks where the risk of scour is of limited magnitude*

rock (and eventual collapse of the revetment). Measures such as groynes and spurs of wood or stone have also been used as hydraulic protection for banks by reducing near-bank velocities and encouraging sediment deposition in the formerly eroding areas (Shields *et al.*, 1995a, 1998a).

10.3 Meeting the challenges?

The challenges for contemporary river channel management (pp. 235–41) have resulted in new approaches to management (p. 245) that involve a revised management framework and new techniques that attempt to work-with-the-river (pp. 279–312) using a design-with-nature philosophy. The ultimate test of these methods is whether they have resulted in significant and sustainable environmental benefit to river channels while also meeting the other management objectives. However, restoration-based management does not inherently involve 'doing good' (Beschta *et al.*, 1994; Kondolf, 1995) but requires an integral commitment to project evaluation (pp. 229–32) against measurable success criteria.

One simple indicator of project design success is whether structural projects remain intact following high flows. In this regard, there have been several notable failures of morphological reconstruction projects that either washed out or abandoned their created channel in flooding shortly after implementation (Deep Run, Maryland, Smith, 1997; Uvas Creek, California, Kondolf *et al.*, 2001; Figure 10.11). In both cases the reconstructed channel was not appropriate to its geomorphological surroundings: both were constructed as regular, single-thread, sinuous channels but (later) catchment historical investigation showed each channel to carry a high sediment load that resulted in a naturally irregular planform in which morphology responds rapidly to flood events. Instream devices can also be vulnerable such as in southwest Oregon where, following moderate flooding, 60% of 161 fish habitat structures on 15 streams were damaged because the structures were inappropriate to the flash flood regime, high sediment loads and erodible bank materials (Frissell and Nawa, 1992). In Moore's Gulch, California, log weirs for fish habitat demonstrated progressive failure in 1993–7 during a period that included a 4- and 35-year flood event (Lassettre, 1997). Even though the cases above could be argued to be examples of poor engineering technique rather than a poor application *per se*, all suggest a failure to meet the challenge of 'managing rivers as ecosystems, incorporating past, present and future conditions' (Section 9.1). Each example does, however, provide a valuable learning experience, a factor that is critical in improving future design (Petroski, 1992).

Indications of project success or failure in terms of environmental performance may be more subtle in comparison. An approach based on a reach-level index of morphological complexity was used to determine the potential habitat diversity in rehabilitated reaches of lowland English streams relative to boundary conditions provided by channelized and reference stream conditions (Skinner *et al.*, 2003), with success attributed to those projects that most closely replicated the character of the reference reaches (Figure 10.12). An alternative is to base the approach on before and after field monitoring. In The Netherlands, the Tongelreep and Keersop channels developed local pool scouring, bank undercutting, coarsening of materials and the formation of depositional bedforms in 2–3 years following construction, changes that can probably be ascribed to post-construction

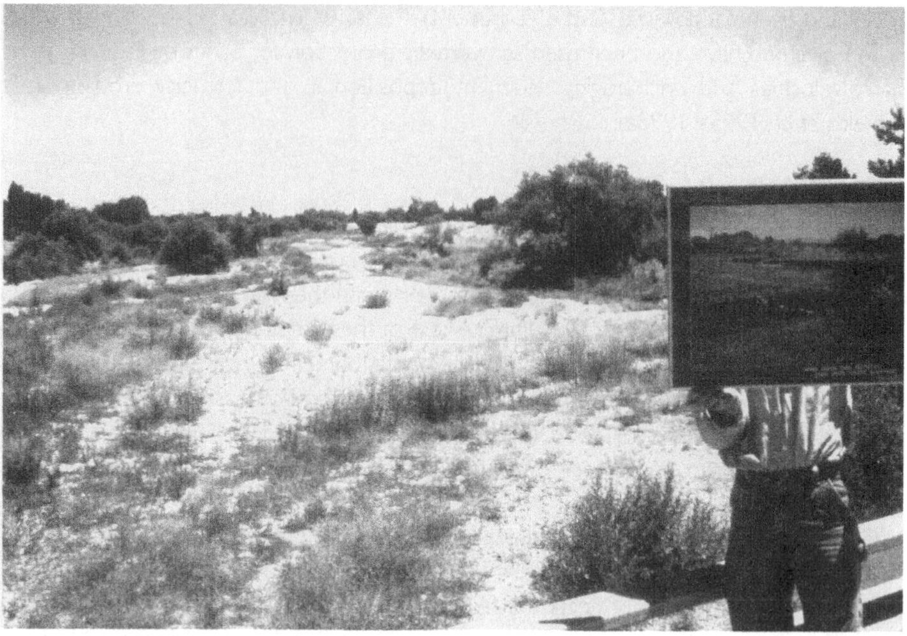

Figure 10.11 *Failure of restoration by morphological reconstruction at Uvas Creek, central California. The completed scheme (November 1995 – see inset photo) was washed out by February 1996 (details in Kondolf et al., 2001) (photograph by G.M. Kondolf)*

adjustment and may not reflect longer-term trends (Wolfert, 2001). In Evan Water, Scotland, two high flow events soon after construction of a sinuous diversion channel caused scour and bank erosion, particle sorting and point bar formation leading to greater morphological diversity without compromising the overall stability of the diversion (Gilvear and Bradley, 1997). In the Red River, Idaho, physical and biological monitoring of meanders reconstructed to recreate historic meadow conditions indicated that, in the short term, channel flow depths and velocities were more suitable for Chinook salmon rearing (relative to habitat suitability curves), the frequency of pools increased, median particle sizes did not change, young Chinook numbers increased but resident salmonid densities did not change, possibly owing to stresses imposed by high summer water temperatures (Clayton, 2002).

Unfortunately, despite the burgeoning interest in restoration-based management approaches, there is a lag in the corresponding output from rigorous post-project appraisals. Various logistical difficulties can be identified. First, there is the extended time span inherent in making 'effects-centred' evaluations rather than the traditional construction-centred auditing of compliance with design intentions. To date, there are few if any monitoring schemes that have extended to the decadal time frame required as a minimum for assessing geomorphological impacts (Downs and Kondolf, 2002). Monitoring of biological populations may require even longer to detect statistically valid population trends because of their inherently high variability (Clayton, 2002) and because construction disturbance may cause a short-term negative impact on flora and fauna (e.g. Biggs *et al.*,

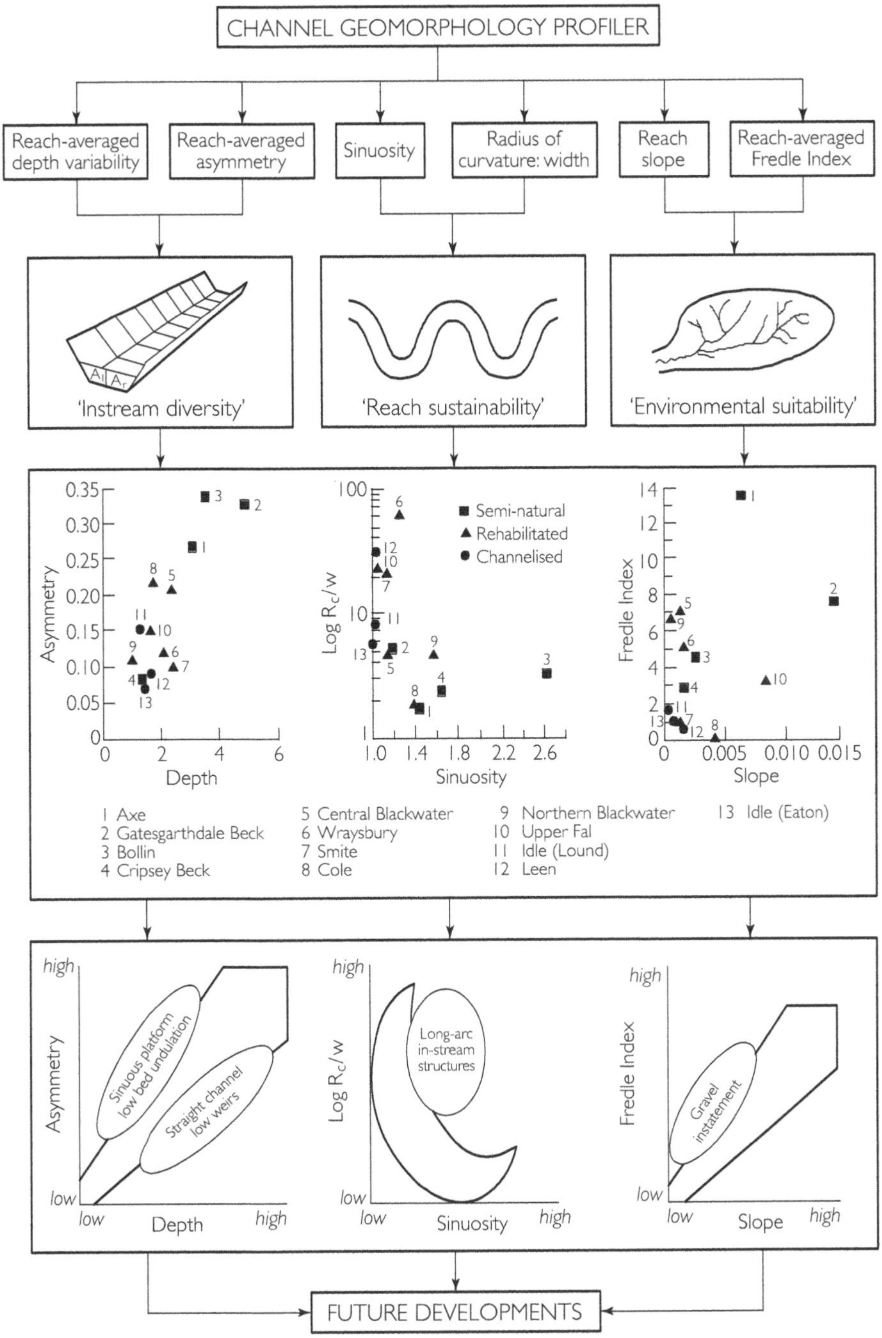

Figure 10.12 *Pilot results from a measure of potential habitat diversity (Channel Geomorphology Profiler) applied to rehabilitated reaches in lowland English streams (Skinner et al., 2003). Three bivariate plots measure instream diversity, reach sustainability and environmental suitability against boundary conditions provided by semi-natural and channelized reaches (based on logic similar to that shown in Figure 7.4)*

1998). Second, the need to evaluate schemes either in their comparative spatial context (i.e. how does this channel compare with others?) and temporal context (i.e. how is this scheme performing relative to the flows experienced since its implementation?) are new concepts (see Section 8.5) without standard approaches. Third, project 'success' in terms of performance and learning experience can be defined in many ways and will depend on the project objectives, making simple interpretation difficult. It is perhaps most appropriate if project findings are considered in terms of their utility under adaptive management (Downs and Kondolf, 2002). As an example, Figure 10.13 illustrates ways in which geomorphological and engineering understanding gained as part of a channel restoration project can be recycled as the basis for improving future practice. A further element in the

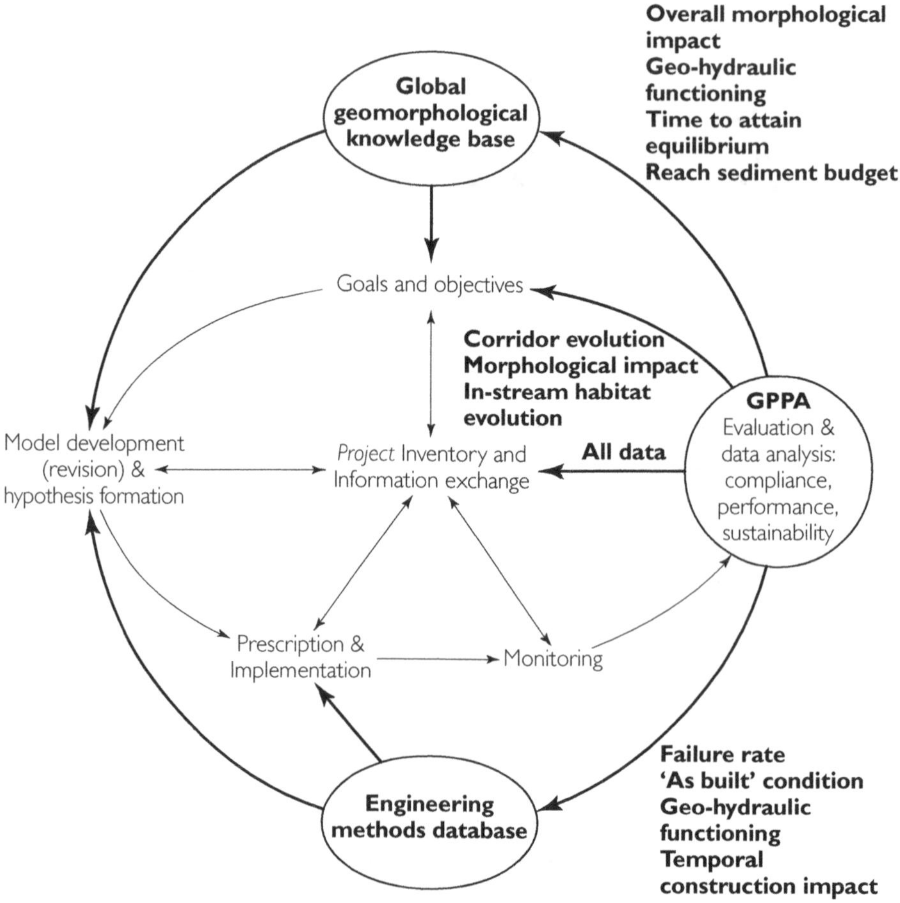

Figure 10.13 *Potential feedback of understanding obtained during geomorphological post-project appraisal undertaken as part of an adaptive management process. Bold lines represent flows of information into global geomorphological knowledge base, engineering methods database and the knowledge base for setting practicable goals and objectives (Downs and Kondolf, 2002)*

definition of project 'success' involves the scheme's economic performance in both market and non-market terms. This is a largely untested aspect of project assessment although Denmark's largest river restoration project involving 2200 ha of the lower Skjern River was subjected to analysis that concluded that, under all but the least favourable economic scenario, the net social benefit of the project was robustly positive (Dubgaard *et al.*, 2002).

In trying to achieve a design-with-nature approach to channel management where rivers are managed as ecosystems and competing management interests are balanced, the practical challenge is to use the planning arena to formulate a vision and specific targets for management, and to use creative management methods that work with the particular river to promote the desired effect. A large majority of the projects implemented under this philosophy have been labelled river restoration although they have actually been exercises in environmental rehabilitation that concurrently meet human needs: true restoration (*senso* Cairns, 1991) would generally require compromising human activities to a level that is either impractical or unpalatable. Also, the variety of methods available for restoration-based management schemes are best accommodated under an adaptive management approach that views projects as experiments from which to learn (Holling, 1978; Walters, 1986), in contrast to the uniformity of methods and assumption of predictive certainty that characterized former approaches. This requires new approaches to project assessment and evaluation. However, as many of these new approaches are still being developed and refined it is difficult to conclude whether they have been successful. It seems probable that most implemented projects have resulted in only minor environmental benefit, partly because design-with-nature approaches have not been followed thoroughly: instead, projects have occurred on very short lengths of river on low budgets and with insufficient planning (Table 9.5). It is also highly probable that many projects have involved the 'creation' (National Research Council, 1992) or 'naturalization' (Rhoads and Herricks, 1996) of alternative channel geometries with new process rates, durations, magnitudes and frequencies that are resulting in previously unknown habitat arrangements (Bradshaw, 1987, 2002; Skinner *et al.*, 2003) and for which the biotic response is not yet clearly understood. Other projects still have been essentially traditional channel management schemes re-cast for political acceptability and which have undoubtedly contributed to a further decline in the 'health' of river channel ecosystems. Despite these concerns, the period of the last 30 years has seen a change in the *approach* to channel management that can be described as revolutionary. This leads to the question (Chapter 11) of what comes next.

V

Revision

This concluding chapter attempts a forward look at millennial methods for river channel management stressing a pluralist focus necessary to cater for the range of riverine environments that have to be managed, and suggesting the rudiments of a river channel management process that we term 'management-with-nature', as a more broadly based successor to the primarily technical concerns of design-with-nature.

11 Towards Sustainable Catchment Hydrosystems: Management with Nature?

11.1 Lessons from late twentieth-century experience

Themes from the first six chapters were collated (Table 7.1) and argued to represent the essential *requirements* from which river channel management was being re-defined in the late twentieth century. Four requirements identified as the fundamental building blocks for the new approach (see details on pp. 172–5) were:

1 new scientific perspectives on river channel management recognizing the challenge of managing the catchment hydrosystem;
2 legislated concern for sustainability in river channel management underpinned by the requirement for environmental assessment;
3 recognition that a revised approach to management was needed in which conservation concerns were integral and which included the re-management of river channels as *river restoration*;
4 development of new technical practices that strive to work with natural river processes in order to achieve 'design-with-nature' solutions.

Elements of these requirements are now evident in several programmes such as the Hydrology for Environment, Life and Policy (HELP) initiative, which grew out of a joint meeting of UNESCO and WMO to assist in creating hydro-socio-ecological management plans. HELP intends to support the development of new datasets and watershed management models to improve the links between hydrology and the needs of society, with the goal of delivering social, economic and environmental benefits to stakeholders through sustainable and appropriate use of water, centred on deploying hydrological science in the support of integrated catchment management (Bonnell and Askew, 2000). Thus, the initiative assists in bridging the normal paradigm differences between advanced experimental research with the more pressing sociologic and ecologic demands for water resource management and policy, by ensuring that (1) scientists are aware of the critical management questions and (2) that policy makers are familiar with the latest scientific theories that could aid management. A further example bridging science and management is provided by the California Bay-Delta Authority, with its programme to resolve the competing interests for water between urban, agricultural and conservation requirements, taking place in the highly impoverished river corridors of the Sacramento and San Joaquin valleys. Grants awarded under the science component of this initiative demand articulated conceptual models of fluvial system operation as the foundation of investigation, the collection of high quality scientific data to underpin management recommendations and use of an adaptive management framework as a basis for iterative river system improvements.

Such developments exemplify how, in just a few decades, environmental awareness has increased significantly and how environmental assessment has become possibly the single most important aspect of the management process, conditioning project goals, influencing the management approach, providing the basis for setting specific resource management and environmental targets for post-project evaluations and for advising on appropriate management methods to achieve the desired targets and goals (Chapter 8). The process of environmental assessment is becoming increasingly sophisticated, partly as the questions asked of the assessment become more specific and challenging, and partly as scientific knowledge of the connections between river processes, form, habitat and biota improves. Even more recently, river restoration, a term encompassing a variety of conservation-based management approaches (Table 9. 2), has become highly popular with the public and with management agencies (Chapter 9). This may reflect both increasing evidence of the nonsustainability of earlier approaches and a shift away from the view of river channel management as the exclusive concern of the civil engineer, being succeeded by multidisciplinary teams that include environmental scientists and public representatives. Analyses required for river restoration make it a potential integrating focus within river basin management that can lead to an improvement in environmental conditions, with reduced risk and the prospect of sustainable benefit for water resources management. Conservation-based approaches imply developing a 'design-with-nature' philosophy that utilizes nonstructural as well as structural methods to work with the river to achieve the desired functions and environmental improvements (Chapter 10). The wide variety of available methods that may be used conjunctively include a more strategic application of channel reinforcement, the use of vegetation either for its structural value or for its living attributes, the use of natural river processes (or the restoration of these processes) to facilitate natural recovery, and the use of corridor and catchment policies regulating land-based actions in a way that will not harm the dynamics of channel processes. Project evaluations so far indicate variable degrees of success, which is inevitable as various methods are trialled. General levels of environmental improvement achieved by restoration is probably fairly modest to date, partly because many schemes have been confined to very short lengths of river, with low budgets and with inadequate planning and institutional support mechanisms.

11.2 Prospect: rudiments of 'river channel management with nature'

Experience teaches us that it doesn't!

Norman MacCaig (*A World of Difference*, 1983: 42)

Setting aside MacCaig's assertion, and assuming that we can learn from the past, one of the defining conclusions from revised approaches to river channel management is that no one solution works for all cases. Technical challenges vary from river to river, according to the degree of impairment to natural environmental functions instigated by land use pressures and previous management actions, thus justifying the need to design-with-nature. Additionally, the guiding vision and targeted objectives for each management scheme should also vary between rivers according to the physical catchment characteristics of the drainage basin (geology, soils, vegetation, topography); its climatic and evolutionary history (Gregory, 2003c); the social, political and economic aspects that reflect prevailing national legislation, institutional jurisdiction and organization; the degree and type of public involvement; and funding available. As such, there are five dimensions for pluralistic river channel management, namely:

1 *the importance of place* – scientifically founded river channel management has evolved to teach us that the strategy devised for any river channel should be constructed with awareness of the particular channel and its spatial environmental context;
2 *the implications of scale* – as the interconnectedness of components of the catchment hydrosystem becomes increasingly understood, the prospect is that for management to be effective, it must look increasingly towards catchment-scale solutions with implications for the methods used;
3 *situation in time* – because there are time lags in river channel response and a long-term legacy that is dependent upon the character and history of the catchment, river channel management should be referenced to the temporal position in its sequence of development, although our imperfect knowledge of catchment hydrosystems means that there will continue to be uncertainty related to the precise character of future channel adjustment – with consequent management implications;
4 *the cultural context* – cultural differences between countries and regions will continue to be the reasons for differential responses to river channel management challenges, including the priority concerns and funding available, and according to the status accorded to environmental management. As such 'Basic assumptions and values associated with divergent cultures should be considered in any efforts towards achieving sustainability' of aquatic systems (Davis *et al.*, 2003);
5 *the political framework* – influences funding available, just as culture does, but the determination to undertake channel management has to be incorporated in the political system and then logically and legally implemented. Legislation for management is therefore critical, and explains why, in Europe, the Water Framework Directive is likely to cause the greatest fundamental change to channel management practices since legislation first required environmental assessment.

Options suitable for river channel management will therefore differ between rivers, and will be time, culture- and politics-dependent, so that achieving a sustainable solution requires not only technical know-how, but also an overall management *process* in which acknowledgement of the benefit of maintaining or restoring natural river processes is an integral component. Achieving these ideals requires more than simply design-with-nature; instead an overall process of *management-with-nature* now seems destined to shape river channel management over the foreseeable future. The rudiments of such management-with-nature relate to knowledge of the river channel environment, and to aspects of management and institutional structure as proposed in Table 11.1, based on issues

Table 11.1 Rudiments of river channel management with nature

Rudiment	**Issues**	**Management component**
Understanding the past and the present	As management objectives have increasingly endeavoured to produce channels that are as natural as possible, questions surrounding the definition of natural (Macnaughten and Urry, 1998) for a particular location have been highlighted. This issue, in combination with the range of permissible and feasible options for a project (derived following the process illustrated in Figure 10.1) serve to reinforce the notion that full 'restoration' is rarely possible or sustainable under contemporary catchment conditions, such that the various terms used to describe conservation-based management alternatives (Table 9.2) are more than a matter *merely* of semantics.	1. Understanding what is nature
	Involves the integration of social, economic and ecosystem knowledge about individual drainage basins and highlights linkages that limit management and assist in identifying knowledge gaps that can be addressed through process-oriented research (National Research Council, 1999). This demands a commitment to long-term data collection, monitoring efforts, dedicated geographic databasing entities, improved catchment modelling capabilities founded on high quality terrain data acquisition and representation, justified methods of data abstraction, improved geomorphological process models of drainage basin operations and investment in the collection of validation data, especially for sediment transport (Downs and Priestnall, 2003). Databases for scientific information and technical guidance are especially necessary for emerging fields in river channel management such as dam removal (The Aspen Institute, 2002; The Heinz Center, 2002).	2. Initiate the collation of scientific geospatial databases
	Concerns the continued quantification of rates of damage and recovery of channel environments (Cairns, 2002) so to enlarge the multidisciplinary database of 'safe, usable' knowledge (Statzner and Sperling, 1993) appropriate to river channel management. This is likely to require a new generation of researchers trained in multidisciplinary 'freshwater environmental studies' that integrate social and environmental science needs related to river management, and a new generation of	3. Commitment to scientific funding

Table 11.1 continued

Rudiment	Issues	Management component
	engineers educated using revised and interdisciplinary teaching curricula (Williams, 2001).	
	Further investigation of channel dynamics and sensitivities, both to learn whether the project performance goals were met, but also to benefit future practice (Downs and Kondolf, 2002). A lack of monitoring threatens to constrain emerging fields such as river restoration (Frissell and Ralph, 1998) and dam removal, and is required, in part, to understand the regional differences between scheme responses, and to indicate technological advances of potential merit (The Heinz Center, 2002). Evaluation is not simply the sum of monitoring results but requires contextualization (Downs, 2001).	4. Commitment to post-project monitoring and evaluation
Incorporating future conditions	As nonstructural solutions are generally preferable in channel management (see Chapter 9), channel systems with sufficient energy and sediment availability should be left untouched or encouraged to restore themselves with minimal human interference. Supporting actions may involve sympathetic land use planning including managed retreat of operations away from the bank edge, efforts at flow deregulation or re-regulation or strategic river maintenance.	5. Learn to manage natural recovery
	The prediction and extrapolation of the combined impact of future human actions in the drainage basin and the likely impacts of climate changes. Models will include those capable of forecasting the impact of flow variability and high flow events in the catchment, and not those that simply predict 'average' conditions. Two particular challenges relate to sediment transport interactions with the channel perimeter and floodplain, and the catchment-scale ecohydrological consequences of regional predictions of climate change	6. Develop improved predictive models
	The prospect is that 'river restoration' efforts, many of which are really exercises in 'creation' (Brookes and Shields, 1996c) or 'naturalization' (Rhoads *et al.*, 1999) are resulting in river environments that never existed before human alteration of river systems and that possess new channel geometries with process rates, durations, magnitudes and frequencies that result in previously unknown habitat arrangements for which the biotic response is not clearly understood.	7. Learn to manage created environments

Table 11.1 continued

Rudiment	**Issues**	**Management component**
Coping with uncertainties: the culture of management	Requires a process-oriented style of management wherein design solutions, based on best available knowledge but still considered as experiments should be monitored and evaluated following implementation. Vital signs of an adaptive management approach are likely to include a move towards soft engineering and nonstructural solutions to achieve a combination of environmental objectives and human-centred objectives (Clark, 2002), acknowledgement that management plans represent the start of the process, not the end, and the routine adoption of independent scientific and technical peer review (National Research Council, 1999).	8. Promote the use of adaptive management
	Uncertainties can reflect the evolving, nonlinear, nature of the biophysical channel hydrosystem (p. 117) and, under properly targeted adaptive management, the surprise of new discoveries (McLain and Lee, 1996) regarding process–form–habitat–biota functions and linkages can be accommodated within the management process.	9. Set attainable and measurable target indicators
	Apart from apparent certainty as the basis for management, and the practice of 'walking away' from a solution once implemented, there is great institutional resistance to accepting that post-implementation hydrosystem change is desirable and of long-term benefit. Therefore, education requires managers to become 'risk tolerant' rather than 'risk adverse' and '... come to confront uncertainty as a core part of their work and philosophy rather than as an unwelcome but hopefully temporary aberration' (Clark, 2002: 347). This is a significant challenge in an increasingly litigious world.	10. Educate the river managers
Rationalizing risk to support decision-making	Channel processes result in either, or both, hazards or assets (p. 120) such that a paradox is created. For the physical element of decision-support, rationalization of risk occurs through understanding river channel sensitivity to change (p. 124) but river channel management decisions will also include biological, socio-economic and political factors all contributing to the prospect that river managers will be required to make decisions outside of the area of their professional expertise (Clark	11. Assessment of the risks involved

Table 11.1 continued

Rudiment	Issues	Management component
	and Richards, 2002): complex mechanisms may be required using fuzzy logic and probabilistic decision-support to prioritize the allocation of funds, especially for restoration purposes (Clark, 2002). The matter is undoubtedly complicated further by society's 'ecological denial' wherein there is an apparent preference for management actions to consist of one enormous risk rather than a series of behavioural adjustments (Cairns, 2002).	
Management with stakeholders	Creating a biophysical vision of the post-management channel environment is recommended by many (e.g. FISRWG, 1998; Naiman *et al.*, 1998; Brierley and Fryirs, 2001; Williams, 2001), but is likely to provide an enduring challenge as it requires a different solution in every drainage basin and depends upon consensus, which may be a challenge both for stakeholders, used only to considering their own viewpoint, and for management institutions, used to a more formal chain of direction. Recent assessment of restoration projects undertaken by the Worldwide Fund for Nature showed that successful projects commonly involved the long-term commitment of project partners and the development of a shared vision of action (Zöckler, 2000).	12. Formulate shared visions of management outcomes
	Involves the creation of stakeholder groups, either as 'friends of the river' type groups organized by environmental activists or the local community, or the creation of 'rivercare' type organizations convened by the managing authority but dependent on local community involvement. This also requires that management institutions interacting with the stakeholder groups have employees adept at interfacing with the local community at regular meetings where information is exchanged. Issues include resolving perceptions on matters of critical importance to river system functioning, including the role of large woody debris as an asset as well as a hazard, exploring public perception of riparian landscapes away from being 'dark and dangerous' (Petts, 2001), and recognizing that channel management decisions have been driven historically by concerns for safety and economy so that solutions that stress nature and aesthetics will be strongly dependent on public support for their implementation (Smits *et al.*, 2001).	13. Encourage stakeholder education

Table 11.1 continued

Rudiment	**Issues**	**Management component**
	Land availability may be one of the most fundamental issues determining whether restoration projects have the breadth and length of river corridor required to achieve meaningful environmental enhancements. Acquisition of land requires creative solutions for obtaining the use (if not the ownership) of riparian land for community benefit and for pursuing sustainable management solutions and, as such, 'sustainable decisions are likely to be holistic and, given the self-sacrifice that is often required of stakeholders, participatory' (Mirighani and Savenije, 1995; Iyer-Raniga and Treloar, 2000).	14. Facilitation of land acquisition
Management as a reflection of institutional structure	Basin-based management demands basin-friendly institutions. Such issues were first raised from concerns regarding the ability of institutions to adopt integrated river basin planning and management (e.g. Mitchell, 1990, and see Chapter 6) and, for catchment-based approaches to channel management, similar tensions arise related to jurisdiction, spatial boundaries and internal structure. Fragmentation of responsibility and lack of integration of federal agency disputes have hindered watershed management to date in the USA (National Research Council, 1999), and it may be that the increasing involvement of nonprofit organizations in watershed management has occurred because they can offer an alternative and more responsive infrastructure than can existing institutions (von Hagen *et al.*, 1998). Successful catchment-based channel management, like IRBM, demands that their governing institutions recognize that drainage basins are the optimal organizing unit for channel management (National Research Council, 1999), are capable of regionally based analysis and funding long-term solutions, and individual agencies consider the catchment implications of their work from multiple perspectives.	15. Ensuring that institutional organization and structures are sufficiently flexible

identified from recent literature. This analysis suggests that there are at least 15 specific management components that need to be adopted, incorporated or improved upon under a management-with-nature scenario. As river channel management develops, these rudiments should become increasingly coherent until they combine to approach a pluralistic blueprint and prospectus for river channel management that centres on sustainable catchment hydrosystems.

Irrespective of other trends, advances in river channel management will continue to depend upon the vision, creativity and energy of individuals and groups of individuals who have a clearly articulated philosophy on which to base the vision they hope to create. In an increasingly 'e'-based world, it seems fitting to plot this progression by considering the changing 'e'-philosophy towards river channel management. Winkley's (1972) initial focus on *experience* and *engineering* judgement to work-with-the-river was succeeded by the four 'e's of best practice advocated by General Hatch when Chief of the US Army Corps of Engineers encompassing *engineering*, *economy*, *efficiency* and *environment* (Thorne *et al.*, 1997b) while, most recently, successful leadership in achieving sustainable freshwater systems is contended to consist of five 'e's, namely: *empathy*, *education*, *empower*, *enable* and *equity* (Holland *et al.*, 2003). We contend that these perspectives fairly represent the rapid evolution that has characterized river channel management in the late twentieth century, including the development of a design-with-nature approach that is still imperfect but now trending towards a holistic, catchment-wide, management process characterized here as management-with-nature: a philosophy that seeks to reconcile objectives for water resources, hazard reduction and conservation under a single conciliatory scheme, but one that provides river managers with numerous challenges for the future.

This book has aimed to provide ingredients for individuals to develop their own philosophy for river channel management. E-based or otherwise, river channel management will continue to be led by concerns for water resource management and for hazard avoidance, with conservation issues (the preservation or restoration of channel assets) playing an increasingly large role in dictating the actions taken. Just as the *river channel* has been perceived in different ways over the last century, it will continue to be perceived in different ways over the present one; *management* is also susceptible to shifts in fashion and perception – we hope that the two will further become more holistically integrated, underpinned by the contributions of recent decades – always bearing in mind the contemporary implications of Robert Bridges (1844–1930) aphorism that 'Man masters nature not by force but by understanding' (Collins Dictionary, 2002: 289).

References

Abbe, T.B., Montgomery, D.R. and Petroff, C., 1997: Design of stable in-channel wood debris structures for bank protection and habitat restoration: an example from the Cowlitz River, WA. In Wang, S.S.Y., Langendoen, E.J. and Shields, F.D., Jr, (eds), *Proceedings of the conference on management of landscapes disturbed by channel incision*. Oxford MS: University of Mississippi, 809–16.

Ackers, P., 1980: Use of sediment transport concepts in stable channel design. *Proceedings of an international workshop on alluvial river problems*. Roorkee, India, paper II.2.

Adams, W.M., 1985: River basin planning in Nigeria. *Applied Geography*, 5: 297–308.

Adams, W.M. and Perrow, M.R., 1999: Scientific and institutional constraints on the restoration of European floodplains. In Marriott, S.B. and Alexander, J. (eds), *Floodplains: interdisciplinary perspectives*. London: Geological Society, 89–97.

Agence de l'Eau Rhin-Meuse, 1996: Outil d'évaluation de la qualitié du milieu physique – synthèse. Metz: Agence de l'Eau Rhin-Meuse.

Aldridge, B.N. and Eychaner, J.H., 1984: Floods of October 1977 in southern Arizona and March 1978 in central Arizona. *US Geological Survey, Water-Supply Paper* 2223, Washington DC.

Allan, J.D., 1995: *Stream ecology: structure and function of running waters*. London: Chapman and Hall.

Allen, J.R.L., 1965: A review of the origin and characteristics of recent alluvial sediments. *Sedimentology*, 5: 89–191.

Allen, J.R.L., 1970: *Physical processes of sedimentation*. London: Allen and Unwin.

Allen, J.R.L., 1977: Changeable rivers: some aspects of their mechanics and sedimentation. In Gregory, K.J. (ed.), *River channel changes*. Chichester: J.Wiley & Sons, 15–45.

Allison, R.J. and Thomas, D.S.G., 1993: The sensitivity of landscapes. In Allison, R.J. and Thomas, D.S.G. (eds), *Landscape sensitivity*. Chichester: J.Wiley & Sons, 1–5.

American Rivers, Friends of the Earth and Trout Unlimited, 1999: Dam removal success stories: restoring rivers through selective removal of dams that don't make sense. Washington DC: American Rivers Publishing.

Amoros, C., Rostan, J.C., Pautou, G. and Bravard, J.P., 1987a: The reversible process concept applied to the environmental management of large river systems. *Environmental Management*, 11: 607–17.

Amoros, C., Roux, A.L. and Reygrobellet, J.L., 1987b: A method for applied ecological studies of fluvial hydrosystems. *Regulated Rivers: Research and Management*, 1: 17–36.

Anderson, M.G. and Bates, P.D., 1994: Initial testing of a two-dimensional finite element model for floodplain inundation. *Proceedings of the Royal Society, Series A*, 444: 149–59.

Anderson, M.G. and Burt, T.P. (eds), 1985: *Hydrological forecasting*. Chichester: J. Wiley and Sons.

Anderson, M.G., Walling, D.E. and Bates, P.D. (eds), 1996: *Floodplain processes*. Chichester: J. Wiley and Sons.

Andrews, E.D., 1980: Effective and bankfull discharges of streams in the Yampa river basin, Colorado and Wyoming. *Journal of Hydrology*, 46: 311–30.

Arnell, N.W., 1996: *Global warming, river flows and water resources*. Chichester: J. Wiley & Sons.

Arnell, N.W. and Reybard, N.S., 1996: The effects of climate change due to global warming on river flows in Great Britain. *Journal of Hydrology*, 183: 397–424.

Ascough, J.C., II, Baffaut, C., Nearing, M.A. and Liu, B.Y., 1997: The WEPP watershed model: I. Hydrology and erosion. *Transactions of the American Society of Agricultural Engineers*, 40: 921–33.

Ashworth, P.J. and Ferguson, R.I., 1987: Interrelationships of channel processes, changes and sediments in a proglacial river. *Geografiska Annaler*, 68A: 361–71.

Aspen Institute, The, 2002: *Dam Removal: a new option for a new century*. Washington DC: The Aspen Institute.

Attfield, R., 1994: Rehabilitating nature and making nature habitable. In Attfield, R. and Belsey, A. (eds), *Philosophy and the natural environment*. Cambridge: Cambridge University Press, 45–57.

Attfield, R., 1999: *The ethics of the global environment*. Edinburgh: Edinburgh University Press.

Attfield, R. and Belsey, A. (eds), 1994: *Philosophy and the natural environment*. Cambridge: Cambridge University Press.

Bachmann, J. and Wurzer, A., 2001: The Danube: river of life. In Nijland, H.J. and Cals, M.J.R. (eds), *River restoration in Europe: practical approaches – proceedings of the conference on river restoration, Wageningen, the Netherlands, 2000*. Lelystad: Institute for Inland Water Management and Waster Water Treatment / RIZA, 85–89.

Bagnold, R.A., 1960: Sediment discharge and stream power: a preliminary announcement. *US Geological Survey, Professional Paper* 421, Washington DC.

Bagnold, R.A., 1966: An approach to the sediment transport problem from general physics. *US Geological Survey, Professional Paper* 422, Washington DC.

Baird, A.J., 1999: Introduction. In Baird, A.J. and Wilby, R.L. (eds), *Eco-hydrology: plants and water in terrestrial and aquatic environments*. London: Routledge, 1–10.

Baker, V.R., 1991: A bright future for old flows. In Starkel, L., Gregory, K.J. and Thornes, J.B. (eds), *Temperate palaeohydrology: fluvial processes in the temperate zone during the last 15,000 years*. Chichester: J. Wiley & Sons, 497–520.

Baker, V.R., 1994: Geomorphological understanding of floods. *Geomorphology*, 10: 139–56.

Baker, V.R., 1995: Global palaeohydrological change. *Quaestiones Geographicae*, Special Issue 4: 27–36.

Baker, V.R., 1996: Discovering Earth's future in its past: palaeohydrology and global environmental change. In Branson, J., Brown, A.G. and Gregory, K.J. (eds), *Global continental changes: the context of palaeohydrology*. London: Geological Society, Special Publication 115, 75–83.

Baker, V.R., 2003: Palaeofloods and extended discharge records. In Gregory, K.J. and Benito, G. (eds), *Palaeohydrology: understanding global change*. Chichester: J. Wiley & Sons, 307–23.

Baker, V.R. and Costa, J.E., 1987: Flood power. In Mayer, L. and Nash, D. (eds), *Catastrophic flooding*. London: Allen and Unwin, 1–24.

Balling, R.C. and Wells, S.G., 1990: Historical rainfall patterns and Arroyo activity within the Zuni River drainage basin, New Mexico. *Annals of the Association of American Geographers*, 80: 603–17.

Barker, G., 1997: *A framework for the future: green networks with multiple uses in and around towns and cities*. English Nature Research Reports Number 256, Peterborough: English Nature.

Barnard, R.S. and Melhorn, W.N., 1982: Morphologic and morphometric response to channelization: the case history of Big Pine Creek Ditch, Benton County, Indiana. In Craig, R.G. and Craft, J.L. (eds), *Applied geomorphology*. London: George Allen & Unwin, 224–39.

Barrett, K.R., 1999: Ecological engineering in water resources: the benefits of collaborating with nature. *Water International*, 24: 182–88.

Bathurst, J.C., 1993: Flow resistance through the channel network. In Bevan, K. and Kirkby, M.J. (eds), *Channel network hydrology*. Chichester: J. Wiley & Sons, 69–98.

Baxter, G., 1961: River utilization and the preservation of migratory fish. *Proceedings of the Institution of Civil Engineers*, 18: 225–44.

Bayley, P.B., 1991: The flood-pulse advantage and restoration of river-floodplain systems. *Regulated Rivers: Research and Management*, 6: 75–86.

Bayley, P.B. and Li, H.W., 1996: Riverine fishes. In Petts, G.E. and Calow, P. (eds), *River biota, diversity and dynamics*. Oxford: Blackwell Science, 92–122.

Beaumont, P., 1978: Man's impact on river systems: a worldwide review. *Area*, 10: 38–41.

Beckinsale, R.P., 1969a: The human use of open channels. In Chorley, R.J. (ed.), *Water, Earth and man*. London: Methuen, 331–43.

Beckinsale, R.P., 1969b: Rivers as political boundaries. In Chorley, R.J. (ed.), *Water, Earth and man*. London: Methuen, 344–55.

Benito, G., Baker, V.R. and Gregory, K.J. (eds), 1998: *Palaeohydrology and environmental change*. Chichester: J. Wiley & Sons.

Bergen, S.D., Bolton, S.M. and Fridley, J.L., 2001: Design principles for ecological engineering. *Ecological Engineering*, 18: 201–10.

Berger, J.J., 1990: Evaluating ecological protection and restoration projects: a holistic approach to the assessment of complex, multi-attribute resource management problems. Unpublished Ph.D. Thesis, University of California, Davis, Davis CA.

Beschta, R.L., Platts, W.S., Kauffman, J.B. and Hill, M.T., 1994: Artificial stream restoration – money well spent or expensive failure? In *Environmental restoration conference*. University Council on Water Resources 1994 Annual Meeting, 2–5 August, Big Sky, MT, 76–104.

Biggs, J., Corfield, A., Grøn, P., Hansen, H.O., Walker, D., Whitfield, M. and Williams, P., 1998: Restoration of the Rivers Brede, Cole and Skerne: joint Danish and British EU-LIFE demonstration project, V – short-term impacts on the conservation value of aquatic macroinvertebrate and macrophyte assemblages. *Aquatic Conservation: Marine and Freshwater Ecosystems*, Special Issue, 8: 241–64.

Billi, P., Hey, R.D., Thorne, C.R. and Tacconi, P. (eds), 1992: *Dynamics of gravel bed rivers.* Chichester: J. Wiley & Sons.

Binder, W., Jürging, P. and Karl, J., 1983: Natural river engineering – characteristics and limitations. *Garten und Landschaft*, 2: 91–94.

Bisson, P.A. and Montgomery, D.R., 1996: Valley segments, stream reaches and channel units. In Hauer, R.F.R. and Lamberti, G.A. (eds), *Methods in stream ecology.* London: Academic Press, 23–52.

Biswas, A.K., 1967: *History of hydrology.* Amsterdam: North Holland Publishing Co.

Blake, W.H., Walling, D.E. and He, Q., 2002: Using cosmogenic beryllium-7 as a tracer in sediment budget investigations. *Geografiska Annaler*, 84A: 89–102.

Bledsoe, B.P. and Watson, C.C., 2000: Observed thresholds of stream ecosystem degradation in urbanizing areas: a process based geomorphic explanation. In Flug, M. and Frevert, D. (eds), *Watershed management 2000: science and engineering technology for the new millenium.* Reston VA: American Society of Civil Engineers.

Bledsoe, B.P. and Watson, C.C., 2001: Effects of urbanization on channel instability. *Journal of the American Water Resources Association*, 37: 255–70.

Blench, T., 1952: Regime theory for self-formed sediment-bearing channels. *Transactions of the American Society of Civil Engineers*, 117: 383–400.

Blench, T., 1957: *Regime behaviour of canals and rivers.* London: Butterworths.

Blench, T., 1969: *Mobile-bed fluviology: a regime theory treatment of canals and rivers for engineers and hydrologists.* Alberta: The University of Alberta Press.

Bolling, D.M., 1994: *How to save a river: a handbook for citizen action.* Washington DC: Island Press.

Bonnell, M. and Askew, A., 2000: *Report of 2nd international conference on climate and water.* Espoo, Finland. Helsinki: Helsinki University of Technology, Water Resources Laboratory.

Boon, P.J., 1992: Essential elements in the case for river conservation. In Boon, P.J., Calow, P. and Petts, G.E. (eds), *River conservation and management.* Chichester: J. Wiley & Sons, 11–23.

Boon, P.J., 1995: The relevance of ecology to the statutory protection of British Rivers. In Harper, D.M. and Ferguson, A.J.D. (eds), *The ecological basis for river management.* Chichester: J. Wiley & Sons, 239–50.

Boon, P.J., 1998: River restoration in five dimensions. *Aquatic Conservation: Marine and Freshwater Ecosystems*, Special Issue, 8: 257–64.

Boon, P.J., Calow, P. and Petts, G.E. (eds), 1992: *River conservation and management.* Chichester: J. Wiley & Sons.

Boon, P.J., Davies, B.R. and Petts, G.E. (eds), 2000: *Global perspectives on river conservation: science, policy and practice.* Chichester: J. Wiley & Sons.

Booth, D.B., 1990: Stream channel incision following drainage basin urbanization. *Water Resources Bulletin*, 26: 407–18.

Born, S.M., Genskow, K.D., Filbert, T.L., Hernandez-Mora, N., Keefer, M.L. and White, K.A., 1998: Socioeconomic and institutional dimensions of dam removals: the Wisconsin experience. *Environmental Management*, 22: 359–70.

Botkin, D.B., Peterson, D.L. and Calhoun, J.M., 2000. *The scientific basis for validation monitoring of salmon for conservation and restoration plans.* Forks WA: University of Washington, Olympic Natural Resources Center.

Bovee, B.D., 1982: *A guide to stream habitat analysis using the instream flow incremental methodology*. Instream Flow Information Paper 12, Fort Collins CO: US Fish and Wildlife Service.

Bovee, K.D., 1996: Perspectives on two-dimensional river habitat models: the PHABSIM experience. In Leclerc, M., Capra, H., Valetin, S., Boudreault, A. and Cote, Y. (eds), *Ecohydraulique 2000: proceedings of the 2nd IAHR symposium of habitat hydraulics*. Quebec: IAHR, 150–62.

Boyd, P., Broderick, T., Cunial, S. and Nagel, F., 1999: Development of community-based river planning on the north coast of New South Wales. In Rutherford, I. and Bartley, R. (eds), *Second Australian stream management conference: the challenge of rehabilitating Australia's streams*. Adelaide: Cooperative Research Centre for Catchment Hydrology, 87–91.

Bradshaw, A.D., 1987: The reclamation of derelict land and the ecology of ecosystems. In Jordan, W.R., Gilpin, M.E. and Aber, J.D. (eds), *Restoration ecology*. Cambridge: Cambridge University Press, 53–74.

Bradshaw, A.D., 2002: Introduction and philosophy. In Perrow, M.R. and Davy, A.J. (eds), *Handbook of ecological restoration, volume 1: principles of restoration*. Cambridge: Cambridge University Press, 3–9.

Braga, G. and Gervasconi, S., 1989: Evolution of the Po River: an example of the application of historic maps. In Petts, G.E., Moller, H. and Roux, A.L. (eds), *Historical change of large alluvial rivers: Western Europe*. Chichester: J. Wiley & Sons, 113–26.

Brandon, T.W. (ed.), 1987a: *River engineering – part I, design principles*. London: Institution of Water and Environmental Management.

Brandon, T.W. (ed.), 1987b: *River engineering – part II, structures and coastal defence works*. London: Institution of Water and Environmental Management.

Brandt, C.J. and Thornes, J.B. (eds), 1996: *Mediterranean desertification and land use*. Chichester: J. Wiley & Sons.

Branson, J., Brown, A.G. and Gregory, K.J. (eds), 1996: *Global continental changes: the context of palaeohydrology*. Special Publication 115, London: Geological Society.

Bravard, J.P. and Gilvear, D.J., 1996: Hydrological and geomorphological structure of hydrosystems. In Petts, G.E. and Amoros, C. (eds), *Fluvial hydrosystems*. London: Chapman and Hall, 98–116.

Bravard, J.P., Kondolf, G.M. and Piégay, H., 1999a: Environmental and societal effects of channel incision and remedial strategies. In Darby, S.E. and Simon, A. (eds), *Incised river channels: processes, forms, engineering and management*. Chichester: J. Wiley & Sons, 303–41.

Bravard, J.-P., Landon, N., Piery, J.-L. and Piégay, H., 1999b: Principles of engineering geomorphology for managing channel erosion and bedload transport, examples from French rivers. *Geomorphology*, 31: 291–311.

Brice, J.C., 1975. Airphoto interpretation of the form and behaviour of alluvial rivers. Unpublished final report to the US Army Research Office. Cited by Shen, H.W. *et al.*, Federal Highway Administration Report FHWARD-80/160, Washington DC.

Brice, J.C., 1981: *Stability of relocated stream channels*. Technical Report FHWA/RD-80/158, Federal Highways Administration, Washington DC: US Department of Transportation.

Brierley, G.J. and Fryirs, K.A., 2000: River styles, a geomorphic approach to catchment characterization: implications for river rehabilitation in Bega catchment, New South Wales, Australia. *Environmental Management*, 25: 661–79.

Brierley, G.J. and Fryirs, K.A., 2001: Creating a catchment-framed biophysical vision for river rehabilitation programs. In Rutherfurd, I., Sheldon, F., Brierley, G. and Kenyon, C. (eds), *Third Australian stream management conference proceedings: the value of healthy rivers*. Brisbane: Cooperative Research Centre for Catchment Hydrology, Volume 1, 59–65.

Brierley, G.J. and Murn, C.P., 1997: European impacts on downstream sediment transfer and bank erosion in Cobargo catchment, New South Wales, Australia. *Catena*, 31: 119–36.

Brierley, G.J. and Stankoviansky, M., 2002: Geomorphic response to land use change. *Earth Surface Processes and Landforms*, 27(4): 339–462.

Brierley, G.J., Fryirs, K., Outhet, D. and Massey, C., 2002: Application of the River styles framework as a basis for river management in New South Wales, Australia. *Applied Geography*, 22: 91–122.

Brinson, M.H. and Rheinhardt, R., 1996: The role of reference wetlands in functional assessment and mitigation. *Ecological Aplications*, 1996: 69–76.

British Columbia Forest Service, 1995: *Inland watershed assessment procedure guidebook*. Forest Practices Code of British Columbia, BC Environment.

British Columbia Forest Service, 1996: *Channel assessment procedure guidebook*. Forest Practices Code of British Columbia, BC Environment.

Brizga, S. and Finlayson, B. (eds), 2000: *River management: the Australasian experience*. Chichester: J. Wiley & Sons.

Brookes, A., 1985: River channelization: traditional engineering methods, physical consequences and alternative practices. *Progress in Physical Geography*, 9: 44–73.

Brookes, A., 1987a: The distribution and management of channelized streams in Denmark. *Regulated rivers : research and management*, 1: 3–16.

Brookes, A., 1987b: Restoring the sinuosity of artificially managed streams. *Environmental Geology and Water Science*, 10: 33–41.

Brookes, A., 1987c: River channel adjustments downstream of channelization works in England and Wales. *Earth Surface Processes and Landforms*, 12: 337–51.

Brookes, A., 1988: *Channelized rivers: perspectives for environmental management*. Chichester: J. Wiley & Sons.

Brookes, A., 1990: Restoration and enhancement of some engineered river channels: some European experiences. *Regulated Rivers: Research and Management*, 5: 45–56.

Brookes, A., 1992: Recovery and restoration of some engineered British river channels. In Boon, P.J., Calow, P. and Petts, G.E. (eds), *River Conservation and Management*. Chichester: J. Wiley & Sons, 337–52.

Brookes, A., 1994: River channel change. In Calow, P. and Petts, G.E. (eds), *The rivers handbook: hydrological and ecological principles*, volume 2. Oxford: Blackwell Scientific, 55–75.

Brookes, A., 1995: River channel restoration: theory and practice. In Gurnell, A. and Petts, G.E. (eds), *Changing river channels*. Chichester: J. Wiley & Sons, 369–88.

Brookes, A., 1996a: Floodplain restoration and rehabilitation. In Anderson, M.G., Walling, D.E. and Bates, P.D. (eds), *Floodplain processes*. Chichester: J. Wiley & Sons, 553–76.

Brookes, A., 1996b: River channel change. In Petts, G.E. and Calow, P. (eds), *River flows and channel forms: selected extracts from the rivers handbook*. Oxford: Blackwell, 221–42.

Brookes, A., 1997: River dynamics of channel maintenance. In Thorne, C.R., Hey, R.D. and Newson, M.D. (eds), *Applied fluvial geomorphology for river engineering and management*. Chichester: J. Wiley & Sons, 293–307.

Brookes, A. and Gregory, K.J., 1988: Channelization, river engineering and geomorphology. In Hooke, J.M. (ed.), *Geomorphology in environmental planning*. Chichester: J. Wiley & Sons, 145–68.

Brookes, A. and Long, H., 1990: Stort Catchment Morphological Survey, Appraisal report and watercourse summaries. Internal report, Reading: National Rivers Authority.

Brookes, A. and Sear, D.A., 1996: Geomorphological principles for restoring channels. In Brookes, A. and Shields, F.D., Jr (eds), *River channel restoration: guiding principles for sustainable projects*. Chichester: J. Wiley & Sons, 75–101.

Brookes, A. and Shields, F.D., Jr, (eds), 1996a: *River channel restoration: guiding principles for sustainable projects*. Chichester: J. Wiley & Sons.

Brookes, A. and Shields, F.D., Jr, 1996b: Perspectives on river channel restoration. In Brookes, A. and Shields, F.D., Jr. (eds), *River channel restoration: guiding principles for sustainable projects*. Chichester: J. Wiley & Sons, 1–19.

Brookes, A. and Shields, F.D., Jr, 1996c: Towards an approach to sustainable river restoration. In Brookes, A. and Shields, F.D., Jr. (eds), *River channel restoration: guiding principles for sustainable projects*. Chichester: J. Wiley & Sons, 387–402.

Brookes, A., Gregory, K.J. and Dawson, F.H., 1983: An assessment of river channelization in England and Wales. *The Science of the Total Environment*, 27: 97–112.

Brooks, A.P. and Brierley, G.J., 2000: Geomorphic responses of lower Bega River to catchment disturbance, 1851–1926. *Geomorphology*, 18: 291–304.

Brown, A.E. and Howell, D.L., 1992: Conservation of rivers in Scotland: legislative and organizational limitations. In Boon, P.J., Calow, P. and Petts, G.E. (eds), *River conservation and management*. Chichester: J. Wiley & Sons, 407–24.

Brown, A.G., 1987: Holocene floodplain sedimentation and the channel response of the Lower Severn, United Kingdom. *Zeitschrift für Geomorphologie, NF*, 31: 293–310.

Brown, A.G., 1995: Holocene channel and floodplain change: a UK perspective. In Gurnell, A.M. and Petts, G.E. (eds), *Changing river channels*. Chichester: J. Wiley & Sons, 43–64.

Brown, A.G., 1996: Floodplain palaeoenvironments. In Anderson, M.G., Walling, D.E. and Bates, P.D. (eds), *Floodplain processes*. Chichester: J. Wiley & Sons, 95–138.

Brown, A.G., 1997: *Alluvial geoarchaeology: floodplain archaeology and environmental change*. Cambridge: Cambridge University Press.

Brown, A.G., 2002: Learning from the past: palaeohydrology and palaeoecology. *Freshwater Biology*, 47: 817–29.

Brown, A.G. and Keough, M., 1992: Palaeochannels, palaeoland-surfaces and the three dimensional reconstruction of floodplain environmental change. In Carling, P.A. and Petts, G.E. (eds), *Lowland floodplain rivers: geomorphological perspectives*. Chichester: J. Wiley & Sons, 185–202.

Brown, A.G. and Quine, T.A., 1999: Fluvial processes and environmental change: an overview. In Brown, A.G. and Quine, T.A. (eds), *Fluvial processes and environmental change*. Chichester: J. Wiley & Sons, 1–27.

Brown, B., 1982: *Mountain in the clouds: a search for the wild salmon*. New York: Touchstone.

Brown, E.M., Ouyang, D., Asher, J. and Bartholk, J.F., 2002: Interactive distributed conservation planning. *Journal of the American Water Resources Association*, 38: 895–903.

Bruns, D.A., Minshall, G.W., Cushing, C.E., Cummins, K.W., Brock, J.T. and Vannote, R.L., 1984: Tributaries as modifiers of the RCC: analysis of polar ordinations and regression models. *Archive für Hydrobiologie*, 99: 208–20.

Brunsden, D., 1993: Barriers to geomorphological change. In Allison, R.J. and Thomas, D.S.G. (eds), *Landscape sensitivity*. Chichester: J. Wiley & Sons, 7–12.

Brunsden, D., 2001: A critical assessment of the sensitivity concept in geomorphology. *Catena*, 42: 99–123.

Brunsden, D., 2002: Geomorphological roulette for engineers and planners: some insights into an old game. *Quarterly Journal of Engineering Geology and Hydrogeology*, 35: 101–42.

Brunsden, D. and Thornes, J.B., 1979: Landscape sensitivity and change. *Transactions of the Institute of British Geographers*, NS4: 463–84.

Brunsden, D., Doornkamp, J.C. and Jones, D.K.C., 1978: Applied geomorphology: a British view. In Embleton, C., Brunsden, D. and Jones, D.K.C. (eds), *Geomorphology: present problems and future prospects*. Oxford: Oxford University Press, 251–62.

Budiansky, S., 1995: *Nature's keepers: the new science of nature management*. London: Weidenfield and Nicholson.

Buijse, A.D., Coops, H., Staras, M., Jans, L.H., Van Geest, G.J., Grift, R.E., Ibelings, B.W., Oosterberg, W. and Roozen, F.C.J.M., 2002: Restoration strategies for river floodplains along large lowlands in Europe. *Freshwater Biology*, 47: 889–907.

Bull, W.B., 1979: Threshold of critical power in streams. *Bulletin of the Geological Society of America*, 90: 453–64.

Bull, W.B., 1988: Floods; degradation and aggradation. In Baker, R., Kochel R.C. and Patton, P.C. (eds), *Flood geomorphology*. New York: J. Wiley & Sons, 158–65.

Burkham, D.E., 1972: Channel changes on the Gila River in Safford valley, Arizona, 1846–1970. *US Geological Survey, Professional Paper* 665G, Washington DC.

Burton, I., Kates, R.W. and White, G.F., 1978: *The environment as hazard*. Oxford: Oxford University Press.

Bushaw-Newton, K.L., Hart, D.D., Pizzuto, J.E., Thomson, J.R., Egan, J., Ashley, J.T., Johnson, T.E., Horwitz, R.J., Keeley, M., Lawrence, J., Charles, D., Gatenby, C., Kreeger, D.A., Nightengale, T., Thomas, R.L. and Velinsky, D.J., 2002: An integrative approach towards understanding ecological responses to dam removal: the Manatawny Creek study. *Journal of the American Water Resources Association*, 38: 1581–end page.

Byrne, R.B., Ingram, B.L., Starratt, S., Malamaud-Roam, F., Collins, J.N. and Conrad, M.E., 2001: Carbon-isotope, diatom, and pollen evidence for Late Holocene salinity in a brackish marsh in the San Francisco Estuary. *Quaternary Research*, 55: 66–76.

Cahn, R., 1978: *Footprints on the planet: the search for an environmental ethic*. New York: Universe Books.

Cairns, J, Jr, 1991: The status of the theoretical and applied science of restoration ecology. *The Environmental Professional*, 13: 186–94.

Cairns, J, Jr, 2000: Setting ecological restoration goals for technical feasibility and scientific validity. *Ecological Engineering*, 15: 171–80.

Cairns, J, Jr, 2002: Rationale for restoration. In Perrow, M.R. and Davy, A.J. (eds), *Handbook of ecological restoration, volume 1: principles of restoration*. Cambridge: Cambridge University Press, 10–23.

Calder, I.R., 1999: *The blue revolution: land use and integrated water resources management*. London: Earthscan.

Calicott, J.B., 1995: The value of ecosystem health. *Environmental Values*, 4: 345–61.

Calow, P. and Petts, G.E. (eds), 1992: *The rivers handbook, hydrological and ecological principles volume 1*. Oxford: Blackwell Science.

Calow, P. and Petts, G.E. (eds), 1994: *The rivers handbook, the science and management of river environments, volume 2*. Oxford: Blackwell Science.

Cals, M.J.R., 1994: Evaluation of restoration project 'Duursche Waarden' 1989 till 1994. In *Publications and reports of the project 'Ecological Rehabilitation of the Rivers Rhine and Meuse'*. Lelystad: RIZA.

Cals, M.J.R. and van Drimmelen, C., 2001: Space for the river in coherence with landscape planning in the Rhine-Meuse delta. In Nijland, H.J. and Cals, M.J.R. (ed.), *River restoration in Europe: practical approaches – proceedings of the conference on river restoration, Wageningen, The Netherlands, 2000*. Lelystad: Institute for Inland Water Management and Waste Water Treatment / RIZA, 189–95.

Cals, M.J.R., Postma, R., Buijse, A.D. and Marteijn, E.C.L., 1998: Habitat restoration along the River Rhine in the Netherlands: putting ideas into practice. *Aquatic Conservation: Marine and Freshwater Ecosystems*, Special Issue, 8: 61–70.

Canter, L.W., 1985: *Environmental impact of water resources projects*. Chelsea MI: Lewis Publishers Inc.

Caraco, D., 2000: Dynamics of urban stream channel enlargement. In Schueler, T.R. and Holland, H.K. (eds), *The practice of watershed protection: techniques for protecting our nation's streams, lakes, rivers and estuaries*. Ellicott City MD: Center for Watershed Protection, 99–104.

Carling, P.A., 1996: In-stream hydraulics and sediment transport. In Petts, G.E. and Calow, P. (eds), *River flows and channel forms: selected extracts from the rivers handbook*. Oxford: Blackwell Science, 160–84.

Carling, P.A. and Petts, G.E. (eds), 1992: *Lowland floodplain rivers: geomorphological perspectives*. Chichester: J. Wiley & Sons.

Carson, R., 1962: *Silent spring*. Boston MA: Houghton Mifflin.

Cavallin, A., Marchetti, M., Panizza, M. and Soldati, M., 1994: The role of geomorphology in environmental impact assessment. *Geomorphology*, 9: 143–53.

Chang, H.H., 1988: *Fluvial processes in river engineering*. New York: J. Wiley & Sons.

Changming, L., 1989: Problems in management of the Yellow River. *Regulated Rivers: Research and Management*, 3: 361–69.

Changming, L., 2000: A remarkable event of human impacts on the ecosystems: the Yellow River drained dry. Paper read to 29th International Geographical Congress, 17th August 2000. Seoul, Korea.

Chen, Z., Yu, L. and Gupta, A., 2001: The Yangtze River: an introduction. *Geomorphology*, 41: 73–75.

Chien, N., 1985: Changes in river regime after the construction of upstream reservoirs. *Earth Surface Processes and Landforms*, 10: 143–60.

Chin, A., 1989: Step pools in stream channels. *Progress in Physical Geography*, 13: 391–407.

Chin, A., 1999: The morphologic structure of step-pools in mountain streams. *Geomorphology*, 27: 191–204.

Chin, A. and Gregory, K.J., 2001: Urbanization and adjustment of ephemeral stream channels. *Annals of the Association of American Geographers*, 91: 595–608.

Chitlale, S.V., 1973: Theories and relationships of river channel patterns. *Journal of Hydrology*, 19: 285–308.

Chorley, R.J. (ed.), 1969: *Water, Earth and man*. London: Methuen.

Chorley, R.J. and Kennedy, B.A., 1971: *Physical geography: a systems approach*. London: Prentice-Hall International Inc.

Chorley, R.J., Schumm, S.A. and Sugden, D.E., 1984: *Geomorphology*. London: Methuen and Sons.

Chow, V.T., 1964: *Handbook of applied hydrology*. New York: McGraw-Hill Inc.

Chow, V.T., Maidment, D.R. and Mays, L.W., 1988: *Applied Hydrology*. McGraw Hill, New York.

Church, M., 1983: Pattern of instability in a wandering gravel bed channel. In Collinson, A.D. and Lewin, J. (eds), *Modern and ancient fluvial systems, special publication of the International Association of Sedimentologists, 6*. Oxford: Blackwell, 169–80.

Church, M., 1992: Channel morphology and typology. In Calow, P. and Petts, G.E. (eds), *The rivers handbook: hydrological and ecological principles, volume 1*. Oxford: Blackwell, 126–43.

Church, M., 2002: Geomorphic thresholds in riverine landscapes. *Freshwater Biology*, 47: 541–57.

Clark, M.J., 2002: Dealing with uncertainty: adaptive approaches to sustainable river management. *Aquatic Conservation: Marine and Freshwater Ecosystems*, 12: 347–63.

Clark, M.J. and Richards, R.J., 2002: Supporting complex decisions for sustainable river management in England and Wales. *Aquatic Conservation: Marine and Freshwater Ecosystems*, 12: 471–83.

Clayton, S.R., 2002: Quantitative evaluation of physical and biological responses to stream restoration. Unpublished Ph.D. Thesis, University of Idaho, Boise ID.

Clifford, N.J., 1993: Formation of riffle-pool sequences: field evidence for autogenic process. *Sedimentary Geology*, 85: 35–51.

Coates, D.R. (ed.), 1976: *Geomorphology and engineering*. London: George Allen and Unwin.

Coleman, J.M., 1969: Brahmaputra River: channel process and sedimentation. *Sedimentary Geology*, 3: 129–239.

Collins Dictionary, 2002: Collins gem quotations. 4th edition. London: HarperCollins.

Collinson, J.D., 1986: Alluvial sediments. In Reading, H.G. (ed.), *Sedimentary environments and facies*. Oxford: Blackwell Scientific, 20–62.

Cooke, R.U. and Doornkamp, J.C., 1990: *Geomorphology in environmental management: a new introduction*. Oxford: Clarendon Press.

Copeland, R.R., 1994: *Application of channel stability methods – case studies*. Technical Report HL-94–11, Vicksburg MS: US Army Corps of Engineers Waterways Experiment Station.

Coppin, N.J. and Richards, I.G., 1990: *Use of vegetation in civil engineering, construction industry research and information association*. London: Butterworth.

Cortner, H.J. and Moote, M.A., 1994: Trends and issues in land and water resources management: setting the agenda for change. *Environmental Management*, 18: 167–83.

Cosgrove, D., 1990: An elemental division: water control and engineered landscapes. In Cosgrove, D. and Petts, G.E. (eds), *Water, engineering and landscape: water control and landscape transformation in the modern period*. London: Belhaven Press, 1–11.

Cosgrove, D. and Petts, G.E., 1990: *Water, engineering and landscape: water control and landscape transformation in the modern period*. London: Belhaven Press.

Coulthard, T.J., Kirkby, M.J. and Macklin, M.G., 1999: Modelling the impacts of Holocene environmental changes in an upland river catchment, using a cellular automation approach. In Brown, A.G. and Quine, T.A. (eds), *Fluvial processes and environmental changes*. Chichester: J. Wiley & Sons, 31–46.

Cowx, I.G. and Welcomme, R.L., 1998: *Rehabilitation of rivers for fish: a study undertaken by the European Inland Fisheries Advisory Commission of FAO*. Oxford: Blackwell / FAO.

Coy, L.B., 1981: Unified river basin management. In North, R.M., Dworsky, L.B. and Allee, D.J. (eds), *Unified river basin management*. Minneapolis MN: American Water Resources Association, 284–92.

Croke, J. and Mockler, S., 2001: Gully initiation and road-to-stream linkage in a forested catchment, southeastern Australia. *Earth Surface Processes and Landforms*, 26: 205–17.

Culp, J.M., Lowell, R.B. and Cash, K.J., 2000: Integrating mesocosm experiments with field and laboratory studies to generate weight-of-evidence risk assessments for large rivers. *Environmental Toxicology and Chemistry*, 19: 1167–73.

Cunningham, G.M., 1986: Total catchment management – resource management for the future. *Journal of Soil Conservation, NSW*, 42: 4–6.

Cushing, C.E., McIntire, C.D., Sedell, J.R., Cummins, K.W., Minshall, G.W., Petersen, R.C. and Vannote, R.L., 1980: Comparative study of physical-chemical variables of streams using mulitvariate analyses. *Archive für Hydrobiologie*, 89: 343–52.

Cushing, C.E., McIntire, C.D., Cummins, K.W., Minshall, G.W., Petersen, R.C., Sedell, J.R. and Vannote, R.L., 1983: Relationships among chemical, physical and biological indices along river continua based on multivariate analyses. *Archiv für Hydrobiologie*, 98: 317–26.

Dalrymple, T., 1960: Flood frequency analyses. *US Geological Survey, Water Supply Paper* 1543A, Washington DC.

Daniels, R.B., 1960: Entrenchment of the willow drainage ditch, Harrison County, Iowa. *American Journal of Science*, 258: 161–76.

Darby, H.C., 1956: The clearing of woodland in Europe. In Thomas, W.L. (ed.), *Man's role in changing the face of the Earth*. Chicago IL: Unversity of Chicago Press, 183–216.

Darby, H.C., 1983: *The changing fenland*. Cambridge: Cambridge University Press.

Darby, S.E. and Simon, A. (eds), 1999: *Incised river channels: processes, forms, engineering and management*. Chichester: J. Wiley & Sons.

Darby, S.E. and Thorne, C.R., 1995: Fluvial maintenance operations in managed alluvial rivers. *Aquatic Conservation: Marine and Freshwater Ecosystems*, 5: 37–54.

Darby, S.E. and Thorne, C.R., 2000: A river runs through it: morphological and landowner sensitivities along the Upper Missouri River, Montana, USA. *Transactions of the Institute of British Geographers*, NS25: 91–107.

Davis, R.J. and Gregory, K.J., 1994: A new distinct mechanism of river bank erosion in a forested catchment. *Journal of Hydrology*, 157: 1–11.

Davis, S.A., Shaffer, L.R. and Edmister, J.H., 2003: Sustainability of aquatic systems and the role of culture and values. In Holland, M.M., Blood, E.R. and Shaffer, L.R. (eds), *Achieving sustainable freshwater systems: a web of connections*. Washington DC: Island Press, 239–49.

Decamps, H., Fortune, M. and Gazelle, F., 1989: Historical change of the Garonne River, southern France. In Petts, G.E., Moller, H. and Roux, A.L. (eds), *Historical change of large alluvial rivers: Western Europe*. Chichester: J. Wiley & Sons, 249–67.

DeGraff, J.V., 1994: The geomorphology of some debris flows in the southern Sierra Nevada, California. *Geomorphology*, 10: 231–52.

Department of the Environment, 1985: *Digest of environmental protection and water statistics*. London: HMSO.

De Waal, L.C., Large, A.R.G. and Wade, P.M. (eds), 1998: *Rehabilitation of rivers: principles and implementation*. Chichester: J. Wiley & Sons.

Dietrich, W.E. and Dunne, T., 1993: The channel head. In Bevan, K. and Kirkby, M.J. (eds), *Channel network hydrology*. Chichester: J. Wiley & Sons, 175–219.

Dietrich, W.E., Dunne, T., Humphrey, N.F. and Reid, L.M., 1982: Construction of sediment budgets for drainage basins. In Swanson, F.J., Janda, R.J., Dunne, T. and Swanston, D.N. (eds), *Sediment budgets and routing in forested drainage basins*. General Technical Report PNW-141, Portland OR: Pacific Northwest Forest and Range Experiment Station, US Forest Service, 5–23.

Diplas, P., 2002: Integrated decision making for watershed management. *Journal of the American Water Resources Association*, 38: 337–40.

Dooge, J.C., 1974: The development of hydrological concepts in Britain and Ireland between 1674 and 1874. *Hydrological Sciences Bulletin*, 19: 279–302.

Doppelt, B., Scurlock, M., Frissell, C. and Karr, J., 1993: *Entering the watershed: a new approach to save America's river ecosystems*. Washington DC: Island Press.

Douglas, I., 2000: Fluvial geomorphology and river management. *Australian Geographical Studies*, 38: 253–62.

Downs, P.W., 1992: Spatial variation in river channel adjustments in south-east England: implications for channel management. Unpublished Ph.D. Thesis, University of Southampton, Southampton.

Downs, P.W., 1994: Characterization of river channel adjustments in the Thames basin, South-East England. *Regulated Rivers: Research and Management*, 9: 151–75.

Downs, P.W., 1995a: Estimating the probability of river channel adjustment. *Earth Surface Processes and Landforms*, 20: 687–705.

Downs, P.W., 1995b: River channel classification for channel management purposes. In Gurnell, A.M. and Petts, G.E. (eds), *Changing river channels*. Chichester: J. Wiley & Sons, 347–65.

Downs, P.W., 2001: Geomorphological evaluation of river restoration schemes: principles, method, monitoring, assessment, evaluation. Progress? In Nijland, H.J. and Cals, M.J.R. (eds), *River restoration in Europe: practical approaches*. Lelystad: Institute for Inland Water Management and Waste Water Treatment / RIZA, 243–49.

Downs, P.W. and Brookes, A., 1994: Developing a standard geomorphological approach for the appraisal of river projects. In Kirby, C. and White, W.R. (eds), *Integrated river basin development*. Chichester: J. Wiley & Sons, 299–310.

Downs, P.W. and Caruso, B.S., 2000: Three streamscapes project: fluvial geomorphology context for rehabilitation opportunities in the Water of Leith, Dunedin, New Zealand. In Nolan, T.J. and Thorne, C.R. (eds), *Gravel bed rivers 2000 CD ROM*. Wellington: Special Publication of the New Zealand Hydrological Society.

Downs, P.W. and Gregory, K.J., 1993: The sensitivity of river channels in the landscape system. In Allison, R.J. and Thomas, D.S.G. (eds), *Landscape sensitivity*. Chichester: J. Wiley & Sons, 15–30.

Downs, P.W. and Gregory, K.J., 1994: Sensitivity analysis and river conservation sites: a drainage basin approach. In O'Halloran, D., Green, C., Harley, M., Stanley, M. and Knill, J. (eds), *Geological and landscape conservation*. London: Geological Society, 139–43.

Downs, P.W. and Gregory, K.J., 1995: The sensitivity of river channels to adjustment. *The Professional Geographer*, 47: 168–75.

Downs, P.W. and Kondolf, G.M., 2002: Post-project appraisals in adaptive management of river channel restoration. *Environmental Management*, 29: 477–96.

Downs, P.W. and Priestnall, G., 2003: Modelling catchment processes. In Kondolf, G.M. and Piegay, H. (eds), *Tools in fluvial geomorphology*. Chichester: J. Wiley & Sons, 205–30.

Downs, P.W. and Simon, A., 2001: Fluvial geomorphological analysis of the recruitment of large woody debris in Yalobusha river network, central Mississippi, USA. *Geomorphology*, 37: 65–91.

Downs, P.W. and Thorne, C.R., 1996: The utility and justification of river reconnaissance surveys. *Transactions of the Institute of British Geographers*, 21: 455–68.

Downs, P.W. and Thorne, C.R., 1998a: Design principles and suitability testing for rehabilitation in a flood defence channel: the River Idle, Nottinghamshire, UK. *Aquatic Conservation: Marine and Freshwater Ecosystems*, 8: 17–38.

Downs, P.W. and Thorne, C.R., 1998b: Planning for rehabilitation in a flood defence channel: the River Idle, UK. In Hansen, H.O. and Madsen, B.L. (eds), *River restoration '96*. Silkeborg: National Environmental Research Institute, 96–103.

Downs, P.W. and Thorne, C.R., 2000: Rehabilitation of a lowland river: reconciling flood defence with habitat diversity and geomorphological sustainability. *Journal of Environmental Management*, 58: 249–68.

Downs, P.W., Gregory, K.J. and Brookes, A., 1991: How integrated is river basin management? *Environmental Management*, 15: 299–309.

Downs, P.W., Skinner, K.S. and Brookes, A., 1997: Developing geomorphic post-project appraisals for environmentally-aligned river channel management. *Water for a changing global community, IAHR XXVII congress*. San Francisco CA: ASCE, 430–35.

Downs, P.W., Skinner, K.S. and Soar, P.J., 1999: Muddy waters: issues in assessing the impact of in-stream structures for river restoration. In Rutherford, I. and Bartley, R. (eds), *Second Australian stream management conference: the challenge of rehabilitating Australia's streams*. Adelaide: Cooperative Research Centre for Catchment Hydrology, 211–17.

Downs, P.W., Skinner, K.S. and Kondolf, G.M., 2002a: Rivers and streams. In Perrow, M.R. and Davy, A.J. (eds), *Handbook of ecological restoration, volume 2: restoration in practice*. Cambridge: Cambridge University Press, 267–96.

Downs, P.W., Sklar, L. and Braudrick, C.A., 2002b: Addressing the uncertainty in prescribing high flows for river restoration. *Eos Transactions AGU*, 83 (47): Abstract H71F-08.

Downward, S.R., 1995: Information from topographic survey. In Gurnell, A.M. and Petts, G.E. (eds), *Changing river channels*. Chichester: J. Wiley & Sons, 303–23.

Doyle, M.W., Harbor, J.M., Rich, C.F. and Spacie, A., 2000a: Examining the effects of urbanization on streams using indicators of geomorphic stability. *Physical Geography*, 21: 155–81.

Doyle, M.W., Stanley, E.H., Luebke, M.A. and Harbor, J.M., 2000b: Dam removal: physical, biological, and societal considerations. *American Society of Civil Engineers joint conference on water resources engineering and water resources planning and management*. Minneapolis MN: ASCE.

Doyle, M.W., Stanley, E.H. and Harbor, J.M., 2002: Geomorphic analogies for assessing probable channel response to dam removal. *Journal of the American Water Resources Association*, 38: 1567–79.

Doyle, M.W., Harbor, J.M. and Stanley, E.H., 2003a: Towards policies and decision-making for dam removal. *Environmental Management*, 31: 453–65.

Doyle, M.W., Stanley, E.H. and Harbor, J.M., 2003b: Channel adjustments following two dam removals in Wisconsin. *Water Resources Research*, 39: 1011, doi: 10.1029/2003WR002038.

Dubgaard, A., Kallesøe, M.F., Petersen, M.L. and Ladenburg, J., 2002: *Cost-benefit analysis of the Skjern River restoration project*. Department of Economics and Natural Resources, Social Science Series 10, Copenhagen: Royal Veterinary and Agricultural University of Copenhagen.

du Boys, P., 1879: Études du régime du Rhone et l'achan exercée par les eaux sur un lit à fond de graviers indéfiniment affoiullable. *Annales des ponts et chaisseés, ser. 5*, 18: 141–95.

Dunne, T. and Leopold, L.B., 1978: *Water in environmental planning*. San Francisco CA: W.H. Freeman.

Dury, G.H., 1966: The concept of grade. In Dury, G.H. (ed.), *Essays in geomorphology*. London: Heinemann, 211–33.

Dynesius, M. and Nilsson, C., 1994: Fragmentation and flow regulation of river systems in the northern third of the world. *Science*, 266: 759.

Ebersole, J.L., Liss, W.J. and Frissell, C.A., 1997: Restoration of stream habitats in the western United States: restoration as re-expression of habitat capacity. *Environmental Management*, 21: 1–14.

Ebisemiju, F.S., 1989: Patterns of stream channel response to urbanization in the humid tropics and their implications for urban land use planning: a case study from southwestern Nigeria. *Applied Geography*, 9: 273–86.

Ebisemiju, F.S., 1991: Some comments on the use of spatial interpolation techniques in studies of man-induced river channel changes. *Applied Geography*, 11: 21–34.

Elliott, R., 1997: *Faking nature: the ethics of environmnental restoration*. London: Routledge.

Elwood, J.W., Newbold, J.D., O'Neill, R.V. and Van Winkle, W., 1983: Resource spiralling: an operational paradigm for analysing lotic ecosystem. In Fontaine, T.D. and Bartell, S.M. (eds), *Dynamics of lotic ecosystems*. Ann Arbor MI: Ann Arbor Science, 3–27.

Endreny, T.A., 2001: A global initiative for hydro-social-ecological watershed research. *Water Resources Impact*, 3: 20–24.

Enzel, Y., Webb, R.H., Ely, L.L., House, P.K. and Baker, V.R., 1993: Palaeoflood evidence for a natural upper bound to flood magnitudes in the Colorado river basin. *Water Resources Research*, 29: 2287–97.

Erskine, W. and Warner, R.F., 1988: Further assessment of flood- and drought-dominated regimes in south eastern Australia. *Australian Geographer*, 29: 257–61.

Erskine, W. and Webb, A., 1999: A protocol for river restoration. In Rutherford, I. and Bartley, R. (eds), *Second Australian stream management conference: the challenge of rehabilitating Australia's streams*. Adelaide: Cooperative Research Centre for Catchment Hydrology, 237–43.

Erskine, W.D., 1990: Hydrogeomorphic effects of river training works: the case of the Allyn River, NSW. *Australian Geographical Studies*, 28: 62–76.

Erskine, W.D., 1998: Environmental impacts of extractive industries on the Hawkesbury-Nepean river, NSW. In Powell, J. (ed.), *The improvers legacy: environmental studies of the Hawkesbury*. Berowra Heights: Deerubbin Press, 49–73.

Erskine, W.D. and Green, D., 2000: Geomorphic effect of extractive industries and their implications for river management. In Finlayson, B. (ed.), *River management: the Australasian experience*. Chichester: J. Wiley & Sons, 123–49.

Erskine, W.D., Terrazzolo, N. and Warner, R.F., 1999: River rehabilitation from the hydrogeomorphic impacts of a large hydro-electric power project: Snowy River, Australia. *Regulated Rivers: Research and Management*, 15: 3–24.

European Community, 2000: Directive 2000/60/EC of the European Parliament and of the Council of 23 October 2000 establishing a framework for Community action in the field of water policy. *Official Journal of the European Communities*, L327: 1–72.

Everest, F.H. and Sedell, J.R., 1984: Evaluating effectiveness of stream rehabilitation projects. *Pacific Northwest stream habitat management workshop*. Arcata CA: American Fisheries Society, Humboldt State University, 246–55.

Eyles, R.J., 1977: Birchams Creek: the transition from a chain of pools to a gully. *Australian Geographer*, 15: 146–57.

Faulkner, D.J., 1998: Spatially variable historical alluviation and channel incision in west central Wisconsin. *Annals of the Association of American Geographers*, 88: 666–85.

Federal Interagency Stream Restoration Working Group (FISRWG), 1998: *Stream corridor restoration: principles, processes and practices*. United States National Engineering Handbook Part 653, Washington DC.

Ferguson, R.I., 1981: Channel form and channel changes. In Lewin, J. (ed.), *British rivers*. London: George Allen & Unwin, 90–125.

Ferguson, R.I., 1984: The threshold between meandering and braiding. In Smith, K.V.A. (ed.), *Channels and channel control structures, proceedings of the first international conference on hydraulic design in water resources engineering*. Berlin: Springer-Verlag, 6.15–6.29.

Field, J., 2001: Channel avulsion on alluvial fans in southern Arizona. *Geomorphology*, 37: 93–104.

Finlayson, B. and Brizga, S., 2000: Introduction. In Brizga, S. and Finlayson, B. (eds), *River management: the Australasian experience*. Chichester: J. Wiley & Sons, 1–10.

Fitzpatrick, F.A. and Knox, J.C., 2000: Spatial and temporal sensitivity of hydrogeomorphic response and recovery to deforestation, agriculture, and floods. *Physical Geography*, 21: 89–108.

Fookes, P.G. and Vaughan, P.R., 1986: *Handbook of engineering geomorphology*. Glasgow: Blackie.

Friberg, N., Kronberg, B., Svendsen, I.M., Hansen, H.O. and Nielsen, M.B., 1994: Restoration of a channelized reach of the River Gelsa, Denmark: effects on the macroinvertebrate community. *Aquatic Conservation: Marine and Freshwater Ecosystems*, 4: 289–96.

Frissell, C.A. and Nawa, R.K., 1992: Incidence and causes of physical failure of artificial structures in streams of Western Oregon and Washington. *North American Journal of Fisheries Management*, 12: 182–97.

Frissell, C.A. and Ralph, S.C., 1998: Stream and watershed restoration. In Naiman, R.J. and Bilby, R.E. (eds), *River ecology and management: lessons for the Pacific Coastal ecosystem*. New York: Springer, 599–624.

Frissell, C.A., Liss, W.J., Warren, C.E. and Hurley, M.D., 1986: A hierarchical framework for stream habitat classification: viewing streams in a watershed context. *Environmental Management*, 10: 199–214.

Frodeman, R., 1995: Radical environmentalism and the political roots of postmodernism – differences that make a difference. In Oelschlaeger, M. (ed.), *Post modern environmental ethics*. Albany NY: State University of New York Press, 121–35.

Fryirs, K. and Brierley, G.J., 1998: The character and age structure of valley fills in upper Wolumba Creek catchment, South Coast, New South Wales. *Earth Surface Processes and Landforms*, 23: 271–87.

Fryirs, K. and Brierley, G.J., 1999: Slope-channel decoupling in Wolumba catchment, N.S.W., Australia: the changing nature of sediment sources following European settlement. *Catena*, 35: 41–63.

Fryirs, K. and Brierley, G.J., 2000: A geomorphic approach to the identification of river recovery potential. *Physical Geography*, 21: 244–77.

Galloway, G.E., 2000: Three centuries of river management along the Mississippi River: engineering and hydrological aspects. In Smits, A.J.M., Nienhuis, P.H. and Leuven, R.S.E.W. (eds), *New approaches to river management*. Leiden: Backhuys, 51–64.

Gameson, A.H.L. and Wheeler, A., 1977: Restoration and recovery of the Thames estuary. In Cairns, J., Dickson, K.L. and Herricks, E.E. (eds), *Recovery and restoration of damaged ecosystems*. Charlottsville VA: University of Virginia Press, 72–101.

Gan, K. and McMahon, T., 1990: Variability of results from the use of PHABSIM in estimating habitat area. *Regulated Rivers: Research and Management*, 5: 233–39.

Garde, R.J. and Ranga Raju, K.G., 1977: *Mechanics of sediment transportation and alluvial stream problems*. New Delhi: Wiley Eastern Limited.

Gardiner, J.L., 1988a: Environmentally sound river engineering: examples from the Thames catchment. *Regulated Rivers: Research and Management*, 2: 445–69.

Gardiner, J.L., 1988b: River Thames flood defence: a strategic initiative. Paper presented to The Institution of Water and Environmental Management, Central Southern Branch Meeting, 29 June 1988.

Gardiner, J.L., 1990: River catchment planning for land drainage, flood defence and the environment. *Journal of the Institution of Water and Environmental Management*, 4: 442–50.

Gardiner, J.L. (ed.), 1991a: *River projects and conservation: a manual for holistic appraisal*. Chichester: J. Wiley & Sons.

Gardiner, J.L., 1991b: Towards sustainable development in river basins: new professional perspectives. Paper presented to International Symposium on effects of watercourse improvements: assessment, methodology, management assistance. 10–12 September 1991, Namur, Belgium.

Gardiner, J.L., 1994: Sustainable development for river catchments. *Journal of the Institution of Water and Environmental Management*, 8: 308–19.

Gardiner, J.L., 1998: Floodplain management in the United Kingdom. In Bailey, R.G., Jose, P.V. and Sherwood, B.R. (eds), *United Kingdom Floodplains*. Otley: Westbury, 17–26.

Gares, P.A., Sherman, D.J. and Nordstrom, K.F., 1994: Geomorphology and natural hazards. *Geomorphology*, 10: 1–18.

Giardino, J.R. and Marston, R.A., 1999: Engineering geomorphology: an overview of changing the face of the earth. *Geomorphology*, 31: 1–11.

Gilbert, G.K., 1877: *Report on the geology of the Henry Mountains*. Washington DC: US Geological Survey, Rocky Mountain Region Report.

Gilvear, D.J., 1999: Fluvial geomorphology and river engineering: future roles utilizing a fluvial hydrosystems framework. *Geomorphology*, 31: 229–45.

Gilvear, D.J. and Bradley, S., 1997: Geomorphological adjustment of a newly engineered upland sinuous gravel-bed river diversion: Evan Water, Scotland. *Regulated Rivers: Research and Management*, 13: 377–89.

Glitz, D., 1983: Atificial channels – the 'ox-bow' lakes of tomorrow: the restoration of the course of the Wandse in Hamberg-Rahlstedt. *Garten und Landschaft*, 2: 109–11.

Gomez, B., Phillips, J.D., Magilligan, F.J. and James, L.A., 1998: Geomorphic and sedimentologic controls on the effectiveness of an extreme flood. *Journal of Geology*, 106: 87–95.

Gordon, J.E., Brazier, V., Thompson, D.B.A. and Horsfield, D., 2001: Geo-ecology and the conservation management of sensitive upland landscapes in Scotland. *Catena*, 42: 323–32.

Gordon, N.D., McMahon, T.A. and Finlayson, B.L., 1992: *Stream hydrology: an introduction for ecologists*. Chichester: J. Wiley & Sons.

Gore, J.A., 1985a: Mechanisms of colonization and habitat enhancement for benthic macroinvertebrates in restored river channels. In Gore, J.A. (ed.), *The restoration of rivers and streams: theories and experience*. Boston MA: Butterworth, 81–101.

Gore, J.A. (ed.), 1985b: *The restoration of rivers and streams: theories and experience*. Boston MA: Butterworth.

Gore, J.A., 1985c: Introduction. In Gore, J.A. (ed.), *The restoration of rivers and streams: theories and experiences*. Boston MA: Butterworth, vii–xii.

Gould, G.A., 1977: Preserving instream flows under the appropriation doctrine – problems and possibilities. In Lamb, B.L. (ed.), *Protecting instream flows under western water law – selected papers*. Instream Flow Information Paper No. 2, FWS/OBS-77/47. Reston VA: U.S. Fish and Wildlife Service.

Graf, W.L., 1975: The impact of suburbanisation on fluvial geomorphology. *Water Resources Research*, 11: 690–92.

Graf, W.L., 1977: The rate law in fluvial geomorphology. *American Journal of Science*, 277: 178–91.

Graf, W.L., 1984: A probabilistic approach to the spatial assessment of river channel instability. *Water Resources Research*, 20: 953–62.

Graf, W.L. (ed.), 1985: *The Colorado River – instability and basin management*. Washington DC: Association of American Geographers.

Graf, W.L., 1988a: Applications of catastrophe theory in fluvial geomorphology. In Anderson, M.G. (ed.), *Modelling geomorphological systems*. Chichester: J. Wiley & Sons, 33–47.

Graf, W.L., 1988b: *Fluvial processes and dryland rivers*. Berlin: Springer Verlag.

Graf, W.L., 1996: Geomorphology and policy for restoration of impounded rivers. In Rhoads, B.L. and Thorn, C.E. (eds), *The scientific nature of geomorphology*. New York: J. Wiley & Sons, 443–73.

Graf, W.L., 1999: Dam nation: a geographic census of American dams and their large-scale hydrologic impacts. *Water Resources Research*, 35: 1305–11.

Graf, W.L., 2000: Locational probability for a dammed, urbanizing stream: Salt River, Arizona, USA. *Environmental Management*, 25: 321–35.

Graf, W.L., 2001: Damage control: restoring the physical integrity of America's rivers. *Annals of the Association of American Geographers*, 91: 1–27.

Gray, D.H. and Sotir, R.B., 1996: *Biotechnical and soil bioengineering for slope stabilization: a practical guide for erosion control*. New York: J. Wiley & Sons.

Green, C.H. and Tunstall, S.M., 1992: The amenity and environmental value of river corridors in Britain. In Boon, P.J., Calow, P. and Petts, G.E. (eds), *River Conservation and Management*. Chichester: J. Wiley & Sons, 425–41.

Greer, C., 1979: *Water management in the Yellow River Basin of China*. Austin TX: University of Texas Press.

Gregory, K.J., 1976a: Changing drainage basins. *Geographical Journal*, 142: 237–47.

Gregory, K.J., 1976b: Lichens and the determination of river channel capacity. *Earth Surface Processes*, 1: 273–86.

Gregory, K.J., 1977a: Channel network and metamorphosis in northern New South Wales. In Gregory, K.J. (ed.), *River Channel Changes*. Chichester: J. Wiley & Sons, 389–410.

Gregory, K.J. (ed.), 1977b: *River channel changes*. Chichester: J. Wiley & Sons.

Gregory, K.J., 1979a: Changes of drainage network composition. *Geology*, 3: 19–28.

Gregory, K.J., 1979b: Hydrogeomorphology: how applied should we become? *Progress in Physical Geography*, 3: 84–101.

Gregory, K.J. (ed.), 1983: *Background to palaeohydrology: a perspective*. Chichester: J. Wiley & Sons.

Gregory, K.J., 1985: *The nature of physical geography*. London: Arnold.

Gregory, K.J., 1987a: Environmental effects of river channel changes. *Regulated Rivers: Research and Management*, 1: 358–63.

Gregory, K.J., 1987b: The hydrogeomorphology of alpine proglacial areas. In Gurnell, A.M. and Clark, M.J. (eds), *Glacio-fluvial sediment transfer*. Chichester: J. Wiley & Sons, 87–107.

Gregory, K.J., 1987c: The power of nature – energetics in physical geography. In Gregory, K.J. (ed.), *Energetics of the physical environment*. Chichester: J. Wiley & Sons, 1–31.

Gregory, K.J., 1992a: Changing physical environment and changing physical geography. *Geography*, 77: 323–34.

Gregory, K.J., 1992b: Vegetation and river channel process interactions. In Boon, P.J., Calow, P. and Petts, G.E. (eds), *River conservation and management*. Chichester: J. Wiley & Sons, 255–69.

Gregory, K.J., 1995: Human activity and palaeohydrology. In Gregory, K.J., Starkel, L. and Baker, V.R. (eds), *Global Continental Palaeohydrology*. Chichester: J. Wiley & Sons, 151–72.

Gregory, K.J. (ed.), 1997: *Fluvial geomorphology of Great Britain.* Geological Conservation Review Series, Joint Nature Conservation Committee. London: Chapman and Hall.

Gregory, K.J., 1998: Applications of palaeohydrology. In Benito, G., Baker, V.R. and Gregory, K.J. (eds), *Palaeohydrology and environmental change*. Chichester: J. Wiley & Sons, 13–26.

Gregory, K.J., 2000: *The changing nature of physical geography*. London: Arnold.

Gregory, K.J., 2002: Urban channel adustments in a management context. *Environmental Management*, 29: 620–33.

Gregory, K.J., 2003a: The limits of wood in world rivers. In Gregory, S.V., Boyer, K.L. and Gurnell, A.M. (eds), *The ecology and management of wood in world rivers*. Bethesda MD: American Fisheries Society, 1–20.

Gregory, K.J., 2003b: Palaeohydrology and river channel management. *Geological Society of India*, in press.

Gregory, K.J., 2003c: Palaeohydrology, environmental change and river channel management. In Gregory, K.J. and Benito, G. (eds), *Palaeohydrology: understanding global change*. Chichester: J. Wiley & Sons, 357–78.

Gregory, K.J. and Benito, G. (eds), 2003a: *Palaeohydrology: understanding global change*. Chichester: J. Wiley & Sons.

Gregory, K.J. and Benito, G., 2003b: Potenial of palaeohydrology in relation to global change. In Gregory, K.J. and Benito, G. (eds), *Palaeohydrology: understanding global change*. Chichester: J. Wiley & Sons, 3–15.

Gregory, K.J. and Brookes, A., 1983: Hydrogeomorphology downstream of bridges. *Applied Geography*, 3: 85–101.

Gregory, K.J. and Chin, A., 2002: Urban stream channel hazards. *Area*, 34: 312–21.

Gregory, K.J. and Davis, R.J., 1992: Coarse woody debris in stream channels in relation to river channel management in wooded areas. *Regulated Rivers: Research and Management*, 7: 117–36.

Gregory, K.J. and Davis, R.J., 1993: The perception of riverscape aesthetics: an example from two Hampshire rivers. *Journal of Environmental Management*, 39: 171–85.

Gregory, K.J. and Gurnell, A.M., 1988: Vegetation and river channel form and process. In Viles, H.A. (ed.), *Biogeomorphology*. Oxford; Blackwell, 11–42.

Gregory, K.J. and Madew, J.R., 1982: Land use changes, flood frequency and channel adjustments. In Hey, R.D., Bathurst, J.C. and Thorne, C.R. (eds), *Gravel bed rivers: fluvial processes, engineering and management*. Chichester: J. Wiley & Sons, 757–82.

Gregory, K.J. and Maizels, J.K., 1991: Morphology and sediments: typological characteristics of fluvial forms and deposits. In Starkel, L., Gregory, K.J. and Thornes, J.B. (eds), *Temperate palaeohydrology*. Chichester: J. Wiley & Sons, 31–59.

Gregory, K.J. and Park, C.C., 1976: Stream channel morphology in northwest Yorkshire. *Revue de Geomorphologie Dynamique*, 25: 63–72.

Gregory, K.J. and Walling, D.E., 1973: *Drainage basin form and process: a geomorphological approach*. Arnold, London.

Gregory, K.J., Gurnell, A.M. and Hill, C.T., 1985: The permanence of debris dams related to river channel processes. *Hydrological Sciences Journal*, 30: 53–74.

Gregory, K.J., Davis, R.J. and Downs, P.W., 1992: Identification of river channel change due to urbanisation. *Applied Geography*, 12: 299–318.

Gregory, K.J., Davis, R.J. and Tooth, S., 1993: Spatial distribution of coarse woody debris dams in the Lymington Basin, Hampshire, UK. *Geomorphology*, 6: 207–24.

Gregory, K.J., Gurnell, A.M., Hill, C.T. and Tooth, S., 1994: Stability of the pool-riffle sequence in changing river channels. *Regulated Rivers: Research and Management*, 9: 35–43.

Gregory, K.J., Starkel, L. and Baker, V.R. (eds), 1995: *Global continental palaeohydrology*. Chichester: J. Wiley & Sons.

Gregory, S.V., Swanson, F.J., McKee, W.A. and Cummins, K.W., 1991: An ecosystem perspective on riparian zones. *BioScience*, 41: 540–51.

Gregory, S.V., Gurnell, A.M., Gregory, K.J., Bolton, S., Medvedeva, L.A., Semenchenco, A., Mahkinov, A.N., Sobota, D., Baurer, J. and Staley, K., 2000: Bibliography: world literature on wood in streams, rivers, estuaries and riparian areas. *International conference on wood in world rivers*. Corvalis OR: Oregon State University.

Griffin, L.M., 1998: *Saving the Marin-Sonoma Coast*. Healdsburg: Sweetwater Springs Press.

Grift, R.E., Buijse, A.D., Van Densen, W.L.T. and Klein-Breteler, J.G.P., 2001: Restoration of the river–floodplain interaction: benefits for the fish community in the River Rhine. *Archive für Hydrobiologie*, 135 (Suppl.): 173–85.

Griggs, G.B. and Paris, L., 1982: Flood control failure: San Lorenzo River, California. *Environmental Management*, 6: 407–19.

Guerrier, Y., 1995: *Values and the environment: a social science perspective*. Chichester: J. Wiley & Sons.

Guglielmini, D., 1697: Della natura di fiumi trattata fiscio matematico. Cited by Lane, E.W., 1955: The importance of fluvial morphology in hydraulic engineering. *Proceedings American Civil Engineers, Journal of the Hydraulics Division*, 81: 1–17.

Guha, R. and Martinez-Alier, J., 1997: *Varieties of environmentalism: essays north and south*. London: Earthscan.

Gurnell, A.M. and Petts, G.E., 2002: Island-dominated landscapes of large floodplain rivers, a European perspective. *Freshwater Biology*, 47: 581–600.

Gurnell, A.M., Angold, P. and Gregory, K.J., 1994: Classification of river corridors: issues to be addressed in developing an operational methodology. *Aquatic Conservation: Marine and Freshwater Ecosystems*, 4: 219–31.

Gurnell, A.M., Peiry, J.-L. and Petts, G.E., 2003: Using historical data in fluvial geomorphology. In Kondolf, G.M. and Piégay, H. (eds), *Tools in fluvial geomorphology*. Chichester: J. Wiley & Sons, 205–30.

Haas, J.W., 1981: Managing natural resources in the 1980's. In North, R.M., Dworsky, L.B. and Allee, D.J. (eds), *Unified river basin management, proceedings symposium*. Minneapolis MN: American Water Resources Association, 55–61.

Halbert, C.L. and Lee, K.N., 1991: Implementing adaptive management. *The Northwest Environmental Journal*, 7: 136–50.

Haltiner, J.P., 1995: Environmentally sensitive approaches to river channel managment. In Thorne, C.R., Abt, S.R., Barends, F.B.J., Maynard, S.T. and Pilarczyk, K.W. (eds), *River, coastal and shoreline protection: erosion contol using riprap and armourstone*. Chichester: J. Wiley & Sons, 545–56.

Haltiner, J.P., Kondolf, G.M. and Williams, P.B., 1996: Restoration approaches in California. In Brookes, A. and Shields, F.D., Jr (eds), *River channel restoration: guiding principles for sustainable projects*. Chichester: J. Wiley & Sons, 291–329.

Handy, R.L., 1972: Alluvial cutoff dating from subsequent growth of a meander. *Geological Society of America Bulletin*, 83: 475–80.

Hansen, H.O., Boon, P.J., Madsen, B.L. and Iversen, T.M. (eds), 1998: River restoration: the physical dimension. *Aquatic Conservation: Marine and Freshwater Ecosystems*, Special Issue, 8 (1): 1–264.

Happ, S.C., Rittenhouse, G. and Dobson, G.C., 1940: *Some principles of accelerated stream and valley sedimentation*. Technical Bulletin 695, US Department of Agriculture.

Hardin, G., 1968: The tragedy of the commons. *Science*, 162: 1243–48.

Harmon, M.E., Franklin, J.F., Swanson, F.J., Sollins, P., Gregory, S.V., Lattin, J.D., Anderson, N.H., Cline, S.P., Aumen, N.G., Sedell, J.R., Lienkamper, G.W., Cromack, K., Jr and Cummins, K.W., 1986: Ecology of coarse woody debris in temperate ecosystems. In MacFadyen, A. and Ford, E.D. (eds), *Advances in ecological research*. New York: Harcourt Brace Jovanovich, 133–302.

Harper, D.M. Ebrahimnezhad, M. and Climente i Cot, F., 1998: Artificial riffles in river rehabilitation: setting the goals and measuring the successes. *Aquatic Conservation: Marine and Freshwater Ecosystems*, 8: 5–16.

Harper, D.M. and Everard, M., 1998: Why should the habitat-level approach underpin holistic river survey and management? *Aquatic Conservation: Marine and Freshwater Ecosystems*, 8: 395–413.

Harper, D.M. and Ferguson, A.J.D. (eds), 1995: *The ecological basis for river management*. Chichester: J. Wiley & Sons.

Harper, D.M., Smith, C.D. and Barham, P.J., 1992: Habitats as the building blocks for river conservation assessment. In Boon, P.J., Calow, P. and Petts, G.E. (eds), *River conservation and management*. Chichester: J. Wiley & Sons, 311–19.

Harper, D.M., Smith, C.D., Barham, P.J. and Howell, R., 1995: The ecological basis for the management of the natural environment. In Harper, D.M. and Ferguson, A.J.D. (eds), *The Ecological basis for river management*. Chichester: J. Wiley & Sons, 219–38.

Harper, D.M., Kemp, J.L., Vogel, B. and Newson, M.D., 2000: Towards the assessment of 'ecological integrity' in running waters of the United Kingdom. *Hydrobiologia*, 422/423: 133–42.

Harvey, A.M., 2001: Coupling between hillslopes and channels in upland fluvial systems: implications for landscape sensitivity, illustrated from the Howgill Fells, northwest England. *Catena*, 42: 225–50.

Harvey, A.M. and Schumm, S.A., 1999: Indus River dynamics and the abandonment of the Mohenjo Daro. In Meadows, A. and Meadows, P.S. (eds), *The Indus River, biodiversity, humankind*. Oxford: Oxford University Press, 333–48.

Harvey, M.D. and Watson, C.C., 1986: Fluvial processes and morphological thresholds in incised channel restoration. *Water Resources Bulletin*, 22: 359–68.

Hasfurther, R., 1985: The use of meander parameters in restoring hydrologic balance to reclaimed stream beds. In Gore, J.A. (ed.), *The restoration of rivers and streams: theories and experience*. Boston MA: Butterworth, 21–40.

Haslam , S.M. and Wolseley, P.A., 1981: *River vegetation: its identification, assessment and management*. Cambridge: Cambridge University Press.

Hauer, R.F.R. and Lamberti, G.A. (eds), 1996: *Methods in stream ecology*. London: Academic Press.

Hawkins, C.P., Kerschner, J.L., Bisson, P.A., Bryant, M.D., Decker, L.M., Gregory, S.V., McCullough, D.A., Overton, C.K., Reeves, G.H., Steedman, R.J. and Young, M.K., 1993: A hierarchical approach to classifying stream habitat features. *Fisheries*, 18: 3–12.

Heaney, J.P., 1993: New directions in water resources planning and management. *Water Resources Update*, No.93, Autumn 1993.

Heathcote, I.W., 1998: *Integrated watershed management: principles and practices*. New York: J. Wiley & Sons.

Heiler, G., Hein, T., Schiemer, F. and Bornette, G., 1995: Hydrological connectivity and flood pulses as the central aspects for the integrity of a river-floodplain system. *Regulated Rivers: Research and Management*, 11: 351–61.

Heinz Center, The, 2002: *Dam removal: science and decision making*. Washington DC: The H. John Heinz III Center for Science Economics and the Environment.

Hemphill, R.W. and Bramley, M.E., 1989: *Protection of river and canal banks: a guide to selection and design*. London: Butterworth.

Henry, C.P., Amoros, C. and Giuliani, Y., 1995: Restoration ecology of riverine wetlands. 2: an example in a former channel of the Rhône River. *Environmental Management*, 19: 903–13.

Heuvelmans, M., 1974: *The river killers*. Harrisburg PA: Stackpole Books.

Hey, R.D., 1982: Gravel-bed rivers: form and processes. In Hey, R.D., Bathurst, J.C. and Thorne, C.R. (eds), *Gravel-bed rivers: fluvial processes, engineering and management*. Chichester: J. Wiley & Sons, 5–13.

Hey, R.D., 1994a: Environmentally sensitive river engineering. In Calow, P. and Petts, G.E. (eds), *The Rivers Handbook: hydrological and ecological principles*. Oxford: Blackwell Scientific, 337–62.

Hey, R.D., 1994b: 'Restoration of gravel-bed rivers: principles and practices'. In Shrubsole, D. (ed.), *'Natural' channel design: perspectives and practice*. Cambridge, Ontario: Canadian Water Resources Association, 157–73.

Hey, R.D., 1997a: River engineering and management in the 21st century. In Thorne, C.R., Hey, R.D. and Newson, M.D. (eds), *Applied fluvial geomorphology for river engineering and management*. Chichester: J. Wiley & Sons, 3–11.

Hey, R.D., 1997b: Stable river morphology. In Thorne, C.R., Hey, R.D. and Newson, M.D. (ed.), *Applied fluvial geomorphology for river engineering and management*. Chichester: J. Wiley & Sons, 223–39.

Hey, R.D. and Heritage, G.L., 1993. *Draft guidelines for the design and restoration of flood alleviation schemes*. R&D Note 154, Bristol: National Rivers Authority.

Hey, R.D. and Thorne, C.R., 1986: Stable channels with mobile gravel beds. *Journal of Hydraulic Engineering*, 112: 671–89.

Hey, R.D. and Winterbottom, A.N., 1990: River engineering in National Parks: the case of the River Wharfe, U.K. *Regulated Rivers: Research and Management*, 5: 35–44.

Hey, R.D., Bathurst, J.C. and Thorne, C.R. (eds), 1982: *Gravel-bed rivers: fluvial processes, engineering and management*. Chichester: J. Wiley & Sons.

Hickin, E.J. and Nanson, G.C., 1975: The character of channel migration on the Beatton River, Northeast British Columbia. *Geological Society of America Bulletin*, 86: 474–87.

Hickin, E.J. and Nanson, G., 1984: Lateral migration rates of river bends. *Journal Hydraulic Engineering, Proceedings American Society Civil Engineers*, 110: 1557–67.

Hill, M.T., Platts, S. and Beschta, R.L., 1991: Ecological and geomorphological concepts for instream and out-of-channel flow requirements. *Rivers*, 2: 198–210.

Hirschboeck, K., 1988: Flood hydroclimatology. In Baker, V.R., Kochel, R.C. and Patton, P.C. (eds), *Flood geomorphology*. Chichester: J. Wiley & Sons, 27–49.

Hobday, S.R., 1952: *Coulson and Forbes on the law of waters, sea, tidal and inland and land drainage*. London: Sweet and Maxwell.

Hodgen, M. 1939. Domesday watermills. *Antiquity* 13, 261–79.

Hodges, F., 1976: Law relating to water. In Rodda, J.C. (ed.), *Facets of hydrology*. London: J. Wiley & Sons, 315–30.

Holland, M.M., Blood, E.R. and Shaffer, L.R., 2003: The state of freshwater resources: conclusions and recommendations. In Holland, M.M., Blood, E.R. and Shaffer, L.R. (eds), *Achieving sustainable freshwater systems: a web of connections*. Washington DC: Island Press, 317–21.

Holling, C.S. (ed.), 1978: *Adaptive environmental assessment and management*. International Institute for Applied Systems Analysis, 3. New York: J. Wiley & Sons.

Holmes, N.T.H. and Nielsen, M.B., 1998: Restoration of the Rivers Brede, Cole and Skerne: joint Danish and British EU-LIFE demonstration project, I – setting up and delivery of the project. *Aquatic Conservation: Marine and Freshwater Ecosystems*: Special Issue, 8: 185–96.

Holmes, N.T.H., Boon, P.J. and Rowell, T.A., 1998: A revised classification for British rivers based on their aquatic plant communities. *Aquatic Conservation: Marine and Freshwater Ecosystems*, 8: 555–78.

Holmes, T.H., 1998: The River Restoration project and its demonstration sites. In de Waal, L.C., Large, A.R.G. and Wade, P.M. (eds), *Rehabilitation of rivers: principles and implementation*. Chichester: J. Wiley & Sons, 133–48.

Hooke, J.M., 1980: Magnitude and distribution of rates of river bank erosion. *Earth Surface Processes*, 5: 143–57.

Hooke, J.M., 1996: River response to decadal scale changes in discharge regime: the Gila River, SE Arizona. In Branson, J., Brown, A.G. and Gregory, K.J. (eds), *Global continental changes: the context of palaeohydrology*. Special Publication 115, London: Geological Society, 191–204.

Hooke, J.M., 1999: Decades of change: contributions of geomorphology to fluvial and coastal engineering and management. *Geomorphology*, 31: 373–89.

Hooke, J.M., 2003: Coarse sediment connectivity in river channel systems: a conceptual framework and methodology. *Geomorphology*, 56: 79–94.

Hooke, J.M. and Kain, R.P.J., 1982: *Historical change in the physical environment: a guide to sources and techniques*. London: Butterworth Scientific.

Hooper, B.P. and Margerum, R.D., 2000: Integrated watershed management for river conservation: perspectives from experiences in Australia and the United States. In Boon, P.J., Davies, B.R. and Petts, G.E. (eds), *Global perspectives on river conservation: science, policy and practice*. Chichester: J. Wiley & Sons, 509–17.

Horton, R.E., 1945: Erosional development of streams and their drainage basins: hydrophysical approach to quantitative morphology. *Bulletin of the Geological Society of America*, 263: 303–12.

House, P.K., Webb, R.H., Baker, V.R. and Levish, D.R. (eds), 2002: *Ancient floods, modern hazards: principles and application of paleoflood hydrology*. Water Science and Application, vol. 5. Washington DC: American Geophysical Union.

Howarth, W., 1992: *Wisdom's law of watercourses*. Crayford: Shaw and Sons Ltd.

Huet , M., 1949: Apercu des relations entre la pente et les populations des eaux courantes. *Schweizer Zeitschrift Hydrolgie*, 11: 332–51.

Huet, M., 1954: Biologie, profils en long en travers des eaux courantes. *Bulletin Francais Pisciculture*, 175: 41–53.

Hughes, N.F., 1999: Dynamics habitat selection theory: linking stream morphology to salmonid population dynamics. *3rd international symposium on ecohydraulics*. 13–16 July 1999. Utah State University Extension, Salt Lake City UT, on CD-ROM.

Hughes, R.M., Larsen, D.P. and Omernik, J.M., 1986: Regional references sites: a method for assessing stream potentials. *Environmental Management*, 10: 629–35.

Hunt, C.E., 2000: New approaches to river management in the United States of America. In Smits, A.J.M., Nienhuis, P.H. and Leuven, R.S.E.W. (eds), *New approaches to river management*. Leiden: Backhuys, 119–39.

Hupp, C.R. and Simon, A., 1991: Bank accretion and the development of vegetated depositional surfaces along modified alluvial channels. *Geomorphology*, 14: 277–95.

Huppert, D. and Kantor, S., 1998: Economic perspectives. In Naiman, R.J. and Bilby, R.E. (eds), *River ecology and management: lessons for the pacific coastal ecosystem*. New York: Springer, 572–95.

Hynes, H.B.N., 1970: *The ecology of running waters*. Liverpool: Liverpool University Press.

Hynes, H.B.N., 1975: The stream and its valley. *Verhandlungen der Internationalen Vereinigung für Theoretische und Angewandte Limnologie*, 19: 1–15.

International Commission on Large Dams, 1988: *World register of large dams: update*. Paris: ICOLD.

International Union for the Conservation of Nature, United Nations Environment Programme and World Wildlife Fund., 1991: *Caring for the Earth: a strategy for sustainable living*. London: Earthscan.

Iversen, T.M., Kronvang, B., Madsen, B.L., Markmann, P. and Nielsen, M.B., 1993: Re-establishment of Danish streams: restoration and maintenance measures. *Aquatic Conservation: Marine and Freshwater Ecosystems*, 3: 73–93.

Iversen, T.M., Madsen, B.L. and Bøgestrand, J., 2000: River conservation in the European Community, including Scandinavia. In Boon, P.J., Davies, B.R. and Petts, G.E. (eds), *Global perspectives on river conservation: science, policy and practice*. Chichester: J. Wiley & Sons, 79–103.

Iyer-Raniga, U. and Treloar, G., 2000: A context for participation in sustainable development. *Environmental Management*, 20: 879–86.

Jackson, L.L., Lopoukline, N. and Hillyard, D., 1995: Ecological restoration – a definition and comments – commentary. *Restoration Ecology*, 3: 71–75.

Jaeggi, M.N.R. and Zarn, B., 1999: Stream channel restoration and erosion control for incised channels in Alpine environments. In Darby, S.E. and Simon, A. (eds), *Incised river channels: processes, forms, engineering and management*. Chichester: J. Wiley & Sons, 343–69.

James, A., 1999: Time and the persistence of alluvium: river engineering, fluvial geomorphology, and mining sediment in California. *Geomorphology*, 31: 265–90.

Jansen, P., Bendegom, L., Berg, J., de Vries, M. and Zanen, A., 1979: *Principles of river engineering – the non-tidal alluvial river*. London: Pitman.

Jasperse, P., 1998: Policy networks and the success of stream rehabilitation projects in the Netherlands. In de Waal, L.C., Large, A.R.G. and Wade, P.M. (eds), *Rehabilitation of rivers: principles and implementation*. Chichester: J. Wiley & Sons, 13–29.

Jeffries, M. and Mills, D., 1990: *Freshwater ecology: principles and applications*. London: Belhaven Press.

Joglekar, D.V., 1956: *Manual on river behaviour control and training*. Publication No. 60, New Dehli: Central Board of Irrigation and Power.

Joglekar, D.V., 1971: *Manual on river behaviour control and training*. Second edition. Publication No. 60, New Dehli: Central Board of Irrigation and Power.

Johannesson, J. and Parker, G., 1989: Linear theory of river meanders. In Ikeda, S. and Parker, G. (eds), *River meandering*. Water Resources Monograph 12, Washington DC: American Geophysical Union, 181–214.

Jordan, W.R., Gilpin, M.E. and Aber, J.D., 1987: Restoration ecology: ecological restoration as a technique for basic research. In Jordan, W.R., Gilpin, M.E. and Aber, J.D. (eds), *Restoration ecology: a synthetic approach to ecological research*. Cambridge: Cambridge University Press, 3–21.

Jungwirth, M., Muhar, S. and Schmutz, S., 2002: Re-establishing and assessing ecological integrity in riverine landscapes. *Freshwater Biology*, 47: 867–87.

Junk, W.J., Bayley, P.B. and Sparks, R.E., 1989: The flood pulse concept in river-floodplain systems. In Dodge: D.P. (ed.), Proceedings of the International Large River Symposium. Special Publication of the *Canadian Journal of Fisheries and Aquatic Sciences*, 49–55.

Kalbermatten, J.M. and Gunnerson, C.A., 1978: Environmental impacts of international engineering practice. In Gunnerson, C.A. and Kalbermatten, J.M. (eds), *Environmental impacts of international civil engineering projects and practices*. New York: American Society of Civil Engineers, 232–54.

Karr, J.R., Allan, J.D. and Benke, A.C., 2000: River conservation in the United States and Canada. In Boon, P.J., Davies, B.R. and Petts, G.E. (eds), *Global perspectives on river conservation: science, policy and practice*. Chichester: J. Wiley & Sons, 3–39.

Katz, E., 1992: The call of the wild – the struggle against domination and the technological fix of nature. *Environmental Ethics*, 14: 265–73.

Keller, E.A., 1975: Channelizaton: a search for a better way. *Geology*, 3: 246–48.

Keller, E.A., 1976: Channelization: environmental, geomorphic and engineering aspects. In Coates, D.R. (ed.), *Geomorphology and engineering*. New York: George Allen & Unwin, 115–41.

Keller, E.A., 1978: Pools, riffles and channelization. *Environmental Geology*, 2: 119–27.

Keller, E.A. and Hoffman, E.K., 1977: Urban streams: sensual blight or amenity? *Journal of Soil and Water Conservation*, 32: 237–42.

Keller, E.A. and MacDonald, A., 1995: River channel change: the role of large woody debris. In Gurnell, A.M. and Petts, G.E. (eds), *Changing river channels*. Chichester: J. Wiley & Sons, 217–35.

Kellerhals, R., Church, M. and Bray, D.I., 1976: Classification and analysis of river processes. *Journal of the Hydraulics Division, Proceedings of the American Society of Civil Engineers*, 102: 813–29.

Kemp, J.L., Harper, D.M. and Crosa, G.A., 1999: Use of 'functional habitats' to link ecology with morphology and hydrology in river rehabilitation. *Aquatic Conservation: Marine and Freshwater Ecosystems*, 9: 159–78.

Kennedy, B.A., 1997: Classics in physical geography revisited: S.A. Schumm and R.W. Lichty, 1965, Time, space and causality in geomorphology. *Progress in Physical Geography*, 21: 419–23.

Kennedy, R.G., 1895: The prevention of silting in irrigation canals. *Proceedings of the Institution of Civil Engineers, London*, 119: Proceedings No. 2826.

Kern, K., 1992: Rehabilitation of streams in south-west Germany. In Boon, P.J., Calow, P. and Petts, G.E. (eds), *River conservation and management*. Chichester: J. Wiley & Sons, 321–35.

Kirby, C., 1979: *Water in Great Britain*. Harmondsworth: Penguin.

Kirby, C. and White, W.R. (eds), 1994: *Integrated river basin development*. Chichester: J. Wiley & Sons.

Kirkby, M.J., 1980: The stream head as a significant geomorphic threshold. In Coates, D.R. and Vitek, A.D. (eds), *Thresholds in geomorphology*. London: Allen and Unwin, 53–73.

Kirkby, M.J., Abrahart, R., McMahon, M.D., Shao, J. and Thornes, J.B., 1998: MEDALUS soil erosion models for global change. *Geomorphology*, 24: 35–49.

Klamath National Forest, 2002: *North Fork Salmon River watersheds and road sediment source inventory and risk assessment*. Draft report. Yreka CA: USDA Forest Service, Pacific Southwest Region.

Klimek, K. and Starkel, L., 1974: History and actual tendency of floodplain development at the border of the Polish Carpathians. *Abhandlungen der Akademie der Wissenschaften in Göttingen, Mathematisch-Physikelische Klass*, 3, 29: 185–96.

Knighton, A.D., 1984: *Fluvial forms and processes*. London: Arnold.

Knighton, A.D., 1998: *Fluvial forms and processes: a new perspective*. London: Arnold.

Knighton, A.D. and Nanson, G.C., 1993: Anastomosis and the continuum of channel patterns. *Earth Surface Processes and Landforms*, 18: 613–25.

Knox, J.C., 1975: Concept of the graded stream. In Melhorn, W.N. and Flemal, R.C. (eds), *Theories of landform development*. Binghamton NY: State University of New York, Publications in Geomorphology, 169–98.

Knox, J.C., 1983: Responses of river systems to Holocene climates. In Wright, H.E., Jr (ed.), *Late Quaternary environments of the United States: volume 2, the Holocene*. Minneapolis MN: University of Minnesota Press, 26–41.

Knox, J.C., 1984: Fluvial responses to small scale climate changes. In Costa, J.E. and Fleischer, P.J. (eds), *Developments and applications of geomorphology*. Berlin: Springer Verlag, 318–42.

Knox, J.C., 1995: Fluvial systems since 20,000 years BP. In Gregory, K.J., Starkel, L. and Baker, V.R. (eds), *Global continental palaeohydrology*. Chichester: J. Wiley & Sons, 87–108.

Knox, J.C., 1999: Long-term episodic changes in magnitudes and frequencies of floods in the Upper Mississippi River valley. In Brown, A.G. and Quine, T.A. (eds), *Fluvial processes and environmental change*. Chichester: J. Wiley & Sons, 255–82.

Knox, J.C., 2000: Sensitivity of modern and Holocene floods to climate change. *Quaternary Science Reviews*, 19: 439–57.

Knox, J.C., 2001: Agricultural influence on landscape sensitivity in the Upper Mississippi River Valley. *Catena*, 42: 192–204.

Kolb, C., 1963: Sediments forming the bed and banks of the lower Mississippi River and their effect on migration. *Sedimentology* 2: 227–34.

Komura, S. and Simons, D.B., 1967: River bed degradation below dams. *Proceedings American Society of Civil Engineers, Journal Hydraulics Division*, 93: 1–14.

Kondolf, G.M., 1994a: Environmental planning in regulation and management of instream gravel mining in California. *Landscape and Urban Planning*, 29: 185–99.

Kondolf, G.M., 1994b: Geomorphic and environmental effects of instream gravel mining. *Landscape and Urban Planning*, 28: 225–43.

Kondolf, G.M., 1994c: Learning from stream restoration projects. In Harris, R., Kattelman, R., Kerner, H. and Woled, J. (eds), *Fifth biennial watershed management*. Askland OR: Water Resources Center Report Number 86, 107–10.

Kondolf, G.M., 1995a: Five elements for effective stream restoration. *Restoration Ecology*, 3: 133–36.

Kondolf, G.M., 1995b: Geomorphological stream channel classification in aquatic habitat restoration: uses and limitations. *Aquatic Conservation: Marine and Freshwater Ecosystems*, 5: 127–41.

Kondolf, G.M., 1997: Hungry water: effects of dams and gravel mining on river channels. *Environmental Management*, 21: 533–51.

Kondolf, G.M., 2000: Assessing salmonid spawning gravel quality. *Transactions of the American Fisheries Society*, 129: 262–81.

Kondolf, G.M., 2001: Historical change to the San Francisco Bay-Delta watershed: implications for ecosystem restoration. In Nijland, H.J. and Cals, M.J.R. (eds), *River restoration in Europe: practical approaches – proceedings of the conference on river restoration, Wageningen, The Netherlands, 2000*. Lelystad: Institute for Inland Water Management and Waste Water Treatment / RIZA, 327–38.

Kondolf, G.M. and Downs, P.W., 1996: Catchment approach to planning channel restoration. In Brookes, A. and Shields, F.D., Jr (eds), *River channel restoration: guiding principles for sustainable projects*. Chichester: J. Wiley & Sons, 129–48.

Kondolf, G.M. and Larson, M., 1995: Historical channel analysis and its application to riparian and aquatic habitat restoration. *Aquatic Conservation: Marine and Freshwater ecosystems*, 5: 109–26.

Kondolf, G.M. and Micheli, E.R., 1995: Evaluating stream restoration projects. *Environmental Management*, 19: 1–15.

Kondolf, G.M. and Swanson, M.L., 1993: Channel adjustments to reservoir construction and gravel extraction along Stony Creek, California. *Environmental Geology*, 21: 256–69.

Kondolf, G.M. and Wilcock, P.R., 1996: The flushing flow problem: defining and evaluating objectives. *Water Resources Research*, 32: 2589–99.

Kondolf, G.M., Kattelmann, R., Embury, M. and Erman, D.C., 1996: Status of riparian habitat.

Sierra Nevada ecosystem project: final report to congress, volume 2, assessment and scientific basis for management options. Davis CA: University of California, Centers for Water and Wildlife Resources, 1009–30.

Kondolf, G.M., Larsen, E.W. and Williams, J.G., 2000: Measuring and modeling the hydraulic environment for assessing instream flows. *North American Journal of Fisheries Management*, 20: 1016–28.

Kondolf, G.M., Smeltzer, M.W. and Railsback, S.F., 2001: Design and performance of a channel reconstruction project in a coastal California gravel-bed stream. *Environmental Management*, 28: 761–76.

Kondrat'yev, N.Y. and Popov, I.V., 1967: Methodological prerequisities for conducting network observations of channel process. *Soviet Hydrology*, 6: 273–97.

Korhonen, I.M., 1996: Riverine systems in international law. *Natural Resources Journal*, 36: 481–520.

Kuhnle, R.A., Alonso, C.V. and Shields, F.D., Jr, 1999: Geometry of scour holes associated with 90 degrees spur dikes. *Journal of Hydraulic Engineering, ASCE*, 125: 972–78.

Kuhnle, R.A., Alonso, C.V. and Shields, F.D., Jr, 2002: Local scour associated with angled spur dikes. *Journal of Hydraulic Engineering, ASCE*, 128: 1087–93.

Lacey, P.M., 1930: Stable channels in alluvium. *Proceedings of the Institution of Civil Engineers*, 229: 259–92.

Ladson, A. and Tilleard, J., 1999: The Herbert River, Queensland, Tropical Australia: community perception and river management. *Australian Geographical Studies*, 87: 471–86.

Ladson, A., Tilleard, J. and Ewing, S., 1999: Successful stream rehabilitation: first set the goals. In Rutherford, I. and Bartley, R. (eds), *Second Australian stream management conference: the challenge of rehabilitating Australia's streams*. Adelaide: Cooperative Research Centre for Catchment Hydrology, 381–87.

Lamb, B.L. and Doerksen, H.R., 1990: Instream water use in the United States – water laws and methods for determining flow requirements. *National water summary 1987 – hydrologic events and water supply and use*. US Geological Survey Water Supply, Paper 2350. Washington DC, 109–16.

Lamplugh, G.W., 1914: Taming of streams. *Geographical Journal*, 43: 651–56.

Länderarbeitsgemeinschaft Wasser (LAWA), 2000. *Gewässerstrukturgütebewertung in der Buundesrepublik Deutschland, Verhahren für kleine und mittelgroße Fließgewässer*. Berlin: LAWA.

Lane, E.W., 1955a: Design of stable channels. *Transactions of the American Society of Civil Engineers*, 120: 1234–79.

Lane, E.W., 1955b: The importance of fluvial morphology in hydraulic engineering. *Proceedings of the American Society of Civil Engineers*, 81: 1–17.

Lane, E.W., 1957. *A study of the shape of channels formed by natural streams flowing in an erodible material*. MRD Sediment Series 9, Missouri River: US Army Engineer Division.

Lane, S.N., 2000: Channel classification. In Thomas, D.S.G. and Goudie, A. (eds), *The Dictionary of physical geography*. Oxford: Blackwell, 78–81.

Large, A.R.G. and Petts, G.E., 1994: Rehabilitation of river margins. In Calow, P. and Petts, G.E. (eds), *The rivers handbook*. Oxford: Blackwell Scientific, 401–18.

Large, A.R.G. and Prach, K., 1999: Plants and water in streams and rivers. In Baird, A.J. and Wilby, R.L. (eds), *Eco-hydrology: plants and water in terrestrial and aquatic environments*. London: Routledge, 237–68.

Larsen, E.W. and Greco, S.E., 2002: Modeling channel management impacts on river migration: a case study of Woodson Bridge State Recreational Area, Sacramento River, California, USA. *Environmental Management*, 30: 209–24.

Lassettre, N.S., 1997: Evaluation of structures to improve steelhead trout (*Oncorhynchus mykiss*) habitat. Unpublished M.S. Thesis, San Jose State University, San Jose, CA.

Lawler, D.M., 1992: Process dominance in bank erosion systems. In Carling, P.A. and Petts, G.E. (eds), *Lowland floodplain rivers: geomorphological perspectives*. Chichester: J. Wiley & Sons, 117–43.

Lawler, D.M., 1993: The measurement of river bank erosion and lateral change: a review. *Earth Surface Processes and Landforms*, 18: 777–821.

Lawler, D.M., Thorne, C.R. and Hooke, J.M., 1997: Bank erosion and instability. In Thorne, C.R., Hey, R.D. and Newson, M.D. (eds), *Applied fluvial geomorphology for river engineering and management*. Chichester: J. Wiley & Sons, 137–72.

Lawson, J.D. and O'Neill, I.C., 1975: Transport of materials in streams. In Chapman, T.G. and Duxin, F.X. (eds), *Prediction in catchment hydrology*. Canberra: Australian Academy of Science, 165–202.

Lecce, S.A., 1997: Nonlinear downstream changes in stream power on Wisconsin's Blue River. *Annals of the Association of American Geographers*, 87: 471–86.

Leclerc, M., Capra, H., Valentin, S., Boudreault, A. and Côté, Y. (eds), 1996: *Ecohydraulics 2000: proceedings 2nd international symposium on habitat hydraulics*. Québec: Institut National de le Recherche Scientique-Eau, Federation Quebecoise pour le Saumon Atlantique, International Association of Hydraulic Engineering and Research.

Leeks, G.J.L., 1992: Impact of plantation forestry on sediment transport processes. In Billi, P., Hey, R.D., Thorne, C.R. and Tacconi, P. (eds), *Dynamics of gravel-bed rivers*. Chichester: J. Wiley & Sons, 651–73.

Leliavsky, S., 1955: *Irrigation and hydraulic design*. London: Chapman and Hall.

Leliavsky, S., 1957: *An introduction to fluvial hydraulics*. London: Constable.

Leliavsky, S., 1965: *River and canal hydraulics*. London: Chapman and Hall.

Lemons, J. (ed.), 1987: *The Environment Professional*, 9. Special focus on environmental ethics.

Lemons, J., 1999: Environmental ethics. In Alexander, D.E. and Fairbridge, R.W. (eds), *Encyclopedia of environmental science*. Dordrecht: Kluwer Academic, 204–206.

Lenhart, C.F., 2000: The vegetation and hydrology of impoundments after dam removal in southern Wisconsin. Unpublished M.S. Thesis, University of Wisconsin.

Leopold, A., 1949: *Sand county almanac*. Oxford: Oxford University Press.

Leopold, B. and Langbein, W.B. 1963: Association and indeterminacy in geomorphology. In Albritton, C.C. (ed.), *The fabric of geology*. Reading MA: Addison-Wesley, 184–92.

Leopold, L.B., 1973: River channel change with time: an example. *Bulletin of the Geological Society of America*, 84: 1845–60.

Leopold, L.B., 1974: *Water: a primer*. San Francisco: W.H. Freeman and Company.

Leopold, L.B., 1977: A reverence for rivers. *Geology*, 5: 429–30.

Leopold, L.B. and Maddock, T., Jr, 1953. The hydraulic geometry of stream channels and some physiographic implications. Washington DC: *US Geological Survey*, Professional Paper 252.

Leopold, L.B. and Maddock, T., Jr, 1954: *The flood control controversy*. New York: The Ronald Press Company.

Leopold, L.B. and Wolman, M.G., 1957. River channel patterns – braided, meandering and straight. Washington DC: *US Geological Survey, Professional Paper 282B*.

Leopold, L.B., Wolman, M.G. and Miller, J.P., 1964: *Fluvial processes in geomorphology*. London: W H Freeman and Company.

Leuven, R.S.E.W. and Poudevigne, I., 2002: Riverine landscape dynamics and ecological risk assessment. *Freshwater Biology*, 47: 845–65.

Leuven, R.S.E.W., Smits, A.J.M. and Nienhuis, P.H., 2000: Introduction. In Smits, A.J.M., Nienhuis, P.H. and Leuven, R.S.E.W. (eds), *New approaches to river management*. Leiden: Backhuys, 329–47.

Lewin, J., 1976: Initiation of bedforms and meanders in coarse-grained sediment. *Bulletin of the Geological Society of America*, 87: 281–85.

Lewin, J., 1977: Channel pattern changes. In Gregory, K.J. (ed.), *River channel changes*. Chichester: J. Wiley & Sons, 167–84.

Lewin, J., 1996: Floodplain construction and erosion. In Petts, G.E. and Calow, P. (eds), *River flows and channel forms: selected extracts from the rivers handbook*. Oxford: Blackwell Science, 203–20.

Lewin, J. and Macklin, M.G., 1987: Metal mining and floodplain sedimentation in Britain. In Gardiner, V. (ed.), International Geomorphology 1986: *Proceedings first international conference on geomorphology*. Chichester: J. Wiley & Sons, 1009–27.

Lienkaemper, G.W. and Swanson, F.J., 1987: Dynamics of large woody debris in streams of old-growth Douglas-fir forests. *Canadian Journal of Forest Research*, 17: 150–56.

Likens, G.E. and Bilby, R.E., 1982: Development, maintenance, and the role of organic debris dams in New England streams. In Swanson, F.J., Janda, R.J., Dunne, T. and Swanston, D.N. (eds), *Sediment budgets and routing in forested drainage basins*. General Technical Report PNW-141, Pacific Northwest Forest and Range Experiment Station. Portland OR: US Forest Service.

Linsley, R.K., 1982: Rainfall-runoff models: an overview. In Singh, V.P. (ed.), *Rainfall-runoff relationship*. Littleton CO: Water Resources Publications, 3–22.

Linsley, R.K., Kohler, M.A. and Paulhaus, J.L.H., 1949: *Applied Hydrology*. New York: McGraw Hill.

Lisle, T.E., Ciu, Y., Parker, G., Pizzuto, J.E. and Dodd, A.M., 2001: The dominance of dispersion in the evolution of bed material in gravel-bed rivers. *Earth Surface Processes and Landforms*, 26: 1409–20.

Lord, W.B., 1982: Unified river basin management in retrospect and prospect. In Allee, D.J., Dworsky, L.B. and North, R.M. (eds), *Unified river basin management – stage II*. Minneapolis MN: American Water Resources Association, 58–67.

Lowe, J.J., 1991: *Radiocarbon dating: recent applications and future potential*. Cambridge: Quaternary Research Association.

Lowell, R.B., Culp, J.M. and Dube, M.G., 2000: A weight-of-evidence approach for Northern River risk assessment: integrating the effects of multiple stressors. *Environmental Toxicology and Chemistry*, 19: 1182–90.

Luce, C.H. and Wemple, B.C., 2001: Introduction to special issue on hydrologic and geomorphic effects of forest roads. *Earth Surface Processes and Landforms*, 26: 111–13.

Lvovitch, M.I., 1973: The global water balance. *US International Hydrological Decade Bulletin*, 23: 28–42.

Macklin, M.G., 1985: Floodplain sedimentation in the upper Axe valley, Mendip, England. *Transactions of the Institute of British Geographers*, 10: 235–44.

Macklin, M.G. and Lewin, J., 1993: Holocene river alluviation in Britain. *Zeitschrift für Geomorphologie Supplementband*, 88: 109–22.

Macklin, M.G. and Lewin, J., 1997: Channel, floodplain and drainage basin response to environmental change. In Thorne, C.R., Hey, R.D. and Newson, M.D. (eds), *Applied fluvial geomorphology for river engineering and management*. Chichester: J. Wiley & Sons, 15–45.

Macnaughten, P. and Urry, J., 1998: *Contested Natures*. London: Sage.

Maddock, T., 1970: Intermediate hydraulics of alluvial channels. *Journal of the Hydraulics Division, American Society of Civil Engineers*, 96: 2309–23.

Magilligan, F.J., 1992: Thresholds and the spatial variability of flood power during extreme floods. *Geomorphology*, 5: 373–90.

Mahoney, J.M. and Rood, S.B., 1993: A model for assessing the effects of altered flow regimes on the recruitment of riparian cottonwoods. In *Riparian management: common threads and shared interests*. Colorado CO: United States Department of Agriculture Forest Service, 227–32.

Mahoney, J.M. and Rood, S.B., 1998: Streamflow requirements for cottonwood seedling recruitment – an integrative model. *Wetlands*, 18: 634–45.

Mai, Q., Zhao, Y. and Pan, X., 1980: Sediment problems of the lower Yellow River. *Proceedings of the international symposium on river sedimentation*. Beijing: Chinese Society of Hydraulic Engineering, Guanghua Press, 397–406.

Mandelbrot, B. and Wallis, J.R., 1968: Noah, Joseph and operational hydrology. *Water Resources Research*, 7: 543–53.

Manning, R., 1891: On the flow of water in open channels and pipes. *Transactions Institution Civil Engineers of Ireland*, 20: 161–207.

Marchand, M. (ed.), 1993: *Floodplain rehabilitation gemenc: main report*. Delft, Lelystad, Budapest: Delft Hydraulics, RIZA / Vituki.

Marchand, M. and Toornstra, F.H., 1986: *Ecological guidelines for river basin development*. Leiden: Centrum voor Milienkunde, Department 28, Rijksuniversiteit, Leiden.

Margerum, R.D., 1996: *Integrated environmental management: a framework for practice*. Discussion paper No. 6. Armidale: Centre for Water Policy Research. University of New England.

Marriott, S.B., 1998: Channel floodplain interactions and sediment deposition on floodplains. In Bailey, R.G., Jose, P.V. and Sherwood, B.R. (eds), *United Kingdom floodplains*. Otley: Westbury, 43–61.

Matlock, M.D., Osborn, G.S., Hession, W.C., Keniner, A.L. and Storm, D.E., 2001: Ecological engineering: a rationale for standardized curriculum and professional certification in the United States. *Ecological Engineering*, 17: 403–409.

Mays, L.W., 2001: *Water resources engineering*. New York: J. Wiley & Sons.

McCaig, N., 1983: *A world of difference*. London: Chatto and Windus.

McCully, P., 1996: *Silenced rivers: the ecology and politics of large dams*. London: Zed Books.

McDonald, A.T. and Kay, D., 1988: *Water resources issues and strategies: themes in resource management*. Essex: Longman.

McEwan, L.J., Brazier, V. and Gordon, J.E., 1997: Evaluating the geomorphology of freshwaters: an assessment of approaches. In Boon, P.J. and Howell, D.L. (eds), *Freshwater quality: defining the indefinable*. Edinburgh: The Stationery Office (HMSO, SNH), 258–81.

McHarg, I.L., 1969: *Design with nature*. New York: Doubleday/Natural History Press.

McHarg, I.L., 1992: *Design with nature*. Chichester: J. Wiley & Sons.

McHarg, I.L. and Steiner, F.R. (eds), 1998: *To heal the Earth: selected writings of Ian L. McHarg*. Washington DC: Island Press.

McKibben, B., 1989: *The end of nature*. New York: Random House.

McLain, R.J. and Lee, R.G., 1996: Adaptive management, promises and pitfalls. *Environmental Management*, 20: 437–48.

Meadows, D.H., Meadows, D.L., Randers, J. and Behrens, W.W., III, 1972: *The limits to growth: a report for the Club of Rome's project on the predicament of mankind*. New York: Universe Books.

Mellquist, P., 1992: River management – objectives and applications. In Boon, P.J., Calow, P. and Petts, G.E. (eds), *River conservation and management*. Chichester: J. Wiley & Sons, 1–8.

Merritts, D.J. and Vincent, K.R., 1989: Geomorphic response of coastal streams to low, intermediate, and high rates of uplift. *Geological Society of America Bulletin*, 100: 1373–88.

Meybeck, M. and Helmer, R., 1989: The quality of rivers: from pristine stage to global pollution. *Palaeogeography, Palaeoclimatology, Palaeoecology (Global and Planetary Change)*, 75: 283–309.

Miall, A.D., 1985: Architectural-element analysis: a new method of facies analysis applied to fluvial deposits. *Earth Science Reviews*, 22: 261–308.

Michaelides, M. and Wainwright, J., 2002: Modelling the effects of hillslope-channel coupling on catchment hydrological response. *Earth Surface Processes and Landforms*, 27: 1441–57.

Miller, T.K., 1988: Stream channel pattern: a threshold model. *Physical Geography*, 9: 373–84.

Mintzberg, H., 1980: Beyond implementation: an analysis of the resistance to policy analysis. *Infor*, 18: 100–38.

Mirighani, M.M.O. and Savenije, H.H.G., 1995: Incorporation of people's participation in planning and implementation of water resources projects. *Physics and Chemistry of the Earth*, 20: 229–36.

Mitchell, B., 1990: Integrated Water Management. In Mitchell, B. (ed.), *Integrated water management: international experiences and perspectives*. London and New York: Belhaven Press, 1–21.

Mitsch, W.J., 1998: Ecological engineering – the 7-year itch. *Ecological Engineering*, 10: 119–30.

Mitsch, W.J. and Jorgensen, S.E., 1989: *Ecological engineering: an introduction to ecotechnology*. New York: J. Wiley & Sons.

Montgomery, D.R. and Buffington, J.M., 1993: *Channel classification, prediction of channel response, and assessment of channel condition*. Washington State Timber/Fish/Wildlife

Agreement. Report TFW-SH10-93-002. Olympia WA: Department of Natural Resources.

Montgomery, D.R. and Buffington, J.R., 1997: Channel-reach morphology in mountain drainage basins. *Geological Society of America Bulletin*, 109: 596–611.

Montgomery, D.R. and Buffington, J.R., 1998: Channel processes, classification and response. In Naiman, R.J. and Bilby, R.E. (eds), *River ecology and management: lessons from the Pacific coastal ecosystem*. New York: Springer, 13–42.

Montgomery, D.R. and Dietrich, W.E., 1994: A physically based model for the topographic control on shallow landsliding. *Water Resources Research*, 30: 1153–71.

Montgomery, D.R. and Macdonald, L.H., 2002: Diagnostic approach to stream channel assessment and monitoring. *Journal of the American Water Resources Association*, 38: 1–16.

Montgomery, D.R., Grant, G.E. and Sullivan, K., 1995: Watershed analysis as a framework for implementing ecosystem management. *Water Resources Bulletin*, 31: 369–85.

Montgomery, D.R., Dietrich, W.E. and Sullivan, K., 1998: The role of GIS in watershed analysis. In Lane, S.N., Richards, K.S. and Chandler, J.H. (eds), *Landform monitoring, modelling and analysis*. Chichester: J. Wiley & Sons, 241–62.

Morisawa, M., 1968: *Streams: their dynamics and morphology*. London: McGraw-Hill.

Morisawa, M., 1985: *Rivers: form and process*. London: Longman.

Mosley, M.P., 1975: Channel changes on the River Bollin, Cheshire, 1872–1973. *The East Midland Geographer*, 6: 185–99.

Mosley, M.P., 1985: River channel inventory, habitat and instream flow assessments. *Progress in Physical Geography*, 9: 494–523.

Mosley, M.P., 1987: The classification and characterisation of rivers. In Richards, K.S. (ed.), *River channels: environment and process*. Oxford: Blackwell, 295–320.

Mosley, M.P., 1992: Land use impacts. In Mosley, M.P. (ed.), *Waters of New Zealand*. Wellington: New Zealand Hydrological Society, 265–84.

Mosley, M.P. and Jowett, I., 1999: River morphology and management in New Zealand. *Progress in Physical Geography*, 23: 541–65.

Moss, B., 1988: *Ecology of fresh waters*. Oxford: Blackwell Science.

Muckleston, K.W., 1990: Integrated water management in the USA. In Mitchell, B. (ed.), *Integrated water management: international experiences and perspectives*. London and New York: Belhaven Press, 22–44.

Muhar, S., Schmutz, S. and Jungwirth, M., 1995: River restoration concepts – goals and perspectives. In Schiemer, E., Zalewski, M. and Thorpe, J.E. (eds), *The importance of aquatic-terrestrial ecotones for freshwater fish*. Dordrecht: Kluwer Academic.

Mulvaney, T.J., 1850: On the use of self-registering rain and flood gauges. *Transactions of the Institution of Civil Engineers of Ireland*, 4: 1–8.

Myers, T.J. and Swanson, S., 1996: Temporal and geomorphic variations in stream stability and morphology: Mahogany Creek, Nevada. *Water Resources Bulletin*, 32: 253–64.

Nace, R., 1974. General evolution of the concept of the hydrological cycle. Paris: United Nations Environment Programme/World Meteorological Organization/International Association of Hydrological Sciences.

Naess, A., 1973: The shallow and the deep, long range ecology movement: a summary. *Inquiry*, 16: 95–100.

Naiman, R.J. (ed.), 1992: *Watershed management: balancing sustainability and environmental change*. New York: Springer.

Naiman, R.J. and Bilby, R.E. (eds), 1998: *River ecology and management: lessons from the Pacific coastal ecoregion*. New York: Springer.

Naiman, R.J. and Décamps, H. (eds), 1990: *The ecology and management of aquatic–terrestrial ecotones*. Paris: UNESCO and Carnforth: The Parthenon Publishing Group.

Naiman, R.J., Lonzarich, D.G., Beechie, T.J. and Ralph, S.C., 1992: General principles of classification and the assessment of conservation potential in rivers. In Boon, P.J., Calow, P. and Petts, G.E. (eds), *River conservation and management*. Chichester: J. Wiley & Sons, 93–119.

Naiman, R.J., Bisson, P.A., Lee, R.G. and Turner, M.G., 1998: Watershed management. In Naiman, R.J. and Bilby, R.E. (eds), *River ecology and management: lessons for the Pacific coastal ecoregion*. New York: Springer, 642–61.

Nanson, G.C., 1986: Episodes of vertical accretion and catastrophic stripping: a model of disequilibrium floodplain development. *Geological Society of America Bulletin*, 97: 1467–75.

Nanson, G.C. and Croke, J.C., 1992: A genetic classification of floodplains. *Geomorphology*, 4: 459–86.

Nanson, G.C. and Erskine, W.D., 1988: Episodic changes of channels and floodplains on coastal rivers in New South Wales. In Warner, R.F. (ed.), *Fluvial geomorphology of Australia*. Sydney: Academic Press, 201–21.

Nanson, G.C. and Young, R.W., 1981: Downstream reduction of rural channel size with contrasting effects in small coastal streams in Southeastern Australia. *Journal of Hydrology*, 52: 239–55.

Nanson, G.C., Barbetti, M. and Taylor, G., 1995: River stabilisation due to changing climate and vegetation during the Late Quaternary in western Tasmania, Australia. *Geomorphology*, 13: 145–58.

Nash, R.F., 1967: *Wilderness and the American mind*. New Haven CT: Yale University Press.

Nash, R.F., 1989: *The rights of nature: a history of environmental ethics*. Madison WI: University of Wisconsin Press.

National Research Council, 1992: *Restoration of aquatic ecosystems: science, technology and public policy*. Washington DC: National Academy Press.

National Research Council, 1999: *New strategies for America's watersheds*. Washington DC: National Academy Press.

National Rivers Authority, 1992. *River corridor surveys: methods and procedures*. Conservation Technical Handbook 1. Bristol: National Rivers Authority.

National Rivers Authority, 1995. *River channel typology for catchment and river management*. R&D 539/NDB/T. Bristol: National Rivers Authority.

Neill, C.R. and Yaremko, E.K., 1988: Regime aspects of flood control channelization. In White, W.R. (ed.), *International conference on river regime*. Chichester: J. Wiley & Sons, Hydraulics Research Limited, 317–29.

Neller, R.J., 1988: Complex channel response to urbanisation in the Dumaresq Creek drainage basin, New South Wales. In Warner, R.F. (ed.), *Fluvial geomorphology of Australia*. Sydney: Academic Press, 323–41.

Neller, R.J., 1989: Induced channel enlargement in small urban catchments, Armidale, New South Wales. *Environmental Geology and Water Science*, 14: 167–71.

Nestler, J., Schneider, T. and Latka, D., 1993: RCHARC: a new method for physical habitat analysis. *Symposium on engineering hydrology*. San Francisco CA: ASCE, 294–99.

Nevins, T.H.F., 1965: River classification with particular reference to New Zealand. In *Fourth New Zealand geography conference*. Dunedin, 83–90.

New South Wales Government, 1986: *Floodplain development manual*. Sydney: New South Wales Government.

Newbold, J.D., 1996: Cycles and spirals of nutrients. In Petts, G.E. and Calow, P. (eds), *River flows and channel forms: selected extracts from The Rivers Handbook*. Oxford: Blackwell Science, 130–59.

Newbold, J.D., Elwood, J.W., O'Neill, R.V. and Van Winkle, W., 1981: Measuring nutrient spiralling in streams. *Canadian Journal of Fisheries and Aquatic Sciences*, 38: 860–63.

Newbury, R. and Gaboury, M., 1993: Exploration and rehabilitation of hydraulic habitats in streams using principles of fluvial behaviour. *Freshwater Biology*, 29: 195–210.

Newson, M.D., 1980: The geomorphological effectiveness of floods – a contribution stimulated by two recent events in mid-Wales. *Earth Surface Processes*, 5: 1–16.

Newson, M.D., 1988: Applied physical geography: the opportunities and constraints of environmental issues revealed by river basin management. *Scottish Geographical Magazine*, 104: 67–71.

Newson, M.D., 1992: *Land, water and development: river basin systems and their sustainable management*. London: Routledge.

Newson, M.D., 1994: Sustainable integrated development and the basin sediment system: guidance from fluvial geomorphology. In Kirby, C. and White, W.R. (eds), *Integrated river basin development*. Chichester: J. Wiley & Sons, 1–10.

Newson, M.D., 1995: Fluvial geomorphology and environmental design. In Gurnell, A.M. and Petts, G.E. (eds), *Changing river channels*. Chichester: J. Wiley & Sons, 413–32.

Newson, M.D., 1997: *Land, water and development: sustainable management of river basin systems*, 2nd Edition. London: Routledge.

Newson, M.D., 2002: Geomorphological concepts and tools for sustainable river ecosystem management. *Aquatic Conservation: Marine and Freshwater Ecosystems*, 12: 365–79.

Newson, M.D. and Lewin, J., 1991: Climatic change, river flow extremes and fluvial erosion – scenarios for England and Wales. *Progress in Physical Geography*, 15: 1–17.

Newson, M.D. and Newson, C.L., 2000: Geomorphology, ecology and river channel habitat: mesoscale approaches to basin-scale challenges. *Progress in Physical Geography*, 24: 195–217.

Newson, M.D., Hey, R.D., Bathurst, J.C., Brookes, A., Carling, P.A., Petts, G.E. and Sear, D.A., 1997: Case studies in the application of geomorphology to river management. In Thorne, C.R., Hey, R.D. and Newson, M.D. (eds), *Applied fluvial geomorphology for river engineering and management*. Chichester: J. Wiley & Sons, 311–63.

Newson, M.D., Clark, M.J., Sear, D.A. and Brookes, A., 1998a: The geomorphological basis for classifying rivers. *Aquatic Conservation: Marine and Freshwater Ecosystems*, 8: 415–30.

Newson, M.D., Harper, D.M., Padmore, C.L., Kemp, J.L. and Vogel, B., 1998b: A cost-effective approach for linking habitats, flow types and species requirements. *Aquatic Conservation: Marine and Freshwater Environments*, 8: 431–46.

Newson, M.D., Gardiner, J.L. and Slater, S., 2000: Planning and managing for the future. In Acreman, M. (ed.), *The hydrology of the UK*. London: Routledge, 244–69.

Newson, M.D., Pitlick, J. and Sear, D.A., 2002: Running water: fluvial geomorphology and river restoration. In Perrow, M.R. and Davy, A.J. (eds), *Handbook of ecological restoration, volume 1: principles of restoration*. Cambridge: Cambridge University Press, 133–52.

Nicholas, A.P., Ashworth, P.J., Kirkby, M.J., Macklin, M.G. and Murray, T., 1995: Sediment slugs: large scale fluctuations in fluvial sediment transport rates and storage volumes. *Progress in Physical Geography*, 19: 500–19.

Nielsen, M.B., 1996: Lowland stream restoration in Denmark. In Brookes, A. and Shields, F.D., Jr (eds), *River channel restoration: guiding principles for sustainable approaches*. Chichester: J. Wiley & Sons, 269–89.

Nienhuis, P.H., Leuven, R.S.E.W. and Ragas, A.M.J. (eds), 1998: *New concepts for sustainable management of river basins*. Leiden: Backhuys.

Nijland, H.J. and Cals, M.J.R., 2001a: Conference considerations, conclusions and recommendations. In Nijland, H.J. and Cals, M.J.R. (eds), *River restoration in Europe: practical approaches – proceedings of the conference on river restoration, Wageningen, the Netherlands, 2000*. Lelystad: Institute for Inland Water Management and Waste Water Treatment/RIZA, 17–30.

Nijland, H.J. and Cals, M.J.R. (eds), 2001b: *River restoration in Europe: practical approaches – proceedings of the conference on river restoration, Wageningen, the Netherlands, 2000*. Lelystad: Institute for Inland Water Management and Waste Water Treatment / RIZA.

Ning Chien, 1985: Changes in river regime after the construction of upstream reservoirs. *Earth Surface Processes and Landforms*, 10: 143–59.

Nixon, M., 1966: Flood regulation and river training. In Thorn, R.B. (ed.), *River engineering and water conservation works*. London: Butterworth, 293–97.

Norwegian Insititute of Technology, 1994: *Proceedings of the 1st international symposium on habitat hydraulics*. Tronheim: Norwegian Insititute of Technology.

Nunnally, N.R. and Keller, E.A., 1979. *Use of fluvial processes to minimise the adverse effects of stream channelization*. University of North Carolina: Water Resources Research Institute Report 114.

O'Connor, K.A., 1995: Watershed management planning: bringing the pieces together. M.S. Thesis, California State Polytechnic University, Pomona CA.

Odum, H.T., 1962: Man in the ecosystem. *Proceedings of the Lockwood Conference on the Suburban Forest and Ecology, Bulletin Connecticut Agricultural Station*, 652: 57–75.

Oelschlaeger, M. (ed.), 1995: *Post modern environmental ethics*. Albany NY: State University of New York Press.

Oglesby, R.T., Carlson, C.A. and McCann, J.A., 1972: *River ecology and man*. New York: Academic Press.

Ohmori, H. and Shimazu, H., 1994: Distribution of hazard types in a drainage basin and its relation to geomorphological setting. *Geomorphology*, 10: 95–106.

Osborne, L.L., Bayley, P.B., Higler, L.W.G., Statzner, B., Triska, F. and Iversen, T.M., 1993: Restoration of lowland streams: an introduction. *Freshwater Biology*, 29: 187–94.

Osman, A.M. and Thorne, C.R., 1988: Riverbank stability analysis I: theory. *Journal of Hydraulic Engineering*, 114: 134–50.

Osterkamp, W.R. and Hedman, E.R., 1982. Perennial streamflow characteristics related to channel geometry characteristics in Missouri River Basin. *US Geological Survey, Professional Paper* 1242, Washington DC.

Osterkamp, W.R., Lane, E.J. and Foster, G.R., 1983. An analytical treatment of channel–morphology relations. *US Geological Survey, Professional Paper* 1288, Washington DC.

Owen, M., 1973: *The Tennessee Valley Authority*. London: Praeger Publishers.

Owens, P.N., Walling, D.E. and Leeks, G.J.L., 1999: Deposition and storage of fine-grained sediment within the main channel system of the River Tweed, Scotland. *Earth Surface Processes and Landforms*, 24: 1061–76.

Padmore, C.L., 1996: Biotopes and their hydraulics: a method for defining the physical component of freshwater quality. In Boon, P.J. and Howell, D.L. (eds), *Freshwater quality: defining the indefinable*. Edinburgh: HMSO, Scottish Natural Heritage, 251–57.

Page, K., Nanson, G.C. and Price, D., 1996: Chronology of Murrumbidgee River palaeochannels on the Riverine Plain, southeastern Australia. *Journal of Quaternary Science*, 11: 311–26.

Paice, C. and Hey, R.D., 1989: Hydraulic control of secondary circulation in meander bends to reduce outer bank erosion. In Albertson, M.L. and Kia, R.A. (eds), *Design of hydraulic structures*. Rotterdam: Balkema, 249–54.

Palissy, B., 1580: *Discours Admirables de la nature de la eaux et fontaines, des pierres, des terres*. Paris.

Palmer, C.G., Peckham, B. and Soltau, F., 2000: The role of legislation in river conservation. In Boon, P.J., Davies, B.R. and Petts, G.E. (eds), *Global perspectives on river conservation: science, policy and practice*. Chichester: J. Wiley & Sons, 475–91.

Palmer, L., 1976: River management criteria for Oregon and Washington. In Coates, D.R. (ed.), *Geomorphology and engineering*. New York: George Allen & Unwin, 329–46.

Panizza, M., 1987: Geomorphological hazard assessment and the analysis of geomorphological risk. In Gardiner, V. (ed.), *International geomorphology 1986: proceedings of the first international conference on geomorphology*. Chichester: J. Wiley & Sons, 225–29.

Pantulu, R., 1983: River basin management. *Ambio*, 12: 109–11.

Pardo, L. and Armitage, P.D., 1997: Species assemblages as descriptors of mesohabitats. *Hydrobiologia*, 344: 111–28.

Park, C.C., 1977a: Man induced changes in stream channel capacity. In Gregory, K.J. (ed.), *River channel changes*. Chichester: J. Wiley & Sons, 121–44.

Park, C.C., 1977b: World-wide variations in hydraulic geometry exponents of stream channels: an analysis and some observations. *Journal of Hydrology*, 33: 133–46.

Parker, D.J. and Penning Rowsell, E.C., 1980: *Water planning in Britain*. London: George Allen & Unwin.

Parker, G. and Andres, D., 1976: Detrimental effects of river channelization. *Proceedings of conference rivers '76*. New York: American Society of Civil Engineers, 1248–66.

Parr, T.W., Sier, A.R.J., Battarbee, R.W., Mackay, A. and Burgess, J., 2003: Detecting environmental change: science and society – perspectives on long-term research and monitoring in the 21st century. *The Science of the Total Environment*, 310: 1–8.

Patrick, D.M., Smith, L.M. and Whitten, C.B., 1982: Methods for studying accelerated fluvial change. In Hey, R.D., Bathurst, J.C. and Thorne, C.R. (eds), *Gravel-bed rivers: fluvial processes, engineering and management*. Chichester: J. Wiley & Sons, 783–816.

Patrick, R., 1982: What is the condition of our surface water? In Environmental Associates (eds), *Proceedings of the national water conference, water quality and the Clean Water Act*. Philadelphia PA: Academy of Natural Sciences.

Pearce, D. (ed.), 1993: *Blueprint 3: measuring sustainable development*. London: Earthscan.

Pearson, W.D., 1992: Historical changes in water quality and fishes in the Ohio River. In *Water Quality in North American River Systems*. Columbus OH: Battelle Press, 207–31.

Pennack, R.W., 1971: Toward a classification of lotic habitats. *Hydrobiolgia*, 38: 321–34.

Perrault, P., 1674: *De l'origine des fontaines*. New York: Hafner, Paris: Trans A La Rocque.

Peters, M.R., Abt, S.R., Watson, C.C., Fischenich, J.C. and Nestler, J.M., 1995: Assessment of restored riverine habitat using RCHARC. *Water Resources Bulletin*, 31: 745–52.

Petersen, M.S., 1986: *River engineering*. Englewood Cliffs, NJ: Prentice Hall.

Petersen, R.C.J., Madsen, B.L., Wilzbach, M.A., Magadza, C.H.D., Paarlberg, A., Kullberg, A. and Cummins, K.W., 1987: Stream management: emerging global similarities. *Ambio*, 16: 166–79.

Petroski, H., 1992: *To engineer is human: the role of failure in successful design*. New York: Vintage Books.

Petts, G.E., 1977: Channel response to flow regulation: the case of the River Derwent, Derbyshire. In Gregory, K.J. (ed.), *River channel changes*. Chichester: J. Wiley & Sons, 145–64.

Petts, G.E., 1979: Complex response of river channel morphology subsequent to reservoir construction. *Progress in Physical Geography*, 3: 329–62.

Petts, G.E., 1982: Channel changes within regulated rivers. In Adlam, B.H., Fenn, C.R. and Morris, L. (eds), *Papers in Earth studies*. Norwich: Geobooks, 117–42.

Petts, G.E., 1984: *Impounded rivers: perspectives for ecological management*. Chichester: J. Wiley & Sons.

Petts, G.E., 1989: Historical analysis of fluvial hydrosystems. In Petts, G.E., Moller, H. and Roux, A.L. (eds), *Historical change of large alluvial rivers: Western Europe*. Chichester: J. Wiley & Sons, 1–18.

Petts, G.E., 1990: Water, engineering and landscape: development, protection and restoration. In Cosgrove, D. and Petts, G.E. (eds), *Water, engineering and landscape: water control and landscape transformation in the modern period*. London: Belhaven Press, 188–208.

Petts, G.E. 1994: Large-scale river regulation. In Roberts, N. (ed.). *The changing global environment*. Oxford: Blackwell, 262–84.

Petts, G.E., 1996: Sustaining the ecological integrity of large floodplain rivers. In Anderson, M.G., Walling, D.E. and Bates, P.D. (eds), *Floodplain processes*. Chichester: J. Wiley & Sons.

Petts, G.E., 1998: Floodplain rivers and their restoration: a European perspective. In Bailey, R.G., Jose, P.V. and Sherwood, B.R. (eds), *United Kingdom floodplains*. Westbury: Otley, 29–41.

Petts, G.E., 1999: River regulation. In Alexander, D.E. and Fairbridge, R.W. (eds), *Encyclopedia of environmental science*. Dordrecht: Kluwer Academic Publishers, 521–28.

Petts, G.E., 2001: Geo-ecological perspectives for the mulitple use of European river systems. In Nijland, H.J. and Cals, M.J.R. (eds), *River Restoration in Europe: practical approaches*. Lelystad: Institute for Inland Water Management and Waste Water Treatment/RIZA, 49–53.

Petts, G.E. and Amoros, C. (eds), 1996a: *Fluvial hydrosystems*. London: Chapman and Hall.

Petts, G.E. and Amoros, C., 1996b: Fluvial systems: a management perspective. In Amoros, C. (ed.), *Fluvial hydrosystems*. London: Chapman and Hall, 263–78.

Petts, G.E. and Bravard, J.E., 1996: A drainage basin perspective. In Petts, G.E. and Amoros, C. (eds), *Fluvial hydrosystems*. London: Chapman and Hall, 13–36.

Petts, G.E. and Calow, P. (eds), 1996: *River flows and channel forms: selected extracts from the Rivers Handbook*. Oxford: Blackwell.

Petts, G.E. and Maddock, I., 1994: Flow allocation for in-river needs. In Calow, P. and Petts, G.E. (eds), *The rivers handbook*. Oxford: Blackwell Scientific Publications, 289–307.

Petts, G.E. and Maddock, I., 1996: Flow allocation for in-river needs. In Petts, G.E. and Calow, P. (eds), *River restoration: selected extracts from the Rivers Handbook*. Oxford: Blackwell Science, 60–79.

Petts, G.E. and Wood, R., 1988: River regulation in the UK. *Regulated Rivers: Research and Management*, 2(3); 199–477.

Petts, G.E., Maddock, I., Bickerton, M. and Ferguson, A.G.D., 1995: Linking hydrology and ecology: the scientific basis for river management. In Harper, D.M. and Ferguson, A.G.D. (eds), *The ecological basis for river management*. Chichester: J. Wiley & Sons, 1–16.

Petts, G.E., Sparks, R. and Campbell, I., 2000: River restoration in developed economies. In Boon, P.J., Davies, B.R. and Petts, G.E. (eds), *Global perspectives on river conservation: science, policy and practice*. Chichester: J. Wiley & Sons, 493–508.

Phillips, J.D., 1992: The end of equilibrium. *Geomorphology*, 5: 195–201.

Phillips, J.D., 1999: Divergence, convergence and self-organization in landscapes. *Annals of the Association of American Geographers*, 89: 466–88.

Phillips, J.D., 2001: Human impacts on the environment:unpredictability and primacy of place. *Physical Geography*, 27: 1–23.

Phillips, J.D., 2003: Sources of nonlinearity and complexity in geomorphic systems. *Progress in Physical Geography*, 27: 1–23.

Pickett, S.T.A. and White, P.S., 1985: Patch dynamics: a synthesis. In Pickett, S.T.A. and White, P.S. (eds), *The ecology of natural disturbance and patch dynamics*. New York: Academic Press, 371–84.

Pickup, G., 1984: Geomorphology in tropical rivers. I. Landforms, hydrology and sedimentation in the Fly and Lower Purai, Papua New Guinea. *Catena Supplement*, 5: 1–17.

Pickup, G., 1986: Fluvial landforms. In Jeans, D.N. (ed.), *Australia: a geography*. Sydney: University Press, 148–79.

Piégay, H., Cuaz, M., Javelle, E. and Mandier, P., 1997: Bank erosion management based on geomorphological, ecological and economic criteria on the Galaure River, France. *Regulated Rivers: Research and Management*, 13: 433–48.

Piégay, H., Thevenet, A., Kondolf, G.M. and Landon, N., 2000: Physical and human factors influencing potential fish habitat distribution along a mountain river, France. *Geografiska Annaler Series A – Physical Geography*, 82A: 121–36.

Pitkethly, A.S., 1990: Integrated water management in England. In Mitchell, B. (ed.), *Integrated water management: international experiences and perspectives*. London: Belhaven, 119–47.
Pizzuto, J., 2002: Effects of dam removal on river form and process. *Bioscience*, 52: 683–91.
Platts, S.W., Megahan, W.F. and Minshall, G.W., 1983. *Methods for evaluating stream, riparian and biotic conditions*. United States Department of Agriculture, Forest Service: General Technical Report INT-138. Washington DC.
Poole, G.C., 2002: Fluvial landscape ecology: addressing uniqueness within the river discontinuum. *Freshwater Biology*, 47: 641–60.
Powell, J.W., 1875: *Exploration of the Colorado River of the West 1869–72*. Cambridge: University of Chicago and Cambridge University Press, reprint.
Prestegaard, K.L., Matherne, A.M., Shane, B., Houghton, K., O'Connell, M. and Katyl, N., 1994: Spatial variations in the magnitude of the 1993 floods, Raccoon River basin, Iowa. *Geomorphology*, 10: 169–82.
Przedwojski, B., Blazejewski, R. and Pilarczyk, K.W., 1995: *River training techniques: fundamentals, design and application*. Rotterdam: A.A. Balkema.
Purseglove, J., 1988: *Taming the flood: a history and natural history of rivers and wetlands*. Oxford: Oxford University Press.
Railsback, S.F. and Harvey, B.C., 2002: Analysis of habitat-selection rules using an individual-based model. *Ecology*, 83: 1817–30.
Railsback, S.F., Lamberson, R.H., Harvey, B.C. and Duffy, W.E., 1999: Movement rules for individual-based models of stream fish. *Ecological Modelling*, 123: 73–89.
Ramsbottom, D.M., Millington, R.J. and Bettess, R., 1992. *National River Authority Thames Region: standards of service, reach specification methodology*. HR Wallingford Report EX2652. Wallingford: Hydraulics Research.
Raven, P.J., Fox, R., Everard, M., Holmes, N.T.H. and Dawson, F.H., 1997: River habitat survey: a new system for classifying rivers according to their habitat quality. In Boon, P.J. and Howell, D.L. (eds), *Freshwater quality: defining the indefinable*. Edinburgh: The Stationery Office (HMSO, SNH), 215–34.
Raven, P.J., Holmes, N.T.H., Dawson, F.H., Everard, M., Fozzard, L. and Rouen, K.J., 1998. *River habitat quality: the physical character of rivers and streams in the United Kingdom and Isle of Man*. River Habitat Survey Report 2. Bristol: Environment Agency.
Raven, P.J., Holmes, N.T.H., Charrier, P., Dawson, F.H., Naura, M. and Boon, P.J., 2002: Towards a harmonized approach for hydromorphological assessment of rivers in Europe: a qualitative comparison of three survey methods. *Aquatic Conservation: Marine and Freshwater Ecosystems*, 12: 405–24.
Ray, J., 1713: *Miscellaneous discourses concerning the dissolution and changes of the world*. Third edition. London: Samuel Smith.
Redclift, M.R., 1995: Values and global environmental change. In Guerrier, Y. (ed.), *Values and the Environment: a social science perspective*. Chichester: J. Wiley & Sons, 7–17.
Reid, I., Bathurst, J.C., Carling, P.A., Walling, D.E. and Webb, B.W., 1997: Sediment erosion, transport and deposition. In Thorne, C.R., Hey, R.D. and Newson, M.D. (eds), *Applied fluvial geomorphology for river engineering and management*. Chichester: J. Wiley & Sons, 95–135.
Reid, L.M., 1998: Cumulative watershed effects and watershed analysis. In Naiman, R.J. and

Bilby, R.E. (eds), *River ecology and management: lessons for the Pacific coastal ecosystem.* New York: Springer, 476–501.

Reid, L.M. and Dunne, T., 1996: *Rapid evaluation of sediment budgets.* Reiskirchen: Catena Verlag.

Reid, L.M. and McCammon, B.P., 1993: *A procedure for watershed analysis.* Portland OR: Interagency Forest Ecosystem Management Assessment Team, USDA Forest Service Region 6.

Reiser, D.W., Ramey, M.P. and Wesche, T.A., 1989: Flushing flows. In Gore, J.A. and Petts, G.E. (eds), *Alternatives in regulated river management.* Baton Rouge FL: CRC Press, 91–135.

Reisner, M., 1986: *Cadillac desert: the American West and its disappearing water.* London: Pengiun Books.

Renssen, H. and Knoop, J.M., 2000: A global river routing network for use in hydrological modelling. *Journal of Hydrology*, 230: 230–43.

Renwick, W.H., 1992: Equilibrium, disequilibrium, and nonequilbrium landforms in the landscape. *Geomorphology*, 5: 265–76.

Rhoads, B.L., 1987: Stream power terminology. *Professional Geographer*, 39: 189–95.

Rhoads, B.L., 1995: Stream power: a unifying theme for urban fluvial geomorphology. In Herricks, E.E. (ed.), *Urban runoff and receiving systems: an interdisciplinary analysis of impact, monitoring and management, proceedings engineering foundation conference August 4–9, 1991. Mount Crested Butte CO.* Boca Ration FL: Lewis Publishers, 91–101.

Rhoads, B.L. and Herricks, E.E., 1996: Naturalization of headwater streams in Illinois: challenges and possibilities. In Brookes, A. and Shields, F.D., Jr (eds), *River channel restoration: guiding principles for sustainable projects.* Chichester: J. Wiley & Sons, 331–67.

Rhoads, B.L., Wilson, D., Urban, M. and Herricks, E.E., 1999: Interactions between scientists and non-scientists in community-based watershed management: emergence of the concept of stream naturalization. *Environmental Management*, 24: 297–308.

Rice, S., 1998: Which tributaries disrupt downstream fining along gravel-bed rivers? *Geomorphology*, 22: 39–56.

Richards, K., 1982: *Rivers: form and process in alluvial channels.* London: Methuen.

Richards, K.S., 1973: Hydraulic geometry and channel roughness – a nonlinear system. *American Journal Science*, 273: 877–96.

Richards, K.S. (ed.), 1987: *River channels environment and process.* Oxford: Blackwell.

Richards, K.S., 1999: The magnitude-frequency concept in fluvial geomorphology: a component of a degenerating research programme? *Zeitschrift für Geomorphologie Supplementband*, 115: 1–18.

Richards, K.S. and Lane, S.N., 1997: Prediction of morphological changes in unstable channels. In Thorne, C.R., Hey, R.D. and Newson, M.D. (eds), *Applied fluvial geomorphology for river engineering and management.* Chichester: J. Wiley & Sons, 269–92.

Ricker, W.E., 1934: An ecological classification of certain Ontario streams. *Ontario Fisheries Research Laboratory Publication*, 49: 1–114.

Riley, A.L., 1989: Overcoming federal water policies: the Wildcat-San Pablo Creeks case. *Environment*, 31: 12 pp.

Riley, A.L., 1998: *Restoring streams in cities: a guide for planners, policy makers and citizens.* Washington DC: Island Press.

River Restoration Centre, 1999: *River restoration: manual of techniques: restoring the River Cole and River Skerne, UK*. Silsoe: River Restoration Centre.

Roberts, C.R., 1989: Flood frequency and urban induced channel change: some British examples. In Beven, K. and Carling, P.A. (eds), *Floods: hydrological, sedimentological and geomorphological implications*. Chichester: J. Wiley & Sons, 57–82.

Rodriguez-Iturbe, I., 2000: Ecohydrology: a hydrologic perspective of climate-soil-vegetation dynamics. *Water Resources Research*, 36: 3–9.

Rodriguez-Iturbe, I. and Rinaldo, A., 1998: *Fractal river basins*. Cambridge: Cambridge University Press.

Roni, P., Beechie, T.J., Bilby, R.E., Leonetti, F.E., Pollock, M.M. and Pess, G.R., 2002: A review of stream restoration techniques and a hierarchical strategy for prioritizing restoration in Pacific northwest watersheds. *North American Journal of Fisheries Management*, 22: 1–20.

Rosgen, D.L., 1994: A classification of rivers. *Catena*, 22: 169–99.

Rosgen, D.L., 1996: *Applied river morphology*. Pagosa Springs CO: Wildland Hydrology.

Rowntree, K.M. and Wadeson, R.A., 1996: Translating channel geomorphology into aquatic habitat: application of the hydraulic biotope concept to an assessment of discharge related habitat changes. In Leclerc, M. (ed.), *Ecohydraulics 2000, proceedings of the 2nd internaitonal symposium of habitat hydraulics*. Quebec: International Association of Hydraulics Research, 342–51.

Rumsby, B.T. and Macklin, M.G., 1996: River response to the last neoglacial (the 'Little Ice Age') in northern, western and central Europe. In Branson, J., Brown, A.G. and Gregory, K.J. (eds), *Global continental changes: the context of palaeohydrology*. London: Geological Society Special Publication 115, 217–33.

Rumsby, B.T. and Macklin, M.G., 1994: Channel and floodplain response to recent abrupt climate change: the Tyne basin, northern England. *Earth Surface Processes and Landforms*, 19: 499–515.

Rust, B.R., 1978: A classification of alluvial river systems. In Miall, A.D. (ed.), *Fluvial sedimentology*. Calgary: Canadian Society of Petroleum Geologists, Memoir 5, 745–55.

Rutherfurd, I., 1999: Sand slugs in a large granite catchment: the Glenelg River, SE Australia. In Mosley, M.P., Harvey, M.D. and Anthony, D. (eds), *Applying geomorphology to environmental management*. Littleton CO: Water Resources Publications.

Rutherfurd, I., 2000: Some human impacts on Australian stream channel morphology. In Brizga, S. and Finlayson, B. (eds), *River management: the Australasian experience*. Chichester: J. Wiley & Sons, 11–49.

Sacramento River Advisory Council, 2000. *Sacramento River conservation area handbook*. Sacramento CA: California Department of Water Resources.

Saha, S.K., 1981: River basin planning as a field of study: design of a course structure for practitioners. In Saha, S.K. and Barrow, C.J. (eds), *River Basin Planning: Theory and Practice*. J. Wiley & Sons, Chichester, 9–40.

Saha, S.K. and Barrow, C.J., 1981: Introduction. In Saha, S.K. and Barrow, C.J. (eds), *River basin planning: theory and practice* Chichester: J. Wiley & Sons, 1–7.

Said, R., 1983: *The River Nile: geology, hydrology and utilization*. Oxford: Pergammon Press.

Schiemer, F., Baumgartner, C. and Tockner, K., 1999: Restoration of floodplain rivers: the Danube Restoration Project. *Regulated Rivers: Research and Management*, 15: 231–44.

Schofield, N.J., Collier, K.J., Quinn, J., Sheldon, F. and Thoms, M.C., 2000: River conservation in Australia and New Zealand. In Boon, P.J., Davies, B.R. and Petts, G.E. (eds), *Global perspectives on river conservation: science, policy and practice*. Chichester: J. Wiley & Sons, 311–33.

Schropp, M.H.I. and Bakker, C., 1998: Secondary channels as a basis for the ecological rehabilitation of Dutch rivers. *Aquatic Conservation: Marine and Freshwater Ecosystems*, 8: 53–59.

Schueler, T.R. and Holland, H.K. (eds), 2000: *The practice of watershed protection*. Elliott City MD: Center for Watershed Protection,

Schulze, R.E., 1997: Impacts of global climate change in a hydrologically vulnerable region: challenges to South African hydrologists. *Progress in Physical Geography*, 21: 113–36.

Schumm, S.A., 1963. *A tentative classification of alluvial river channels*. US Geological Survey, Circular 477: Washington DC.

Schumm, S.A., 1969: River metamorphosis. *Journal of the Hydraulics Division, ASCE*, HY1: 255–63.

Schumm, S.A., 1973: Geomorphic thresholds and complex response of drainage systems. In Morisawa, M. (ed.), *Fluvial geomorphology*. New York: SUNY Binghamton, Publications in Geomorphology, 299–309.

Schumm, S.A., 1977a: Applied fluvial geomorphology. In Hails, J.R. (ed.), *Applied geomorphology*. Amsterdam: Elsevier, 119–56.

Schumm, S.A., 1977b: *The fluvial system*. New York: Wiley-Interscience.

Schumm, S.A., 1979: Geomorphic thresholds: the concept and its applications. *Transactions of the Institute of British Geographers*, NS4: 485–515.

Schumm, S.A., 1981: Evolution and response of the fluvial system, sedimentologic implications. *Society of Economic Paleontologists and Mineralogists, Special Publication* 31: 19–29.

Schumm, S.A., 1985a: Explanation and extrapolation in geomorphology: seven reasons for geologic uncertainty. *Transactions Japanese Geomorphological Union*, 6: 1–18.

Schumm, S.A., 1985b: Patterns of alluvial rivers. *Annual Review of Earth and Planetary Sciences*, 13: 5–27.

Schumm, S.A., 1988: Geomorphic hazards – problems and predictions. *Zeitschrift für Geomorphologie*, NF 67: 17–24.

Schumm, S.A., 1991: *To interpret the Earth: ten ways to be wrong*. Cambridge: Cambridge University Press.

Schumm, S.A., 1994: Erroneous perceptions of fluvial hazards. *Geomorphology*, 10: 129–38.

Schumm, S.A., 1999: Causes and controls of channel incision. In Darby, S.E. and Simon, A. (eds), *Incised river channels: processes, forms, engineering and management*. Chichester: J. Wiley & Sons, 19–33.

Schumm, S.A. and Lichty, R.W., 1963: Channel widening and floodplain construction along the Cimarron river in south-western Kansas. *US Geological Survey, Professional Paper* 352D: Washington DC.

Schumm, S.A. and Lichty, R.W., 1965: Time, space and causality in geomorphology. *American Journal of Science*, 263: 110–19.

Schumm, S.A. and Rea, D.K., 1995: Sediment yield from disturbed earth systems. *Geology*, 23: 391–94.

Schumm, S.A., Harvey, M.D. and Watson, C.C., 1984: *Incised channels: morphology, dynamics and control*. Littleton CO: Water Resources Publications.

Schumm, S.A., Mosley, M.P. and Weaver, W.E., 1987: *Experimental fluvial geomorphology*. New York: J. Wiley & Sons.

Schumm, S.A., Dumont, J.F. and Holbrook, J.M., 2000: *Active tectonics and alluvial rivers*. Cambridge: Cambridge University Press.

Sear, D.A., 1994: River restoration and geomorphology. *Aquatic Conservation: Marine and Freshwater Ecosystems*, 4: 169–77.

Sear, D.A. and Newson, M.D., 2003: Environmental change in river channels: a neglected element. Towards geomorphological typologies, standards and monitoring. *The Science of the Total Environment*, 310: 17–23.

Sear, D.A., Darby, S.E., Thorne, C.R. and Brookes, A., 1994: Geomorphological approach to stream stabilization and restoration: case study of the Mimmshall Brook, Hertfordshire, UK. *Regulated Rivers: Research and Management*, 9: 205–23.

Sear, D.A., Newson, M.D. and Brookes, A., 1995: Sediment-related river maintenance: the role of fluvial geomorphology. *Earth Surface Processes and Landforms*, 20: 629–47.

Sedell, J.R., Bisson, P.A., Swanson, F.J. and and others, 1988: What we know about large trees that fall into streams and rivers. In Maser, C., Tarrant, R.F., Trappe, J.M. and others (eds), *From the forest to the sea: a story of fallen trees*. Pacific Northwest Research Station, General Technical Report PNW-229. Portland OR: USDA Forest Service, 47–82.

Sedell, J.R., Richey, J.E. and Swanson, F.J., 1989: The river continuum concept: a basis for the expected behaviour of very large rivers? In Dodge, D.P. (ed.), Proceedings of the international large river symposium. Special Publication of the *Canadian Journal of Fisheries and Aquatic Sciences*, 106: 49–55.

Sellin, R.H.J., Giles, A. and Van Beesten, D.P., 1990: Post-implementation appraisal of a two-stage channel in the River Roding, Essex. *Journal of the Institution of Water and Environmental Management*, 4: 119–30.

Shannon, M.A., 1998: Social organizations and institutions. In Naiman, R.J. and Bilby, R.E. (eds), *River ecology and management: lessons for the Pacific coastal ecosystem*. New York: Springer, 529–52.

Sharpin, M., Barter, S. and Csanki, S., 1999: Stormwater planning in New South Wales, Australia. In Joliffe, I.B. and Ball, J.E. (eds), *8th international conference on urban storm drainage*. Barton A.C.T.: Institution of Engineers, Australia, 2006–14.

Shaw, E.M., 1994: *Hydrology in Practice*. Third Edition. London: Chapman and Hall.

Shen, H.W., 1973: *Environmental impact on rivers (river mechanics III)*. Fort Collins CO: H.W. Shen.

Shen, H.W., Schumm, S.A., Nelson, J.D., Doehring, D.O., Skinner, M.M. and Smith, G.L., 1981: *Methods for assessment of stream-related hazards to highways and bridges*. Technical Report for US Department of Transportation FHWA/RD-80/160. Fort Collins CO: Colorado State University.

Shen, H.W., Tabios, G. and Harder, J.A., 1994: Kissimmee river restoration study. *Journal of Water Resources Planning and Management*, 120: 330–49.

Sherrard, J.J. and Erskine, W.D., 1991: Complex response of sand bed streams to upstream impoundment. *Regulated Rivers: Research and Management*, 6: 53–70.

Shields, A., 1936. *Anwendung der ahnlichkeitsmechanil und der turbulenzforschung suf die geschiebebewegung*. Berlin: Milleilungen der Preussischen Versuchsanstlt fur Wasserbau.

Shields, F.D., Jr, 1982. *Environmental features for flood control channels*. Environmental and Water Quality Operational Studies Technical Report E-82–7. Vicksburg MS: US Army Engineer Waterways Experiment Station.

Shields, F.D., Jr, 1996: Hydraulic and hydrologic stability. In Brookes, A. and Shields, F.D., Jr (eds), *River channel restoration: guiding principles for sustainable projects*. Chichester: J. Wiley & Sons, 24–74.

Shields, F.D., Jr, and Abt, S.R., 1989: Sediment deposition in cutoff meander bends and implications for effective management. *Regulated Rivers: Research and Management*, 4: 381–96.

Shields, F.D., Jr, and Hoover, J.J., 1991: Effects of channel restabilization on habitat diversity, Twentymile Creek, Mississippi. *Regulated Rivers: Research and Management*, 6: 163–81.

Shields, F.D., Jr, Knight, S.S. and Cooper, C.M., 1995a: Incised stream physical habitat restoration with stone weirs. *Regulated Rivers : Research and Management*, 10: 181–98.

Shields, F.D., Jr, Knight, S.S. and Cooper, C.M., 1995b: Rehabilitation of watersheds with incising channels. *Water Resources Bulletin*, 31: 971–81.

Shields, F.D., Jr, Knight, S.S. and Cooper, C.M., 1998a: Addition of spurs to stone toe protection for warmwater fish habitat rehabilitation. *Journal of the American Water Resources Association*, 34: 1427–36.

Shields, F.D., Jr, Knight, S.S. and Cooper, C.M., 1998b: Rehabilitation of aquatic habitats in warmwater streams damaged by channel incision in Mississippi. *Hydrobiologica*, 382: 63–86.

Shorter, A.H., 1971: *Paper making in the British Isles: an historical and geographical study*. Newton Abbott; David and Charles.

Showers, K.B., 2000: Popular participation in river conservation. In Boon, P.J., Davies, B.R. and Petts, G.E. (eds), *Global perspectives on river conservation: science, policy and practice*. Chichester: J. Wiley & Sons, 459–74.

Shuman, J.R., 1995: Environmental considerations for assessing dam removal alternatives for river restoration. *Regulated Rivers: Research and Management*, 11: 249–61.

Siggers, G.B., Bates, P.D., Anderson, M.G., Walling, D.E. and He, Q., 1999: A preliminary investigation of the integration of modelled floodplain hydraulics with estimates of overbank floodplain sedimentation derived from Pb-210 and Cs-137 measurements. *Earth Surface Processes and Landforms*, 24: 211–31.

Simon, A., 1989: A model of channel response in disturbed alluvial channels. *Earth Surface Processes and Landforms*, 14: 11–26.

Simon, A. and Collison, A.J.C., 2002: Quantifying the mechanical and hydrologic effects of riparian vegetation on streambank stability. *Earth Surface Processes and Landforms*, 27: 527–46.

Simon, A. and Darby, S.E., 1999: The nature and significance of incised river channels. In Darby, S.E. and Simon, A. (eds), *Incised rivers: processes, forms, engineering, management*. Chichester: J. Wiley & Sons, 3–18.

Simon, A. and Downs, P.W., 1995: An interdisciplinary approach to evaluation of potential instability in alluvial channels. *Geomorphology*, 12: 215–32.

Simon, A. and Thomas, R.E., 2002: Processes and form of an unstable alluvial system with resistant, cohesive, streambeds. *Earth Surface Processes and Landforms*, 27: 699–715.

Simon, A., Outlaw, G.S. and Thomas, R., 1989: Evaluation, modeling, and mapping of potential bridge scour, West Tennessee. *Proceedings of the national bridge scour symposium.* Washington DC: Federal Highways Administration, US Department of Transportation, 112–39.

Simon, A., Curini, A., Darby, S.E. and Langendoen, E.J., 1999: Streambank mechanics and the role of bank and near-bank processes in incised rivers. In Darby, S.E. and Simon, A. (eds), *Incised rivers: processes, forms, engineering, management.* Chichester: J. Wiley & Sons, 123–52.

Simons, H.E.J., Bakker, C., Schropp, M.H.I., Jans, L.H. and Kok, F.R., 2001: Man-made secondary channels along the River Rhine (the Netherlands): results of post-project monitoring. *Regulated Rivers: Research and Management*, 17: 473–91.

Sinha, R. and Jain, V., 1998: Flood hazards of North Bihar rivers, Indo-Gangetic plains. *Memoir Geological Society of India*, 41: 27–52.

Skinner, K.S., 1999: Geomorphological post-project appraisal of river rehabilitation schemes in England. Unpublished Ph.D. Thesis, University of Nottingham, Nottingham.

Skinner, K.S., Downs, P.W. and Brookes, A., 1999: Assessing the success of river rehabilitation schemes. *Water resources into the new millenium: past accomplishments, new challenges.* Seattle WA: ASCE, on CD-ROM.

Skinner, K.S., Downs, P.W. and Brookes, A., 2004: The channel geomorphology profiler: piloting a technique for aiding post-project appraisals of river rehabilitation schemes. Unpublished.

Smart, P.L. and Francis, P.D. (eds), 1991: *Quaternary dating methods – a users guide.* Technical guide no. 4. Cambridge: Quaternary Research Association.

Smil, V., 1979: *China's energy: achievements, problems, prospects.* New York: Praeger Publishers.

Smith, B., 1994: The River Medway project: an example of community participation in integrated river management. In Kirby, C. and White, W.R. (eds), *Integrated river basin development.* Chichester: J. Wiley & Sons, 377–87.

Smith, C.L., Gilden, J., Steel, B.S. and Mrakovcich, K., 1998: Sailing the shoals of adaptive management: the case of salmon in the Pacific Northwest. *Environmental Management*, 22: 671–81.

Smith, D.I., 1998: *Water in Australia: resources and management.* Melbourne: Oxford University Press.

Smith, K., 1992: *Environmental hazards: assessing risk and reducing disaster.* London: Routledge.

Smith, K., 1998: *Frank Lloyd Wright: America's master architect.* New York: Abbeville Press.

Smith, S., 1997: Changes in the hydraulic and morphological characteristics of a relocated stream channel. Unpublished M.S. Thesis, Annapolis MD: University of Maryland.

Smits, A.J.M., Nienhuis, P.H. and Leuven, R.S.E.W. (eds), 2000: *New approaches to river management.* Leiden: Backhuys.

Smits, A.J.M., Cals, M.J.R. and Drost, H.J., 2001: Evolution of European river basin management. In Nijland, H.J. and Cals, M.J.R. (eds), *River restoration in Europe: practical approaches.*

Lelystad: Institute for Inland Water Management and Waste Water Treatment/RIZA, 41–48.

Soar, P.J. and Thorne, C.R., 2001: *Channel restoration design for meandering rivers*. Coastal and Hydraulics Laboratory ERDC/CHL CR-01–1, Vicksburg MS: US Army Engineer Research and Development Center.

Soar, P.J., Thorne, C.R., Downs, P.W. and Copeland, R.R., 1998: Geomorphic engineering for river restoration design. In Hayes, D.F. (ed.), *Engineering approaches to ecosystem restoration, proceeding of the wetlands engineering and river restoration conference*. Denver CO: ASCE, on CD-ROM.

Somner, T.R., Nobriga, M.L., Harrell, W.C., Batham, W. and Kimmerer, W.J., 2001: Floodplain rearing of juvenile chinook salmon: evidence of enhanced growth and survival. *Canadian Journal of Aquatic Sciences*, 58: 325–33.

Sparks, R.E., 1995: Need for ecosystem management of large rivers and their floodplains. *BioScience*, 45: 168–82.

Stanford, J.A., 1996: Landscapes and catchment basins. In Hauer, R.F.R. and Lamberti, G.A. (eds), *Methods in stream ecology*. London: Academic Press, 3–22.

Stanford, J.A. and Ward, J.V., 1986: Fish of the Colorado system. In Davies, B.R. and Walker, K.R. (eds), *The ecology of river systems*. Dordrecht: Dr W. Junk, 385–402.

Stanford, J.A. and Ward, J.V., 1992: Management of aquatic resources in large catchments: recognizing interaction between ecosystem connectivity and environmental disturbance. In Naiman, R.J. (ed.), *Watershed Management*. New York: Springer Verlag, 91–124.

Stanford, J.A. and Ward, J.V., 1993: An ecosystem perspective of alluvial rivers: connectivity and the hyporheic corridor. *Journal of the North American Benthological Society*, 12: 48–60.

Stanley, E.H. and Doyle, M.W., 2002: A geomorphic perspective on nutrient retention following dam removal. *BioScience*, 52: 693–701.

Starkel, L., 1983: The reflection of hydrologic changes in the fluvial environment on the temperate zone during the last 15,000 years. In Gregory, K.J. (ed.), *Background to palaeohydrology*. Chichester: J. Wiley & Sons, 213–36.

Starkel, L., 1991: The Vistula River valley: a case study for central Europe. In Starkel, L., Gregory, K.J. and Thornes, J.B. (eds), *Temperate palaeohydrology: fluvial processes in the temperate zone over the last 15,000 years*. Chichester: J. Wiley & Sons, 171–88.

Starkel, L., Gregory, K.J. and Thornes, J.B. (eds), 1991: *Temperate palaeohydrology: fluvial processes in the temperate zone during the last 15,000 years*. Chichester: J. Wiley & Sons.

Statzner, B. and Higler, B., 1985: Questions and comments of the river continuum concept. *Canadian Journal of Fisheries and Aquatic Sciences*, 42: 1038–44.

Statzner, B. and Sperling, F., 1993: Potential contribution of system-specific knowledge (SSK) to stream management decisions: ecological and economic aspects. *Freshwater Biology*, 29: 313–42.

Statzner, B., Gore, J.A. and Resh, V.H., 1988: Hydraulic stream ecology: observed patterns and potential applications. *Journal of the North American Benthological Society*, 7: 307–60.

Stephenson, D., 1858: *The principles and practice of canal and river engineering*. Edinburgh: A&C Black.

Stillwater Sciences, 2002: *Merced River Corridor restoration plan*. Berkeley CA: Stillwater Sciences.

Strahler, A.N., 1952: Dynamic basis of geomorphology. *Geological Society of America Bulletin*, 63: 923–37.

Strahler, A.N., 1956: The nature of induced erosion and aggradation. In Thomas, W.L. (ed.), *Man's role in changing the face of the Earth*. Chicago: Unversity of Chicago Press, 621–38.

Strahler, A.N., 1992: Quantitative/dynamic geomorphology at Columbia 1945–60: a retrospective. *Progress in Physical Geography*, 16: 65–84.

Street, E., 1981: The role of electricity in the Tennessee Valley Authority. In Saha, S.K. and Barrow, C.J. (eds), *River basin planning: theory and practice*. Chichester: J. Wiley & Sons, 233–52.

Szilagy, J., 1932: Flood control on the Tisza River. *Military Engineer*, 24: 632.

Taylor, M.P. and Lewin, J., 1997: Non-synchronous response of adjacent floodplain systems to Holocene environmental change. *Geomorphology*, 18: 251–64.

Taylor, P., 1986: *Respect for nature*. Princeton NJ: Princeton University Press.

Thomas, D.F. and Goudie, A. (eds), 2000: *The dictionary of physical geography*. Third edition, Oxford: Blackwell.

Thomas, M.F., 2001: Landscape sensitivity in time and space – an introduction. *Catena*, 42: 83–98.

Thompson, A., 1986: Secondary flows and the pool-riffle unit: a case study of the processes of meander development. *Earth Surface Processes and Landforms*, 11: 631–42.

Thoms, M.C. and Walker, K.F., 1993: Channel change associated with two adjacent weirs on a regulated lowland alluvial river. *Regulated Rivers: Research and Management*, 8: 271–84.

Thomson, J.R., Taylor, M.P., Fryirs, K.A. and Brierley, G.J., 2001: A geomorphological framework for river characterization and habitat assessment. *Aquatic Conservation: Marine and Freshwater Management*, 11: 373–89.

Thorne, C.R., 1992: Bend scour and bank erosion on the meandering Red River, Louisiana. In Carling, P.A. and Petts, G.E. (eds), *Lowland floodplain rivers: geomorphological perspectives*. Chichester: J. Wiley & Sons, 95–115.

Thorne, C.R., 1993: *Guidelines for the use of stream reconnaisance record sheets in the field*. Contract Report HL-93–2. Vicksburg VA: Waterways Experimental Station, US Army Corps of Engineers.

Thorne, C.R., 1997: Channel types and morphological classification. In Thorne, C.R., Hey, R.D. and Newson, M.D. (eds), *Applied fluvial geomorphology in river engineering management*. Chichester: J. Wiley & Sons, 175–222.

Thorne, C.R., 1998: *Stream reconnaissance handbook: geomorphological investigation and analysis of river channels*. Chichester: J. Wiley & Sons.

Thorne, C.R., 1999: Bank processes and channel evolution in the incised rivers of north-central Mississippi. In Darby, S.E. and Simon, A. (eds), *Incised rivers: processes, forms, engineering, management*. Chichester: J. Wiley & Sons, 97–121.

Thorne, C.R., 2002: Geomorphic analysis of large alluvial rivers. *Geomorphology*, 44: 203–219.

Thorne, C.R., Bathurst, J.C. and Hey, R.D. (eds), 1987: *Sediment transport in gravel-bed rivers*. Chichester: J. Wiley & Sons.

Thorne, C.R., Allen, R.G. and Simon, A., 1996a: Geomorphological river channel reconnaissance for river analysis, engineering and management. *Transactions of the Institute of the British Geographers*, 21: 469–83.

Thorne, C.R., Reed, S. and Doornkamp, J.C., 1996b: *A procedure for assessing river bank erosion problems and solutions*. R&D Report 28. Bristol: National Rivers Authority.

Thorne, C.R., Hey, R.D. and Newson, M.D., 1997a: *Applied fluvial geomorphology for river engineering and management*. Chichester: J. Wiley & Sons.

Thorne, C.R., Newson, M.D. and Hey, R.D., 1997b: Application of applied fluvial geomorphology: problems and potential. In Thorne, C.R., Hey, R.D. and Newson, M.D. (eds), *Applied fluvial geomorphology in river engineering management*. Chichester: J. Wiley & Sons, 365–70.

Thorne, C.R., Downs, P.W., Newson, M.D., Clark, M.J. and Sear, D.A., 1998. *River geomorphology: a practical guide*. National Centre for Risk Analysis and Options Appraisal Guidance Note 18. London: Environment Agency.

Tockner, K., Malard, F. and Ward, J.V., 2000: An extension of the flood pulse concept. *Hydrological Processes*, 14: 2861–83.

Tockner, K., Ward, J.V., Edwards, P.J. and Kollmann, J., 2002: Riverine landscapes: an introduction. *Freshwater Biology*, 47: 497–500.

Tooth, S. and Nanson, G.C., 2000: Equilibrium and nonequilibrium conditions in dryland rivers. *Progress in Physical Geography*, 21: 183–211.

Toth, L.A., 1996: Restoring the hydrogeomorphology of the channelized Kissimmee River. In Brookes, A. and Shields, F.D., Jr (eds), *River channel restoration: guiding principles for sustainable projects*. Chichester: J. Wiley & Sons, 371–83.

Toth, L.A., Obeysekera, T.B., Perkins, W.A. and Loftin, M.K., 1993: Flow regulation and restoration of Florida's Kissimmee River. *Regulated Rivers: Research and Management*, 8: 155–66.

Toth, L.A., Melvin, S.L., Arrington, D.A. and Chamberlain, J., 1998: Managed hydrologic manipulations of the channelised Kissimmee River: implications for restoration. *BioScience*, 8: 757–64.

Towner Coston, H.E., Penetlow, F.T.K. and Butcher, R.W., 1936: *River management: the making, care and development of salmon and trout rivers*. London: Lonsdale.

Townsend, C.R., 1989: The patch dynamics concept of stream community ecology. *Journal of the North American Benthological Society*, 8: 36–50.

Tuan, Y.F., 1968: *The hydrological cycle and the wisdom of God: a theme in geoteleology*. Toronto: University of Toronto Press.

Tuan, Y.F., 1993: *Passing strange and wonderful: aesthetics, nature and culture*. Washington DC: Island Press,

Tung, Y.K., 1985: Channel scouring potential using logistic analysis. *Journal of Hydraulic Engineering, Proceedings of the American Society of Civil Engineers*, 111: 194–201.

Tunstall, S.M., Penning-Rowsell, E.C., Tapsell, S.M. and Eden, S.E., 2000: River restoration: public attitudes and expectations. *Journal of the Chartered Institution of Water and Environmental Management*, 14: 363–70.

Turner, R.K., 1993a: Postscript: future prospects. In Turner, R.K. (ed.), *Sustainable environmental economics and management: principles and practice*. London: Belhaven Press, 383–86.

Turner, R.K., 1993b: Sustainability: principles and practice. In Turner: R.K. (ed.), *Sustainable environmental economics and management: principles and practice*. London: Belhaven Press, 3–36.

Udvardy, M.D.F., 1981: The riddle of dispersal: dispersal theories and how they affect vicariance biogeography. In Nelson, G. and Rosen, D.E. (eds), *Vicariance biogeography: a critique*. New York: Columbia University Press, 6–29.

Uitto, J.I. and Wolf, A.T., 2002: Water wars? Geographical perspectives: an introduction. *Geographical Journal*, 168: 289–92.

Underwood, A.J., 1994: Spatial and temporal problems with monitoring. In Calow, P. and Petts, G.E. (eds), *The rivers handbook*. Oxford: Blackwell Scientific, 101–23.

United Nations, 1970. *Integrated river basin development: report of a panel of experts*. New York: U.N. Department of Economic and Social Affairs.

United Nations Economic Commission for Europe, 1993. *Protecting water resources and catchment ecosystems*. Water Series 1 ECE/ENVWA/31. Geneva: UNECE.

United Nations World Commission on Environment and Development, 1987: *Our common future*. Oxford: Oxford University Press.

United States Department of Agriculture, Forest Service, 2001: *Forest plan monitoring and evaluation report – 2001*. Coeur d'Alene ID: Idaho Panhandle National Forests.

United States Environmental Protection Agency, 1993: *Reference guide to pollution prevention resources*. Washington DC: USEPA Office of Pollution prevention and Toxics.

United States Fish & Wildlife Service / Hoopa Valley Tribe, 1999: *Trinity River flow evaluation, final report*. Arcata: US Fish & Wildlife Service and Hoopa Valley Tribe.

Vallentine, H.R., 1967: *Water in the service of man*. Harmondsworth: Penguin Books.

Van den Berg, J.H., 1995: Prediction of alluvial channel pattern of perennial rivers. *Geomorphology*, 12: 259–79.

Van Rijn, L.C., 1987: *Mathematical modelling of morphological processes in the case of suspended sediment transport*. Delft: Delft Hydraulics. Delft Hydraulics Communication, No. 382.

Van Velson, R., 1979: Effects of livestock grazing upon rainbow trout in Otter Creek. In Cope, O.B. (ed.), *Forum – grazing and riparian/stream ecosystems*. Vienna VA: Trout Unlimited, 53–55.

Vannote, R.L., Minshall, G.W., Cummins, K.W., Sedell, J.R. and Cushing, C.E., 1980: The river continuum concept. *Canadian Journal of Fisheries and Aquatic Science*, 37: 130–37.

Vaughn, D.M., 1990: Flood dynamics of a concrete-lined urban stream in Kansas City, Missouri. *Earth Surface Processes and Landforms*, 15: 525–37.

Viessman, W. and Welty, C., 1985: *Water management technology and institutions*. New York: Harper Row.

Vischer, D., 1989: Impact of 18th and 19th century river training works: three case studies from Switzerland. In Petts, G.E., Moller, H. and Roux, A.L. (eds), *Historical change of large alluvial rivers: Western Europe*. Chichester: J. Wiley & Sons, 19–40.

Vivash, R., Ottosen, O., Janes, M. and Sørensen, H.V., 1998: Restoration of the Rivers Brede, Cole and Skerne: joint Danish and British EU-LIFE demonstration project, II – the river restoration works and other related practical aspects. *Aquatic Conservation: Marine and Freshwater Ecosystems*, 8: 197–208.

Volker, A. and Henry, J.C., 1988: *Side effects of water resources management, 172*. Wallingford: I.A.H.S. Press.

Von Hagen, B., Beebe, S., Schoonmaker, P. and Kellogg, E., 1998: Nonprofit organizations and watershed management. In Naiman, R.J. and Bilby, R.E. (eds), *River ecology and management: lessons for the Pacific coastal ecosystem*. New York: Springer, 625–41.

Wade, P.M., Large, A.R.G. and de Waal, L.C., 1998: Rehabilitation of degraded river habitat: an introduction. In de Waal, L.C., Large, A.R.G. and Wade, P.M. (eds), *Rehabilitation of rivers: principles and implementation*. Chichester: J. Wiley & Sons, 1–10.

Walling, D.E., 1981: Yellow River that never runs clear. *Geographical Magazine*, 53: 568–75.

Walling, D.E., 1999: Linking land use, erosion and sediment yields in river basins. *Hydrobiologia*, 410: 223–40.

Walters, C.J., 1986: *Adaptive management of renewable resources*. New York: Macmillan.

Ward, D., Holmes, N. and Jose, P. (eds), 1994: *The new rivers and wildlife handbook*. Sandy: The Royal Society for the Protection of Birds/National Rivers Authority/Royal Society for Nature Conservation.

Ward, J.V., 1989: The four-dimensional nature of the lotic ecosystem. *Journal of the North American Benthological Society*, 8: 2–8.

Ward, J.V., 1998: Riverine landscapes: biodiversity patterns, disturbance regimes and aquatic conservation. *Biological Conservation*, 83: 269–78.

Ward, J.V. and Stanford, J.A., 1983: The serial discontinuity concept of lotic ecosystems. In Fontaine, T.D. and Bartell, S.M. (eds), *Dynamics of lotic ecosystems*. Ann Arbor MI: Ann Arbor Science, 29–41.

Ward, J.V. and Stanford, J.A., 1987: The ecology of regulated streams: past accomplishments and directions for future research. In Craig, J.F. and Kemper, J.B. (eds), *Regulated streams: advances in ecology*. New York: Plenum Press, 391–409.

Ward, J.V. and Stanford, J.A., 1995: Ecological connectivity in alluvial river ecosystems and its disruption by flow regulation. *Regulated Rivers: Research and Management*, 11: 105–19.

Ward, J.V., Tockner, K., Edwards, P.J., Kollman, J., Bretschko, G., Gurnell, A.M., Petts, G.E. and Rossaro, B., 1999a: A reference river system for the Alps: the Fiume Tagliamento. *Regulated Rivers: Research and Management*, 15: 63–75.

Ward, J.V., Tockner, K. and Schiemer, F., 1999b: Biodiversity of floodplain river ecosystems: ecotones and connectivity. *Regulated Rivers: Research and Management*, 15: 125–39.

Ward, J.V., Tockner, K., Arscott, D.B. and Claret, C., 2002: Riverscape landscape diversity. *Freshwater Biology*, 47: 517–39.

Ward, R.C., 1967: *Principles of hydrology*. London: McGraw Hill.

Ward, R.C., 1978: *Floods: a geographical perspective*. London: Macmillan.

Ward, R.C., 1982: *The fountains of the deep and the windows of heaven*. Hull: University of Hull Press.

Warne, A.G., Toth, L.A. and White, W.A., 2000: Drainage-basin-scale geomorphic analysis to determine reference conditions for ecologic restoration: Kissimmee River, Florida. *Geological Society of America Bulletin*, 112: 884–99.

Warner, R.F., 1987: Spatial adjustment to temporal variations in flood regime in some Australian rivers. In Richards, K.S. (ed.), *River channels, environment and process*. Oxford: Blackwell, 14–40.

Warner, R.F., 1994: A theory of channel and floodplain responses to alternating regimes and its application to actual adjustments in the Hawkesbury River, Australia. In Kirkby: M.J. (ed.), *Process models and theoretical geomorphology.* Chichester: J. Wiley & Sons, 173–200.

Warner, R.F., 2000: Gross channel changes along the Durance River, southern France, over the last 100 years using cartographic data. *Regulated Rivers: Research and Management,* 16: 141–58.

Water Victoria, 1990: *The environmental condition of Victorian streams.* Melbourne: Department of Water Resources.

Wathern, P., 1988: An introductory guide to EIA. In Wathern, P. (ed.), *Environmental impact assessment: theory and practice.* London: Unwin Hyman, 3–30.

Watson, C.C. and Biedenharn, D.S., 1999: Design and effectiveness of grade control structures in incised river channels of North Mississippi, USA. In Darby, S.E. and Simon, A. (eds), *Incised rivers: processes, forms, engineering, management.* Chichester: J. Wiley & Sons, 395–422.

Watts, H. and Fargher, J., 1999: Monitoring and evaluation: tokenism or the potential for achievements in streamway management. In Rutherford, I. and Bartley, R. (eds), *Second Australian stream management conference: the challenge of rehabilitating Australia's streams.* Adelaide: Cooperative Research Centre for Catchment Hydrology, 679–82.

Webb, R.H., Smith, S.S. and McCord, V.A.S., 1991: *Historic channel change of Kanab Creek, Southern Utah and Northern Arizona.* Monograph 9. Grand Canyon AZ: Grand Canyon Natural History Association.

Wharton, G., 1992: Flood estimation from channel size: guidelines for using the channel geometry method. *Applied Geography,* 12: 339–59.

Wharton, G., 1994: Progress in the use of drainage network indices for rainfall–runoff modelling and runoff prediction. *Progress in Physical Geography,* 18: 539–57.

Wharton, G., 1995: Information from channel geometry–discharge relations. In Gurnell, A.M. and Petts, G.E. (eds), *Changing river channels.* Chichester: J. Wiley & Sons, 325–46.

Wharton, G., Arnell, N.W., Gregory, K.J. and Gurnell, A.M., 1989: River discharge estimated from channel dimensions. *Journal of Hydrology,* 106: 365–76.

White, C.M., 1940: Equilibrium grains on the bed of a stream. *Proceedings of the Royal Society of London,* 174A: 322–34.

White, G.F., 1957: A perspective of river basin development. *Law and Contemporary Problems,* 22: 157–87.

White, G.F., 1958: *Changes in urban occupance of flood plains in the United States.* University of Chicago, Department of Geography Research Paper.

White, G.F. (ed.), 1974: *Natural hazards: local, national, global.* New York: Oxford University Press.

White, R.J., 1975: Trout population responses to streamflow fluctuation and habitat management in Big Rock-A-Cri-Creek, Wisconsin. *Verhandlungen der internationalen Vereingung fur theoretische und angewandte Limnologie,* 19: 2469–77.

White, W.R., Bettess, R. and Paris, E., 1982: Analytical approach to river regime. *Journal of the Hydraulics Division, ASCE,* 108: 1179–93.

Whitlow, J.R. and Gregory, K.J., 1989: Changes in urban stream channels in Zimbabwe. *Regulated Rivers: Research and Management,* 4: 27–42.

Whittaker, J.G., 1987: Sediment transport in step-pool streams. In Thorne, C.R., Bathurst,

J.C. and Hey, R.D. (eds), *Sediment transport in gravel-bed rivers.* Chichester: J. Wiley & Sons, 545–79.

Whittaker, J. and Haeggi, M., 1986: 'Blockschwellen'. Mitteilungen der Versuchsanstalt für Wasserbau, Hydrologie und Glaziologie, Nr. 91, ander Eidgenossischen Technischen Hochschüte Zurich.

Whitton, B.A., 1975: *River Ecology.* Oxford: Blackwell.

Wilby, R.A. and Wigley, T.M.L., 1997: Downscaling general circulation model output: a review of methods and limitations. *Progress in Physical Geography,* 21: 530–48.

Williams, G.P., 1978: Bankfull discharge of rivers. *Water Resources Research,* 14: 1141–54.

Williams, G.P. and Wolman, M.G., 1984. Downstream effects of dams on alluvial rivers. *US Geological Survey, Professional Paper* 1286, Washington DC.

Williams, P.B., 2001: River engineering versus river restoration. In Hayes, D.F. (ed.), *ASCE wetlands engineering & river restoration conference 2001.* Reno NV. 27–31 August 2001. Washington DC: American Society of Civil Engineers.

Winkley, B.R., 1972: *River regulation with the aid of nature.* International Commission on Irrigation and Drainage: factors affecting river training and floodplain regulation, 43–57.

Winkley, B.R., 1982: Response of the Lower Mississippi River to river training and re-alignment. In Hey, R.D., Bathurst, J.C. and Thorne, C.R. (eds), *Gravel bed rivers. Fluvial processes, engineering and management.* Chichester: J. Wiley & Sons, 659–81.

Winpenny, J., 1994: *Managing water as an economic resource.* London: Routledge.

Winter, B.D., 1990: *A brief review of dam removal efforts in Washington, Oregon, Idaho and California.* NOAA Technical Memorandum NMFS F/NWR-28. Washington DC: NOAA National Marine Fisheries Service.

Wittfogel, K.A., 1956: The hydraulic civilizations. In Thomas, W.L. (ed.), *Man's role in changing the face of the Earth.* Chicago IL: Unversity of Chicago Press, 152–64.

Wohl, E.E., 2000: *Mountain rivers.* Water Resources Monograph 14. Washington DC: American Geophyical Union.

Wolfert, H.P., 2001: *Geomorphological change and river rehabilitation: case studies on lowland fluvial systems in the Netherlands.* Alterra Scientific Contributions, 6. Wageningen: Alterrra Green World Research.

Wolman, M.G., 1967: A cycle of sedimentation and erosion in urban river channels. *Geografiska Annaler,* 49A: 385–95.

Wolman, M.G. and Gerson, R., 1978: Relative scales of time and effectiveness of climate in watershed geomorphology. *Earth Surface Processes,* 3: 189–208.

Wolman, M.G. and Leopold, L.B., 1957: River flood plains: some observations on their formation. *US Geological Survey, Professional Paper* 271, Washington DC.

Wolman, M.G. and Miller, J.P., 1960: Magnitude and frequency of forces in geomorphic processes. *Journal of Geology,* 68: 54–74.

Woolhouse, C., 1994: Catchment management plans: current successes and future opportunities. In Kirby, C. and White, W.R. (eds), *Integrated river basin development.* Chichester: J. Wiley & Sons, 463–74.

Woolrych, H.W., 1851: *A treatise on the law of waters and of sewers.* Second edition. London: Saunders and Benning.

Wooster, J., 2002: Geomorphic responses following dam removal: a flume study. Unpublished M.S.Thesis, University of California, Davis CA.

Wright, J.F., Armitage, P.D. and Furse, M.T., 1984: A preliminary classification of running water sites in Great Britain based on macro-invertebrates species and the prediction of community type using environmental data. *Freshwater Biology*, 14: 221–56.

Wunderlich, R.C., Winter, B.D. and Meyer, J.H., 1994: Restoration of the Elwha River ecosystem. *Fisheries*, 19: 11–19.

Wyzga, B., 1991: Present day downcutting of the Raba river channel (Western Carpathians, Poland) and its environment effects. *Catena*, 18: 556–66.

Zalewski, M., 2000: Ecohydrology – the scientific background to use of ecosystem properties as management tools: towards sustainability of water resources. *Ecological Engineering*, 16: 1–8.

Zimmerman, R.C., Goodlett, J.C. and Comer, G.H., 1967: *The influence of vegetation on channel form of small streams*. International Association of Scientific Hydrology: Symposium River Morphology, Publication 75. IAHS, 255–75.

Zöckler, C., 2000. *Wise use of floodplains: review of river restoration projects in a number of European countries – a representative sample of WWF projects*. Cambridge: WWF European Freshwater Programme.

Index

Note: Page numbers in *italics* refer to Figures, those in **bold** refer to tables

Lightning Source UK Ltd.
Milton Keynes UK
UKOW06f0727170415

249825UK00002B/29/P

9 780340 759691